100 Years Series

A HUNDRED YEARS OF CHEMISTRY

A HUNDRED YEARS
OF CHEMISTRY

by
ALEXANDER FINDLAY
Emeritus Professor of Chemistry in the University of Aberdeen

Third Edition Revised by
TREVOR I. WILLIAMS

GERALD DUCKWORTH & CO LTD
3 Henrietta Street, London, W.C.2

First published 1937
Second Edition, 1948
Reprinted, 1955
Third Edition, 1965
Reprinted, 1968

© 1965 by Alexander Findlay

SBN 7156. 0168. 7

Printed in Great Britain by
Western Printing Services Limited, Bristol

PREFACE TO THE FIRST EDITION

FOLLOWING on the period of vigorous resurgence inaugurated by Lavoisier, Dalton, Berzelius and Avogadro, modern chemistry began to develop rapidly; and during the past hundred years it has grown into the marvellous edifice of chemical science which we know at the present day. To trace, in broad outlines, how, from the time more especially of Liebig and Wöhler at the beginning of the eighteen-thirties, chemists have studied the properties and transformations of matter and have ascertained the laws according to which these transformations take place; how they have unravelled the inner structure of and the arrangement of the atoms in the molecules of many of nature's most important compounds, and have succeeded in building up these compounds in the laboratory; and how, as the result of great experimental skill and philosophic insight, they have gained ever clearer views of the fundamental nature of matter, is the aim of the present work. From the dawn of civilization, moreover, man has sought to turn his knowledge to practical account in order to increase the comfort and to improve the amenities of everyday life, and an attempt has therefore been made, in this survey of a hundred years of chemistry, to indicate how some, at least, of the discoveries of the scientific laboratory have found application to the advancement of industry and the enhancement of the social and economic well-being of man.

While this survey necessarily deals with advances in our knowledge of matter, it is not forgotten that the erection of the present-day edifice of chemistry is a great human achievement. Short biographical sketches of those, now dead, who have contributed most conspicuously to the work of building have therefore been collected together in an appendix.

I wish here to express my thanks to my colleague, Dr. R. B. Strathdee, for his assistance in reading the proof-sheets.

14 The Chanonry   A.F.
   Old Aberdeen
     April 1937

PREFACE TO THE SECOND EDITION

IN the ten years which have elapsed since the publication of the first edition great advances have been made in chemical theory, in chemical discovery and in the applications of the new knowledge to practical purposes. In the domain of chemical theory the most important advances relate to the constitution of the atomic nucleus and to the artificial transmutation of elements; and we have been the witnesses of one of the greatest achievements of the human mind, the utilization of atomic or nuclear energy. That this great achievement should have found its first use in human and material destruction by an atom bomb is one of those tragedies which are a constant challenge to man's moral nature but which do not diminish the greatness of the achievement or make more difficult the application of man's knowledge to beneficent ends.

Great also have been the advances in our knowledge of human nutrition, in our power of combating disease and in the production of materials which minister to man's comfort and material well-being. To all these advances, based on chemical research, it has been sought to direct attention in this second edition, and the opportunity has also been taken of correcting a few misprints and errors which had passed unnoticed in the first edition.

Beckenham   A.F.
September 1947

PREFACE TO THE THIRD EDITION

IN inviting me to collaborate with him in revising this work, Professor Findlay both paid me a compliment and set me a problem. It is certainly a privilege to have been associated with a distinguished chemist who took up his first academic appointment as long ago as 1900 and yet remains active and alert and keenly interested in current chemical developments. He himself has first-hand knowledge of well over half the long period now spanned; by comparison, I am a novice. On the practical side, the rapidly accelerating progress of chemistry in so many different fields, and increasing dependence on the methods of physics, meant that the amount of new material to be considered was disproportionately large in proportion to the extra span of years to be covered. The difficulties in incorporating it without altering the character of what is essentially an historical work are obvious.

Rather than alter the whole balance of the book by trying to be comprehensive with regard to recent developments, I have contented myself with seeking to indicate some of the main lines of advance and the more important results obtained. Inevitably, much has had to be omitted and the reader seeking fuller information must consult larger and more specialized works. So far as inorganic chemistry is concerned, where the mere number of different elements to be considered makes comprehensive treatment within the present compass especially difficult, we would particularly recommend *Modern Aspects of Inorganic Chemistry* (third edition, 1960) by H. J. Eméleus and J. S. Anderson.

It is hoped that this revised edition will serve the same purpose as its predecessors, to show that modern chemistry is firmly rooted in the past and to present the subject as the product of a process of continuous evolution.

July 1963   TREVOR I. WILLIAMS

7

CONTENTS

CHAPTER ONE

THE HISTORICAL BACKGROUND

ALTHOUGH it may be that in the history of scientific knowledge, as in the social and political history of a people, development is continuous—all that happens flowing from that which has been into that which is to be—the rate of progress is by no means uniform. Periods of rapid and vigorous development alternate with periods of slower and feebler growth. There are epochs in history, there are generations in the life of a science; and each of these generations is generally found to begin with the conception of an hypothesis and the birth of a theory which gives not only a new and more adequate interpretation of the phenomena and laws derived from experimental inquiry, but gives also a new outlook to the student and, to the investigator, a new aim and direction to his endeavours.

Such an epoch in the history of chemistry was inaugurated in the last quarter of the eighteenth century; and, during the first quarter of the nineteenth century, the lines of future development were being drawn and some of the main foundations on which the chemistry of that century was built were being laid. In order, therefore, that we may rightly appreciate how the science of chemistry has grown and developed during the past hundred years, a glance, at least, must be cast over the preceding period into the heritage of which the chemists of the eighteen-thirties entered, for it was a period of vigorous resurgence and a period in which chemical science was inspired with a new life and vision.

In spite of the teaching of Robert Boyle (1627–91) in *The Sceptical Chymist*, chemists could not, all at once, free their minds from the spirit of medieval thought; and, owing to the fact that until towards the close of the eighteenth century chemistry was largely a descriptive and qualitative science, the minds of chemists were dominated, for nearly a century, by the *phlogiston theory*; an erroneous, although doubtless useful, interpretation of the phenomena of combustion and other chemical processes.

During the second half of the eighteenth century, however, the founders of "pneumatic chemistry," or the chemistry of gases— Joseph Black (1728–99); Henry Cavendish (1731–1810); Joseph Priestley (1733–1804); Carl Scheele (1742–86)—were, all unconsciously, preparing the way for the overthrow of this theory. The end came in 1778 when the brilliant French chemist, Antoine Laurent Lavoisier (1743–94), as a result of his classical experiments on the calcination of tin[1] and of mercury and on the decomposition by heat of the calx of mercury,[2] was able to set forth his interpretation of the phenomena of combustion in air, an interpretation which may be summarized in the words: *The process of combustion is a chemical combination of the combustible substance with the oxygen[3] present in the air, the process being accompanied by emission of heat and it may be also of light.* Through the replacement of the phlogiston theory by one "more conformable with the laws of nature and involving less forced explanations and fewer contradictions,"[4] Lavoisier gave to the study of chemistry a new life, a new direction and a wider outlook.

Lavoisier's success in finding the true interpretation of combustion was mainly due to his recognition of the supreme importance of investigating chemical processes quantitatively, and to his adoption of the balance as an essential aid in the interpretation of chemical phenomena. Lavoisier was not the first to employ the balance in the study of chemical processes, but he was the first fully to realize and to make others realize the importance of the balance as an indispensable instrument of chemical investigation. By his use of the balance he was able to dispel the erroneous belief that water is converted into earth on being boiled in a glass vessel, an achievement comparable in its importance with the disproof by Louis Pasteur in 1860 of the doctrine of spontaneous generation; and by his enunciation, in 1785, of the law of conservation of mass, Lavoisier inaugurated the quantitative epoch in chemistry.[5]

---

[1] Lavoisier, *Mém. Acad. Sci.*, 1774, p. 351; *Œuvres*, II, 105.

[2] First described in the *Traité élémentaire de Chimie*, 1789.

[3] This name, derived from the Greek ὀξύς (oxys) = acid, and γεννάω (gennaö) = I produce, was given to the "dephlogisticated air" of Scheele and Priestley by Lavoisier, in the erroneous belief that this element is an essential constituent of acids.

[4] *Œuvres*, II, 233.

[5] On this law which, like all laws founded on experiment, is endowed with the possibility of error, but which was shown in 1908 by H. Landolt (1831–1910), of the University of Berlin, and in 1912 by J. J. Manley (1863–1946), of the University of Oxford, to be accurate within one part in one hundred million, the whole edifice of quantitative chemistry is based.

Ever since the time of Boyle it had been tacitly accepted that the composition of a compound is definite and invariable. In the opening years of the nineteenth century, however, the validity of this assumption, which clearly is of fundamental importance, was challenged and became the subject of controversy and of experimental investigation.

In his *Recherches sur les lois de l'affinité* (1798), which was later greatly expanded and published under the title *Essai de Statique Chimique* (1803), Claude Louis Berthollet (1748–1822), one of the most eminent and honoured chemists of his day and a companion of Napoleon on his expedition to Egypt in 1799, showed that the direction in which a reaction takes place may be altered by variation in the proportions of the substances involved. As a result of faulty experiments, however, and of confusing homogeneous mixtures (solutions) with compounds, Berthollet was misled into drawing the erroneous conclusion that not only the direction of chemical change but also the composition of the product of reaction is influenced by the relative amounts of the reacting substances. This conclusion was opposed to the existing belief in the fixity of composition of compounds and was vigorously contested by the French chemist, Joseph Louis Proust (1754–1826), at that time Professor of Chemistry in the University of Madrid.[1] Berthollet's contention was successfully refuted; and the law of fixed proportions, established on a basis of experiment, was accepted as one of the fundamental laws of chemistry. It serves not only to define a compound but to distinguish it from a homogeneous mixture.

With the introduction of a certain measure of order into chemistry and into the rapidly growing body of facts relating to the quantitative composition of chemical compounds, the time appeared to be ripe for the enunciation of the atomic theory by John Dalton (1766–1844), a man of philosophic insight rather than an accurate experimenter. This theory,[2] which was based on definite conceptions regarding the constitution of matter, enabled one to interpret

[1] See, more especially, Berthollet, *J. de Physique*, 1805, **60**, 284, 347; **61**, 352; Proust, *ibid.*, 1806, **63**, 364.

[2] The theory was put forward by Dalton in a lecture in 1803, and made more widely known by Thomas Thomson in his *System of Chemistry*, 1807, and by Dalton in *A New System of Chemical Philosophy*, 1808. See Roscoe and Harden, *New View of Dalton's Atomic Theory*. An atomic theory was also advanced by William Higgins (1763–1825): for a critical assessment of claims made on behalf of Higgins see Wheeler, *Endeavour*, 1952, XI, 47. See also, Wheeler and Partington, *The Life and Work of William Higgins*.

and co-ordinate the laws of chemical combination in a manner which until then had not been possible.

The most important advance made by Dalton in the development of the hypothesis of the atomic constitution of matter was the introduction of the quantitative factor; and since, according to the atomic theory, the relative weights of the atoms, the so-called *atomic weights*, are fundamental units of chemical science, the determination of these atomic weights was clearly a matter of importance for the verification and for the application of the theory. This fact was recognized not only by Dalton but also, with greater insight, by the Swedish chemist, Jöns Jacob Berzelius (1779–1848), perhaps the most eminent chemical philosopher and accurate analyst of his time. By the large number of analyses which he carried out and by the acumen and philosophic insight which he exhibited in the interpretation of his results, he contributed more to the development of the atomic theory and to the setting up of accurate values of the atomic weights than did any other worker of the time. Of his contributions, moreover, to the development of the atomic theory and the advancement of chemical science, not the least valuable was the introduction of a chemical symbolism which, with slight modification, is in use at the present day.[1] By giving to his symbols a quantitative meaning—the symbol of an element representing one atomic proportion by weight—it was possible "to show briefly and clearly the number of elementary atoms in each compound and, after the determination of their relative weights, present the results of each analysis in a simple and easily retained manner." This symbolism was speedily adopted on the Continent but, in England, only after some considerable time.

Besides the difficulties of accurate quantitative analysis, which skill and the perfecting of the balance could overcome, the determination of atomic weights encountered another difficulty which seemed to be insuperable. By analysis one can determine the *equivalents* or proportions by weight in which the elements combine with a definite proportion, say 8 parts by weight, of oxygen; and if it could be assumed that all compounds are formed by the combination of only one atom of each element, the matter would be very simple. The value of the equivalent would then be the same as the value of the atomic weight with the atomic weight of oxygen

[1] *Annals of Philosophy*, 1813, **2**, 359; 1814, **3**, 51. An alphabetical symbolism had been used in a qualitative manner as early as 1802 by Thomas Thomson, Professor of Chemistry at Glasgow.

equal to 8. But nature does not work to so simple a plan. It was already well known that an element may form more than one compound with oxygen, and so have more than one equivalent; and since the number of elementary atoms present in the compound atom, or smallest particle of the compound, was unknown, there seemed to be no way of deciding which equivalent, or what multiple of the equivalent, should be taken as representing the atomic weight of the element. Chemists sought to help themselves out of the difficulty by applying various rules based on assumptions and analogies, but while Dalton was guided by one set of rules, Berzelius, with much greater experimental skill and scientific insight, was guided by another set which, in a number of cases, led to different values of atomic weights. Confusion reigned, and some chemists lost all interest in the atomic theory, as they found in it no help in the solution of their practical problems. Others contented themselves with the experimentally determinable values of the equivalents, and they made the confusion all the greater by speaking of equivalents and combining proportions as if they were synonymous with atomic weights. It was not until after 1858, as we shall learn more fully later, that agreement was attained regarding atomic weight values and the formulae of compounds.

And yet, as early as 1811, a signpost had been set up and a path had been opened which, if it had been followed, would have led chemists to the desired goal.

The energies of post-Lavoisierian chemists had mainly been bent to the elucidation of the laws of chemical combination by weight; but the investigations of Joseph Louis Gay-Lussac (1778–1850), whom Sir Humphry Davy, writing in 1813, placed at the head of the living chemists of France, led to the discovery, in 1809, of the *law of combination of gases by volume*, which may be stated in the words: *Gases combine with one another by volume in the ratio of whole numbers, and generally small whole numbers; and the volume of the product, if gaseous, bears a simple ratio to the volumes of the reacting gases.*[1]

This law of combination by volume could not be explained directly by Dalton's atomic theory, which interpreted only the laws of combination by weight; and in order to reconcile the laws of combination by weight and volume, the assumption was made by Berzelius and others that equal volumes of gases contain the same number of atoms, the term atom being applied to the smallest

[1] *Mém. Soc. d'Arceuil*, 1809, **2**, 206.

particle of a compound as well as to the smallest particle of an element. In the case of the combination of one volume of hydrogen with one volume of chlorine to form two volumes of hydrogen chloride, however, the assumption made by Berzelius led to the conclusion that the atoms of hydrogen and of chlorine must undergo division, a conclusion which, from the point of view of the atomic theory, was a *reductio ad absurdum*.

This serious difficulty was overcome in 1811 by the Italian physicist, Amedeo Avogadro (1776–1856), Professor of Physics at Turin, in a very simple manner. Avogadro drew a distinction between an *atom*, or the smallest particle of an element which can take part in a chemical reaction, and a *molecule*, or the smallest particle of a substance, whether elementary or compound, which can exist free in a gas; and he assumed that, even in the case of elements, these free molecules may consist, not of single atoms, but of groups of two or more like atoms. By making this assumption, Avogadro showed that the laws of combination by weight and volume can be reconciled if it be postulated that equal volumes of gases, elementary or compound, when under the same conditions, contain the same number of *molecules*.[1] This postulate, moreover, by pointing a way, as we shall learn later, to the determination of relative molecular weights and by thereby giving a means of ascertaining the number of atoms in a molecule, provided a basis for defining the atomic weight of an element. How this could be done was, however, not made clear by Avogadro.

The hypothesis of Avogadro, unfortunately, was presented prematurely to men who were unable to grasp its implications or to free their minds from preconceived ideas, and also before knowledge was sufficiently advanced to give to the hypothesis experimental confirmation. And so, although it was by this hypothesis that the atomic theory was enabled to survive and to become of value in chemistry, its importance was not recognized by contemporary chemists. It was not till 1858 that the importance of Avogadro's hypothesis was fully realized, and that it came to be recognized as the corner-stone of chemical science.[2]

A new and interesting chapter in the history of chemistry was opened in 1800 by the invention of the voltaic cell by Alessandro Volta (1745–1827),[3] an invention which gave to chemists a powerful

[1] Similar conclusions were reached by Ampère (*Ann. Chim.*, 1814, **90**, 43) a later date.

[2] See A. N. Meldrum, *Avogadro and Dalton*, 1904.

[3] *Phil. Trans.*, 1800, p. 403.

and valuable instrument of investigation and of decomposition; and the chemical effects produced by the electric current—the decomposition of water by Nicholson and Carlisle,[1] the decomposition of dissolved salts by Berzelius and Hisinger,[2] and more especially the decomposition of the caustic alkalis by Davy[3]—created quite a sensation. Until 1807 the caustic alkalis, caustic potash and caustic soda, had been regarded as elements, but in that year, Humphry Davy (1778–1829), who first gave to electrochemistry a solid experimental foundation, succeeded, by passng the electric current through the molten alkalis, in decomposing them with liberation of the hitherto unknown metals, potassium and sodium. Soon after, Davy succeeded also in isolating, in a similar manner, the metals calcium, strontium, barium and magnesium.

The metal potassium, owing to its chemical reactivity, was later used for the decomposition of compounds and the isolation of other elements. Thus boron was obtained from boric anhydride (by Davy and by Gay-Lussac and Thenard), silicon from the fluoride (Gay-Lussac and Thenard), and aluminium from the chloride (Wöhler), by heating the compounds with potassium.

The electrochemical experiments of Davy were continued and extended by his even more famous successor at the Royal Institution, Michael Faraday (1791–1867), who in 1834 not only introduced the exceedingly apt terms, electrolyte, electrode, anode, cathode, ion, anion, cation, which are still in use, but discovered the two important laws of electrolysis which go by his name.[4]

Apart from the spectacular discovery of the alkali and alkaline earth metals, the view expressed by Davy that chemical affinity is of the nature of electrical attraction,[5] and the electrochemical theory of compounds put forward by Berzelius[6] in 1819, are direct results of the electrochemical investigations of the early nineteenth century. This theory, which exercised not a little influence on

---

[1] *Nicholson's Journal*, 1801, **4**, 185.

[2] *Ann. Chim.*, 1804, **51**, 172. Wilhelm Hising (1766–1852), who after being raised to the rank of nobility assumed the name of von Hisinger, was an eminent mineralogist and member of a wealthy Swedish family.

[3] *Phil. Trans.*, 1808, **98**, 1.

[4] *Phil. Trans.*, 1834, p. 77; *Experimental Researches on Electricity*, 1831–8. The terms electrode, electrolyte and electrolyse were suggested to Faraday by Dr. Whitlock Nicholl, a medical practitioner in London; and the terms anode, cathode, ion, anion, cation were suggested by William Whewell, a "portentous encyclopaedist", who became Master of Trinity College, Cambridge. (*Notes and Records of the Royal Society of London*, 1961, **16**, 187.)

[5] *Phil. Trans.*, 1807, p. 39; 1826, p. 383.

[6] *Essai sur la théorie des proportions chimiques et sur l'influence chimique de l'électricité.*

chemical thought during the first half of the century, assumed that the elementary atoms bear a predominantly positive or negative charge of electricity; and Berzelius grouped the elements into two classes, electropositive and electronegative, according to the pole at which they were liberated during electrolysis.

Even when two elements combined, the compound formed might still have excess of positive or negative electricity; and combination between oppositely charged compounds could, therefore, also take place. On this basis Berzelius explained the formation of salts by the combination of two oxides, the oxide of a metal or basic oxide, which was electropositive, and the oxide of a non-metal or acid oxide, which was electronegative. The salt, barium sulphate, for example, was thus regarded as being formed by the combination of barium oxide, BaO, and the acid oxide, SO_3, giving $BaO.SO_3$. In all cases, compound formation was regarded as due to the combination of oppositely charged particles.

The constitution of salts for which an explanation was given by the theory of Berzelius was that due to Lavoisier, in whose scheme of chemistry oxygen formed the central element, to which, also, the acid properties of a substance were attributed. This view, however, became untenable (but was not immediately abandoned) when it was shown by Gay-Lussac and Thenard in 1809 that muriatic acid gas contains no oxygen,[1] and when Davy,[2] in 1810, showed that the oxymuriatic acid obtained by Scheele is an element, to which he gave the name chlorine. The salt, sodium chloride, moreover, is formed by the direct combination of sodium and chlorine, and therefore contains no oxygen. In 1815 Davy foreshadowed the present view that hydrogen is the essential element in acids and that acids are salts of hydrogen,[3] a view more clearly defined by Justus von Liebig (1803–73), the great German chemist and pioneer of chemical education, in 1838.[4] The full understanding of the association between hydrogen and acid properties, however, did not come till the eighties of last century.

Until nearly the end of the seventeenth century the interest of chemists had been directed almost entirely to the substances occurring in the non-living mineral world, but in the *Cours de Chymie* (1675), written by the French physician and chemist,

---

[1] *Mém. Soc. d'Arceuil*, II, 339.   [2] *Phil. Trans.*, 1811, p.1.
[3] *Phil. Trans.*, 1815, p. 203. The same view was held by Dulong, who did not, however, publish his researches (see Reports by Cuvier, *Mém. de l'Institut.* 1813–1815, p. cxcviii).
[4] *Annalen*, 1838, **26**, 170.

Nicolas Lémery (1645–1715), the Arabic classification of substances into the three "kingdoms"—animal, vegetable and mineral —was adopted, and substances occurring in plants and animals were brought within the range of chemical interest and investigation. During the eighteenth century a considerable number of compounds, derived mainly from vegetable but, to some extent, also from animal sources, were isolated and their properties studied. Little or no attempt, however, had been made to study these substances in a systematic manner; their composition was not known and they were classified according to origin and general characters rather than according to chemical relationships. On account of their association with the living animal or vegetable organism, that is, with "organized" matter, the products of the animal and vegetable worlds were termed *organic*, in order to distinguish them from the *inorganic* substances which make up, or which can be formed from, inanimate mineral matter.

Although it had been shown by Lavoisier that organic compounds are composed, for the most part, of the elements carbon, hydrogen and oxygen, together sometimes with nitrogen and, less frequently, phosphorus and sulphur—all substances belonging to the mineral "kingdom"—chemists held the view that the power of building up these elements into the complex compounds occurring in the animal or vegetable organism is the sole prerogative of a special *vital force* or vital energy, inherent in the living cell. Moreover, although Berzelius was able to show, in 1814, that the composition of organic compounds follows the laws of fixed and multiple proportions and can be represented by formulae (based on the atomic theory), as in the case of inorganic compounds,[1] he held the view that while inorganic compounds could be prepared artificially in the laboratory, organic compounds could be produced only in the living organism.

In 1828, however, the gulf which seemed to separate the inorganic from the organic compounds was bridged by Friedrich Wöhler (1800–82), when he discovered that the salt, ammonium cyanate, was transformed by heat into urea, a compound which occurs in urine and which had hitherto been known only as a product of animal metabolism.[2] "I must tell you," wrote Wöhler

[1] *Annals of Philosophy*, 1814, **4**, 323, 401; 1815, **5**, 93, 174; *Ann. Chim.*, 1815, **94**, 1, 170, 296; **95**, 51.
[2] It may be pointed out that, in 1826, Henry Hennell, of Apothecaries' Hall, had effected a synthesis of alcohol, which, however, was not at that time recognized as a product of vital activity (*Phil. Trans.*, 1826, p. 240). By absorbing ethylene in

to Berzelius, "that I can make urea without the need of kidneys or of any animal whatever"; and in these words we have the announcement of the birth of modern organic chemistry. This production of an organic compound from inorganic material remained, for a number of years, the only one of its kind, and it was not till about the middle of the century, when the synthesis of other organic compounds began to be successfully carried out, that belief in a vital force and in the existence of a fundamental difference between inorganic and organic compounds was abandoned. The classification of compounds into inorganic and organic is still retained as a matter of convenience and in order to facilitate the mastery of the very large number of organic compounds which are now known and which possess characteristics less frequently met with in the case of inorganic compounds. The connotation of the term "organic" has, however, undergone a change.

Before order could be brought into the study of organic substances, it was necessary to devise means whereby the composition of the compounds could be quantitatively ascertained. Lavoisier was the first to attempt such analyses by burning the substances in oxygen or by oxidizing them by means of red lead, and thereafter determining the amount of carbon dioxide and water produced. The method of combustion was improved in 1810 by Gay-Lussac and Louis Jacques Thenard (1777–1857),[1] in 1815 by Berzelius[2] and, more especially, in 1831 by Liebig,[3] whose method is, with slight modification, widely used at the present day.

In 1815 Gay-Lussac showed that the organic acid, prussic acid, first obtained by Scheele in 1782, is the hydrogen compound of a *radical* or group of elements, CN, which passes unchanged from compound to compound and behaves like an element, giving rise to compounds which are analogous to the chlorides. A still more striking example of such behaviour was revealed by the researches carried out by Liebig, in collaboration with Wöhler,[4] on the radical of benzoic acid, in the course of which a series of compounds was obtained all of which could be regarded as containing the radical,

sulphuric acid he succeeded in isolating sulphovinic acid, and he also showed that when this is distilled with dilute sulphuric acid, alcohol is obtained.

[1] *Recherches physico-chimiques*, 1811, II, 265. The spelling of this chemist's name with an é—Thénard—which is very widespread, is erroneous and was not used by Thenard himself. It may be that, so far as England is concerned, the obituary notice of this chemist published by the Royal Society is responsible for the misspelling of the name. The erroneous spelling is also given in Poggendorff's *Literarisch-biographisches Handwörterbuch*.

[2] *Annals of Philosophy*, 1814, **4**, 330, 401.   [3] *Ann. Physik*, 1831, **21**, 1.

[4] *Annalen*, 1832, **3**, 249.

benzoyl, C_7H_5O.[1] Thus, oil of bitter almonds was represented as the hydrogen compound of the benzoyl radical, $C_7H_5O.H$; benzoic acid, which is formed by the oxidation of this compound, was represented as $C_7H_5O.OH$; and the chloride and cyanide as $C_7H_5O \cdot Cl$ and $C_7H_5O.CN$ respectively. As a result of these discoveries and views, organic chemistry was defined as the *chemistry of the compound radicals*, and it came to be the task of the chemist not only to determine the ultimate composition of an organic compound, but also to ascertain the radicals which were regarded as actually existing in the compound.[2] This quest of the radicals led to a great increase of knowledge regarding organic compounds.

The importance of ascertaining the *constitution* as well as the composition of an organic compound was emphasized in a remarkable manner by the discovery by Liebig in 1823 that the two compounds, silver fulminate and silver cyanate, which have entirely different properties, have the same composition.[3] This fact, which was contrary to what was at that time accepted as axiomatic, namely, that substances having the same composition have the same properties, was thought at first to be due to an error; but the discovery by Faraday in 1825 of a hydrocarbon (now known as butylene), having the same empirical composition as ethylene, the transformation, in 1828, by Wöhler of ammonium cyanate into urea, and, finally, the recognition by Berzelius in 1830 that racemic acid has the same composition as tartaric acid, convinced even Berzelius of the existence of compounds which have the same composition but different properties. For such compounds he, with his accustomed facility for felicitous nomenclature, suggested the name *isomeric*.[4] Berzelius later recognized two special cases of this general phenomenon of isomerism, namely *metamerism* and *polymerism*.

The occurrence of isomerism was, as Gay-Lussac suggested, to be explained as due to differences in the manner in which the constituent elements are combined in the molecule; and as the number of organic compounds increased, the phenomenon of

---

[1] Liebig and Wöhler represented the radical by the formula, $C_{14}H_{10}O_2$.
[2] See Dumas and Liebig, *Compt. rend.*, 1837, **5**, 567.
[3] *Ann. Chim.*, 1823, **24**, 264; Wöhler, *ibid.*, 1824, **27**, 196.
[4] From the Greek ἴσος (isos) equal, and μέρος (meros), a part (*Ann. Physik*, 1830, **19**, 326). The term metameric was applied to compounds which have the same composition and the same molecular weight; the term polymeric to compounds having the same percentage composition but different molecular weight. The term metameric is now generally restricted to compounds of the same chemical type. The term *allotropy* was applied in 1841 to the existence of elements in different forms (*Jahresbericht*, 1841, II, 13).

isomerism was more and more frequently observed. The study of the *constitution* of organic compounds which was thereby imposed on chemists led, during the following decades, to a growing insight into the architecture of molecules and to the discovery of the processes of synthetic chemistry which are among the most remarkable and wonderful achievements of chemical science.

CHAPTER TWO

THE DEVELOPMENT OF
ORGANIC CHEMISTRY, 1835–65

"ORGANIC chemistry just now is enough to drive one mad. It gives me the impression of a primeval tropical forest, full of the most remarkable things, a monstrous and boundless thicket, with no way of escape, into which one may well dread to enter." These words, written by Wöhler[1] to Berzelius in a letter dated January 28th, 1835, show how great was felt to be the need for bringing order into the chaotic mass of organic compounds. The work of Wöhler and Liebig, referred to in the preceding chapter, which made it possible to give a simple explanation of the relations between a large series of compounds, laid the first sound foundation of a theory of organic compounds and was hailed with enthusiasm by Berzelius, then at the height of his fame and influence. "The conclusions which you have drawn from the investigation of bitter-almond oil," wrote Berzelius to Liebig and Wöhler,[2] "are certainly the most important which have so far been reached in the domain of vegetable chemistry, and give promise of shedding an unexpected light over this part of the science. . . . The facts which you have set forth inspire such reflections that they may be regarded as the dawn of a new day in vegetable chemistry."

Although the interpretation that Liebig and Wöhler placed on the experimental facts had, later, to give place to other more adequate interpretations, we may regard their work as marking the beginning of a period, extending over more than a quarter of a century, during which organic chemistry underwent a remarkable development and chemists were led, slowly and painfully, through experimental investigation and heated controversy, to clearer views regarding molecular structure and constitution. These two men, Liebig and Wöhler, pioneers in the development of organic chemistry, form a twin constellation in the chemical firmament. Men

[1] Berzelius–Wöhler, *Briefwechsel.*   [2] *Annalen*, 1832, 3, 284.

of unlike talents, the one was the complement of the other, and by their common work they achieved an unexampled success. History has, indeed, justified the words addressed in 1871 by Liebig to Wöhler,[1] in one of his last letters: "When we are dead . . . the bonds which united us in life will always hold us together in the memory of men as a not frequent example of two men who loyally, without envy or malice, contended and strove in the same domain and yet remained closely united in friendship." The name of the one conjures up a picture of the other.

The enthusiasm with which Berzelius had accepted the views of Liebig and Wöhler regarding the benzoyl radical soon grew cold as he realized that to accept electronegative oxygen as a constituent of an electropositive radical was irreconcilable with his dualistic electrochemical system. He could not, moreover, assent to a view according to which oxygen lost the unique and dominating position which it had held since the time of Lavoisier and became absorbed into a group or radical in equal partnership with other elements. Such a view betokened a revolution in scientific thought comparable with the overthrow of a monarchy and the setting up of a republic of equality and fraternity. Berzelius, therefore, challenged[2] the interpretation put forward by Liebig and Wöhler, and insisted that oil of bitter almonds and its derivatives must be regarded as compounds not of the oxygenated radical, $C_{14}H_{10}O_2(C_7H_5O)$, but of the hydrocarbon residue,[3] $C_{14}H_{10}(C_7H_5)$.

Support for the view that organic compounds contain stable groups of elements or radicals was obtained from a consideration of alcohol (spirit of wine) and its derivatives. Dumas and Polydore Boullay, fils, had, in 1828, regarded these as compounds of the radical C_2H_4, olefiant gas or ethylene,[4] alcohol being represented as C_2H_4,H_2O; ether as $2C_2H_4,H_2O$; hydrochloric ether (ethyl chloride) as C_2H_4,HCl. Ethylene was then erroneously regarded as analogous to ammonia, and the compound ethers were regarded as salts analogous to those formed by ammonia and formulated, for example, as NH_3,HCl. In 1832, Berzelius[5] agreed with this interpretation and called the radical C_2H_4 (or, as he wrote it, C_4H_8),

[1] Liebig–Wöhler, Briefwechsel.
[2] Ann. Physik, 1833, 28, 626.
[3] Called by Berzelius picramyl—from the Greek πικρός (bitter) and άμυγδαλῆ (almond).
[4] Ann. Chim., 1828, 37, 15. The authors, using the atomic weight C = 6, wrote the formula of ethylene C_4H_4.
[5] Annalen, 1832, 3, 282. See also Robert John Kane (Professor of Chemistry to Apothecaries' Hall), Dublin Jour. Med. Chem. Sci., 1833, 2, 348.

etherin; but in the following year he adopted the view,[1] also put forward by Kane, that, just as the salts of ammonia are to be regarded as containing the radical ammonium, NH_4, so the compound ethers derived from alcohol should be regarded as containing the radical[2] C_4H_{10} $(2C_2H_4 + H_2 = 2C_2H_5)$. He put forward the view, also, that alcohol and ether are the oxides of two different radicals, C_2H_6 and C_4H_{10} respectively; but Liebig[3] showed that the facts could best be interpreted on the assumption that the same radical C_4H_{10} (now C_2H_5), which he called *ethyl*, is present in both alcohol and ether, the latter being the oxide of ethyl, $C_4H_{10}O$, and alcohol the hydrate of ether, $C_4H_{10}O,H_2O$. Muriatic ether (ethyl chloride) was represented as $C_4H_{10}Cl_2(C_2H_5Cl)$.

It is perhaps worthy of passing remark, as showing how completely Avogadro's hypothesis was ignored, that Liebig found no difficulty in assuming a higher atomic (molecular) weight for alcohol than for ether, although the latter has the greater vapour density.

The theory of radicals was supported, in 1834, by the work of Dumas and Péligot[4] on an alcohol (now known as methyl alcohol), contained in the aqueous distillate of wood, from which it was deduced that this alcohol and its derivatives could be regarded as derived from a radical *methylene*, CH_2, or a radical *methyl*, CH_3. The formulation of these compounds, therefore, was analogous to that of ethyl alcohol and its derivatives.

Since, on oxidation, wood spirit yields formic acid and spirit of wine yields acetic acid, it seemed, on analogy with benzoic acid, that those acids must contain the radicals CHO and C_2H_3O. Later,[5] Liebig abandoned these oxygenated radicals, so abhorrent to Berzelius, and adopted the radical C_2H_3 (or, as he wrote it, C_4H_6) as the fundamental radical of ethyl alcohol and its derivatives. From this radical, which he called *acetyl*,[6] one can derive ethylene (etherin) and ethyl simply by the addition of hydrogen. Thus one has: acetyl = C_4H_6; etherin (C_4H_8) = $C_4H_6 + H_2$; ethyl (C_4H_{10}) = $C_4H_6 + H_4$; ether $(C_4H_{10}O)$ = $C_4H_6 + H_4 + O$. Two important derivatives of alcohol, namely, aldehyde and acetic acid, were formulated as C_4H_6O,H_2O and as $C_4H_6O_3,H_2O$ respectively.

[1] *Ann. Physik*, 1833, **28**, 626.   [2] Called by Kane (*loc. cit.*) *ethereum*.
[3] *Annalen*, 1834, **9**, 1.
[4] *Annalen*, 1835, **15**, 1; *Mém. Acad. Sci.*, 1838, **15**, 557. Eugène Melchior Péligot (1811–90) became Professor of Chemistry at the Conservatoire des Arts et Métiers, Paris.
[5] *Annalen*, 1839, **30**, 139.
[6] This name is now applied to the radical $CH_3.CO$.

Whatever differences in detail might exist between the views of the three great leaders, Berzelius, Liebig and Dumas,[1] all were convinced of the existence of groups of elements or radicals[2] as constituents of organic compounds, and the theory was a great stimulus to investigation. In a joint communication on the *Present State of Organic Chemistry*, published in 1837, Dumas and Liebig[3] gave enthusiastic expression to the conviction that the great problem had been solved of how nature builds up three or four elements into the manifold and diverse compounds occurring in plants and animals. "Nature has followed a course as simple as it is unexpected, for with the elements she has built up compounds which have all the properties of the elements themselves. And therein lies the whole secret of organic chemistry. . . . In inorganic chemistry the radicals are simple; in organic chemistry the radicals are compound, that is all the difference." That was the conclusion arrived at as the result of years of painstaking experiment and of keen controversy; and a call was given to the energetic young workers of all countries who were animated by a love of science to join in building up a natural classification of organic compounds by the study and determination of their constituent radicals.

In this work of investigating the constituent radicals of organic compounds, the most outstanding success was achieved by Bunsen in his investigation of the cacodyl compounds,[4] in which he proved the presence of the radical, $C_4H_{12}As_2$ (C_2H_6As). By heating the chloride, C_2H_6AsCl with zinc, Bunsen obtained the spontaneously inflammable free cacodyl, $C_4H_{12}As_2$—now written $(CH_3)_2As.As(CH_3)_2$.

With the work of Bunsen the culminating point of the old radical theory with its idea of invariable atomic complexes was reached; but even before he penned, with Liebig, the report on the achievements of the radical theory to which reference has been made, Dumas was discovering facts which were destined to destroy the Berzelius system of dualism on which the radical theory was partly based and into which it had been fitted.

Dumas' investigations had their origin in a trivial circumstance. During a reception held by Charles X at the Tuileries, the candles

---

[1] The arguments for and against the different views regarding the constitution of ether and the compound ethers are summarized by Liebig, *Annalen*, 1836, **19**, 270.

[2] Defined by Liebig, *Annalen*, 1838, **25**, 3.

[3] *Compt. rend.*, 1837, **5**, 567.

[4] *Annalen*, 1839, **31**, 175; 1841, **37**, 1; 1842, **42**, 14; 1843, **46**, 1.

burned with a smoky flame and emitted an irritating vapour. Alexandre Brongniart, Director of the Sèvres Porcelain Works, requested Dumas, his son-in-law, to investigate this behaviour, and it was soon ascertained that the choking fumes were due to hydrochloric acid gas derived from the chlorine which had been used to decolorize the wax.[1] Dumas found, moreover, that the chlorine was not present merely as impurity but was in combination with the wax; and on studying the action of chlorine on oil of turpentine and other substances, and also the action of bromine and of iodine, he was led to the conclusion that chlorine, bromine and iodine have the power of abstracting hydrogen from certain substances and of replacing it, volume for volume, or atom for atom. To this law or *theory of substitution*, as he called it, Dumas, in 1834, gave the name of *metalepsy*.[2] Such cases of substitution, he pointed out, had previously been observed by Gay-Lussac in the case of hydrogen cyanide,[3] by Wöhler and Liebig in the case of oil of bitter almonds, and by Faraday[4] in the action of chlorine on ethylene.

The conversion of alcohol into acetic acid by oxidation was also regarded by Dumas as a case of substitution.[5] Thus, representing alcohol by $C_8H_8,2H_2O$ (C = 6), Dumas wrote,

$$C_8H_8.2H_2O + 4O = C_8H_4O_2.2H_2O + 2H_2O$$

or, using the value 12 for the atomic weight of carbon,

$$C_2H_4.H_2O + 2O = C_2H_2O.H_2O + H_2O$$
Acetic acid

In this case one volume of hydrogen is substituted by half a volume of oxygen.

In his statement of the rules of substitution, Dumas realized that hydrogen is substituted by halogen or oxygen in equivalent amounts, and that whereas one volume of hydrogen is equivalent to one volume of chlorine, bromine or iodine, it is equivalent to only half a volume of oxygen. A difference between atom and equivalent is thus indistinctly realized.

As a result of the views put forward by Dumas, the further investigation of the phenomena of substitution was taken up with great energy, more especially in France; and in 1835 Laurent, a

---

[1] Hofmann, *Berichte*, 1884, **17**, *Referate*, p. 667.
[2] From μετάληψις (metalēpsis), exchange. See *Mém. Acad. Sci.*, 1838, **15**, 548; *Ann. Chim.*, 1834, **56**, 113.
[3] *Ann. Chim.*, 1815, **95**, 210.   [4] *Phil. Trans.*, 1821, p. 47.
[5] *Ann. Chim.*, 1834, **56**, 143.

pupil of Dumas, converted Dumas' theory into a real theory of substitution by making the important addition that when a compound undergoes chlorination, the chlorine takes the place and, as it were, plays the part of the hydrogen which is removed. The chlorinated product, therefore, is not essentially different in character from the original substance.[1]

This doctrine, which was so directly opposed to the dualistic electrochemical theory, drew upon its author the indignation of Berzelius.[2] "An element so eminently electronegative as chlorine could never enter an organic radical: such an idea is contrary to the first principles of chemistry. Its electronegative nature and its powerful affinities would prevent it from entering except as an element in a combination peculiar to itself." Dumas bowed to the storm, asserted that the rules of substitution which he had given were merely empirical, and repudiated responsibility for the views of Laurent,[3] who, however, not only defended them,[4] with, perhaps, somewhat aggressive self-assertiveness, but sought to develop them into a system, the so-called *nucleus theory*.[5] This theory, although not generally accepted, was an advance on the older radical theory in regarding the radical not as an unalterable group of atoms but as being capable of change through substitution without loss of its characteristic properties. Soon the force of facts proved too strong also for Dumas; and when he discovered, through his own experiments, the strong resemblance between trichloracetic acid and acetic acid, he surrendered to the views of Laurent.

Dumas now abandoned entirely the dualistic electrochemical theory of Berzelius, which he regarded as erroneous and harmful to the development of organic chemistry, and he put forward a unitary theory, his *theory of types*,[6] in which a compound was regarded as a uniform whole, not formed by the union of two parts. "Every

---

[1] *Thèse de Docteur*, 1837; *Methods of Chemistry*, trans. by Odling; *Ann. Chim.*, 1836, **61**, 125; 1836, **63**, 377; *Compt. rend.*, 1840, **10**, 413.
[2] *Ann. Chim.*, 1838, **67**, 309; *Compt. rend.*, 1838, **6**, 633. Berzelius wrongly attributed the theory to Dumas.
[3] *Compt. rend.*, 1838, 6, 647, 695.
[4] *Ann. Chim.*, 1837, **66**, 326; *Compt. rend.*, 1840, **10**, 413.
[5] *Methods of Chemistry*; *Compt. rend.*, 1840, **10**, 412. Laurent's nuclei were all hydrocarbons, and by substituting the hydrogen by chlorine, oxygen, etc., radicals were obtained which possessed, in general, the characteristic properties of the nucleus. These radicals could then unite with other atoms or molecules (oxygen, hydrogen chloride, water, etc.) giving rise to acids, hydrochlorides, etc. In this way one could show the generic relationships between a whole series of compounds derived from the same nucleus. The theory was used by Gmelin as a basis of classification of organic compounds.
[6] *Annalen*, 1839, **32**, 101; 1840, **33**, 179; *Compt. rend.*, 1839, **8**, 609.

chemical compound forms a complete whole and cannot, therefore, consist of two parts. Its chemical character is dependent primarily on the arrangement and number of its atoms and in a lesser degree on their chemical nature." From his study of the products of substitution, Dumas concluded that "there exist in organic chemistry certain types which remain unchanged even when the hydrogen which they contain is replaced by equal volumes of chlorine, bromine or iodine." Besides these chemical types, Dumas later assumed, as suggested by Regnault, the existence of mechanical types, or substances having the same number of elementary atoms and produced by substitution, but having essentially different properties.[1]

Although Liebig and his followers in Germany did not accept the narrow definition of a radical adopted by Berzelius and had no objection to assuming the presence of oxygen or electronegative atoms within a radical, they, for some time longer, still adhered to the dualistic theory; and when Dumas, now obsessed by his theory of substitution and types, took up the extreme position that every element of a compound, even carbon, might be replaced by another element without alteration of type, he was violently opposed by Liebig and was laughed to scorn by Wöhler, writing in French over the pseudonym of S. C. H. Windler (Swindler).[2] The rapidly growing body of facts, however, was on the side of the adherents of the substitution theory, and Berzelius alone remained, until his death in 1848, to fight for the lost cause of electrochemical dualism.[3] Even when, in 1842, Louis Henri Frédéric Melsens (1814–86), Professor of Chemistry in Brussels, showed that chloracetic acid could be reconverted to acetic acid by potassium amalgam,[4] Berzelius, unable to abandon a theory on which his whole system of chemistry was based,[5] still strove to transfer to organic

[1] *Annalen*, 1840, 33, 259; *Compt. rend.*, 1840, 10, 149.
[2] *Annalen*, 1840, 33, 308. The skit on Dumas' views was written by Wöhler to Berzelius and was not intended for publication. It was inserted in the *Annalen* by Liebig. In a letter to Liebig, dated March 29th, 1840, Wöhler wrote: "I had not the remotest idea that the skit on the substitution theory would be printed. It was really intended only for Berzelius who derives pleasure from such *allotria* in letters." Wöhler also points out that Liebig should in any case have represented the author as being a Frenchman, such as Ch. Arlatan, instead of a German.
[3] In 1845 Liebig stated that the work of Hofmann on substituted anilines furnished him with decisive proof that "the chemical character of a compound depends in no way on the nature of the elements of which it is composed, as the electrochemical theory requires, but depends solely on the position occupied by these elements." (*Annalen*, 1845, 53, 1.)
[4] *Ann. Chim.*, 1842, 10, 233.
[5] In words of almost tragic prophecy, Berzelius, in 1827, wrote, "The habit of an opinion often leads to the complete conviction of its truth; it hides the weaker parts and makes us incapable of accepting the proofs against it."

compounds, to which it was inherently inapplicable, the theory of electrochemical dualism for which, in the case of inorganic salts, there was some experimental justification, and which, as we shall see, was revived at a much later date and in a modified form. Thus, adopting, but with a different signification, the term *copula* introduced by Gerhardt, Berzelius formulated acetic acid as a coupled oxalic acid, $C_2H_3 + C_2O_3$, with the copula[1] C_2H_3; and chloracetic acid, similarly, was represented by the formula, $C_2Cl_3 + C_2O_3$. While thus preserving a dualistic constitution, Berzelius, by granting that the hydrogen of the copula could be replaced by chlorine without essential alteration of the chemical properties, conceded, no doubt unwittingly, the claims of the substitution theory against which he had previously so strenuously striven. His change of position is difficult to account for,[2] unless it be that under the spell of oxygen worship he conceived that the properties of the acids are dependent solely on the group C_2O_3.

A new aspect was given to the substitution-type theory by Gerhardt, who explained the formation of organic compounds as due to the "pairing" or "copulation" of residues.[3] Compounds, it was thought, were produced by the elimination of hydrogen from one of the reacting substances along with oxygen (or hydroxyl), chlorine, etc., from the other reacting substance and the joining up of the resulting residues. Thus, the formation of sulphobenzene and of benzamide can be represented by the equations,

$$2C_6H_6 + SO_3 = H_2O + (C_6H_5)_2SO_2$$
$$\text{and } C_7H_5OCl + NH_3 = HCl + (C_7H_5O)NH_2$$

In this *theory of residues* the residues were regarded as imaginary substances, different from the free atomic groups of the same composition. Thus, the residue SO_2 in sulphobenzene was not

---

[1] By *copula* Berzelius seems to have meant a neutral hydrocarbon group which could combine with an electropositive or an electronegative group.

[2] Berzelius' mind had lost its flexibility and was unable to adapt itself to the outlook demanded by the new discoveries. Even in 1839 Liebig wrote to Wöhler: "[Berzelius] fights for a lost cause and, quite against his nature, with the pen only . . . It is horrible to think that a time must come when we can believe only in the past as true, and regard the future as mere froth and fraud; when, with all our might, we must cling to that which exists, because we have no longer the power to keep pace; and yet the wheel of time cannot stand still." (Liebig-Wöhler, *Briefwechsel*, I, 147.) It may be pointed out that the researches of Graham on the phosphoric acids (*Phil. Trans.*, 1833, p. 253), and of Liebig on the constitution of organic acids (*Annalen*, 1838, 26, 113), which confirmed the views of Davy and of Dulong that acids are compounds of hydrogen in which the hydrogen can be replaced by metals, were equally opposed to the dualistic theory of Berzelius.

[3] *Ann. Chim.*, 1839, 72, 184.

regarded as sulphur dioxide existing as such in the sulphobenzene; and even in one and the same compound, different residues might be regarded as "copulated," according to the method of preparation, etc., The breakaway of the younger French chemists from the older conceptions of dualism and of pre-existing radicals which could be isolated was complete. Instead of the dualistic formulae of Berzelius and the German chemists, unitary, empirical formulae were now employed.

It was pointed out in the preceding chapter that during the first half of the nineteenth century no general agreement was reached regarding the atomic weights of the elements; and, since the hypothesis of Avogadro was ignored, no agreement about the molecular weight of a compound was possible. This want of agreement with regard to the chemical unit delayed the development of organic chemistry, for it is clear that little real progress in deciding the constitution of a compound, or the arrangement of the atoms within the molecule, could be made so long as no agreement existed as to the number of atoms in the molecule. It was, therefore, one of the greatest services to chemistry on the part of Gerhardt and Laurent that they strove, and with a certain measure of success, to bring about uniformity in the formulation of organic compounds, by reviving the hypothesis of Avogadro and by distinguishing between equivalents, atoms and molecules.

This question was taken up in 1842 by Gerhardt[1] who, from a study of reactions between organic compounds involving the elimination of inorganic compounds—water, hydrogen chloride, ammonia, etc.—showed that there was a want of concordance in the formulation of organic and inorganic compounds. This discordance, which had been brought about by the application of the dualistic theory to organic compounds and by the use of erroneous atomic weights, could be removed by halving the prevailing formulae of organic compounds and by adopting the atomic weights of Berzelius for hydrogen (1), carbon (12), oxygen (16), nitrogen (14), sulphur (32). Thus, acetic acid was given the formula $C_2H_4O_2$ in place of the formula $C_8H_8O_4(C = 6)$, or $C_4H_8O_4(C = 12)$.

Although Gerhardt clarified the position by using for volatile compounds formulae which had a common basis and which indicated those quantities of the compounds which, in the gaseous state, occupy two volumes, the volume of one atom of hydrogen

[1] *J. pr. Chem.*, 1842, **27**, 439; 1843, **28**, 34, 65; **30**, 1; *Précis de Chimie Organique.*

being unity, it remained for Laurent to make clear the distinction between equivalents, atoms and molecules,[1] by attaching to these terms meanings similar to those given to them at the present day. Like Avogadro[2] and Ampère, Laurent regarded the molecules of hydrogen, of oxygen, of chlorine, etc., as consisting of two atoms and of forming "homogeneous compounds" which could then give rise to "heterogeneous compounds" by double decomposition— (HH) + (ClCl) = (HCl) + (HCl)—as had long before been pointed out by Avogadro.

Through the work of Gerhardt and Laurent a sound basis was given for the determination of atomic weights and for the correct formulation of organic compounds, but their views were still without sufficient experimental support and failed to obtain general acceptance. Even Gerhardt, in his *Traité de Chimie Organique*, rather weakly continued to use the old formulae "in order the better to exemplify how irrational they are and to leave to time the justification of a reform which chemists have not yet generally adopted." In Laurent's case it was the concluding tragedy of a somewhat tragic life that an early death deprived him of the credit of establishing the postulate of Avogadro as the basis of molecular and atomic weights.

Soon after the middle of the nineteenth century a notable advance in the system of classification of organic compounds was made with the introduction of the *new type theory* by Gerhardt[3] in 1852, as a result, more especially, of the work of Wurtz, Hofmann and Williamson.

In 1849 Wurtz[4] had discovered the "compound ammonias" or amines, and in the following year Hofmann not only prepared the amines by the action of ammonia on the halogen compounds of the hydrocarbon (alkyl) radicals, methyl, ethyl, etc., but also obtained the alkyl derivatives of aniline.[5] It thus became clear that these compounds must be regarded as derived from ammonia or from aniline by the substitution of one or more hydrogen atoms by alkyl radicals, and could therefore be represented by the formulae:

---

[1] *Ann, Chim.*, 1846, **18**, 266.
[2] Laurent, however, does not mention Avogadro.
[3] *Ann. Chim.*, 1853, **37**, 331.
[4] *Compt. rend.*, 1849, **28**, 223; *Annalen*, 1849, **71**, 326.
[5] *Phil. Trans.*, 1850, p. 93; *Annalen*, 1850, **74**, 117; See also Wurtz, *Ann. Chim.*, 1850, **30**, 498.

$$\left.\begin{array}{l}H\\H\\H\end{array}\right\}N \quad \left.\begin{array}{l}C_2H_5\\H\\H\end{array}\right\}N \quad \left.\begin{array}{l}C_2H_5\\C_2H_5\\H\end{array}\right\}N \quad \left.\begin{array}{l}C_2H_5\\C_2H_5\\C_2H_5\end{array}\right\}N \quad \left.\begin{array}{l}C_6H_5\\H\\H\end{array}\right\}N \quad \left.\begin{array}{l}C_6H_5\\C_2H_5\\H\end{array}\right\}N$$

Ammonia   Ethylamine   Diethylamine Triethylamine   Aniline   Ethylaniline

The ammonia "type" was thus created.

The idea that organic compounds could be represented in accordance with the type or pattern of simple inorganic compounds, by replacing one or more hydrogen atoms by radicals containing carbon, was advanced a step further by Williamson, who prepared ether by the action of ethyl iodide on potassium ethylate[1] and thereby showed that alcohol must be represented by the formula C_2H_5OH, and not by the formulae given to it by Liebig, $C_4H_{10}O,H_2O$. Alcohol and ether, therefore, could be regarded as belonging to the water type[2] as represented by the formulae:

$$\left.\begin{array}{l}H\\H\end{array}\right\}O \quad \left.\begin{array}{l}C_2H_5\\H\end{array}\right\}O \quad \left.\begin{array}{l}C_2H_5\\C_2H_5\end{array}\right\}O \quad \left.\begin{array}{l}C_2H_5\\CH_3\end{array}\right\}O$$

Water   Alcohol   Ether   Methylethyl ether

The question of the formula and molecular weight of alcohol and of ether, about which difference of opinion had long existed, was thereby settled. Moreover, acetic acid and the other acids of that series can also be represented as belonging to the water type, with the formula $\left.\begin{array}{l}C_2H_3O\\H\end{array}\right\}O$ for acetic acid. This formula obviously indicates the possibility of obtaining a compound in which the second hydrogen atom of water is also replaced by C_2H_3O; that is, $\left.\begin{array}{l}C_2H_3O\\C_2H_3O\end{array}\right\}O$. This and other similar compounds, known at first as anhydrous acids and later as *acid anhydrides*, were obtained in 1852 by Gerhardt.[3] "The method here employed of stating the rational constitution of bodies by comparison with water, seems to me," wrote Williamson, "to be susceptible of great extension; and I have no hesitation in saying that its introduction will be of service in simplifying our ideas, by establishing a uniform standard of comparison by which bodies may be judged of." The method

---

[1] *J. Chem. Soc.*, 1852, **4**, 106, 229. See also G. Chancel, *Compt. rend.*, 1850, **31**, 521.

[2] As early as 1846 Laurent had suggested that alcohol and ether might be regarded as substitution products of water (*Ann. Chim.*, 1846, **18**, 266). Guided by his views regarding the formula of ether, Williamson was able to explain the formation of that compound from alcohol and sulphuric acid—a reaction which had been labelled by Mitscherlich and by Berzelius as "catalytic."

[3] *Ann. Chim.*, 1853, **37**, 285; *Traité de Chimie Organique*, IV, 672.

was extended by Gerhardt who generalized the idea of types due to Wurtz, Hofmann and Williamson, as well as to the American chemist, Thomas Sterry Hunt[1] (1826-92), and thus set up the so-called *new type theory*, a blend of the older radical and type theories of Dumas, as a means of classifying organic compounds. To the ammonia and water types, Gerhardt added the hydrogen and hydrogen chloride types, which included the hydrocarbons and the chlorides, iodides, cyanides, etc. Thus:

| Hydrogen | Ethane | Butane | Hydrogen chloride | Ethyl chloride |

To these simple types Williamson added *multiple types e.g.*, double and triple water types, and these were developed by William Odling (1829-1921), Professor of Chemistry in the University of Oxford.[2] On the pattern of these types there could be formulated polyhydric alcohols and polybasic acids. At a later time, *mixed types* and also the *marsh gas type* were introduced by Kekulé.[3]

Besides his classification according to types, Gerhardt made use of the conception of *homologous series* which he had already utilized[4] for classification purposes in 1843, and which had been fore-shadowed by J. H. W. Schiel[5] in the case of the alcohols and by Dumas[6] in the case of the fatty acids. In a homologous series the molecule of each member contains one carbon atom and two hydrogen atoms more than the preceding member.

The theory of types was undoubtedly useful as a means of classifying and schematizing the rapidly increasing number of organic compounds, and it facilitated thereby the study of organic chemistry. It did little, however, to solve the really important problem which faced organic chemists, the problem of the constitution of compounds or the arrangement of the atoms within the molecule. To Gerhardt, who held the view that this problem could not be solved, the formulae according to types represented merely methods of formation or of decomposition; and different formulae might be given to the same compound in order to represent its behaviour in

[1] *Amer. J. Sci.*, 1848, **5**, 265; **6**, 173; 1849, **7**, 399; **8**, 89; *Compt. rend.*, 1861, **52**, 247. Hunt's work was not known in Europe until after Gerhardt's death.
[2] *J. Chem. Soc.*, 1855, **7**, 1.
[3] *Annalen*, 1857, **104**, 129. See also Odling, *loc. cit.*
[4] *Précis de Chimie Organique.*
[5] *Annalen*, 1842, **43**, 107.
[6] *Annalen*, 1843, **45**, 330; *Compt. rend.*, 1842, **15**, 935.

different reactions. Nevertheless, owing to this very fact, Gerhardt was led a little way, at least, towards a recognition and representation of constitution, more especially in connection with the fatty acids. Thus, whereas the behaviour of these acids, in most of their reactions, *e.g.*, formation of salts, esters, etc., is in accordance

with the formulae, $\left.\begin{array}{c} C_2H_3O \\ H \end{array}\right\}O, \left.\begin{array}{c} C_3H_5O \\ H \end{array}\right\}O$, etc., other reactions in-

dicate that the group C_2H_3O, for example, is composed of the group CH_3 and the group CO. Thus, it was shown by H. Kolbe and E. Frankland that methyl cyanide is converted, under the action of caustic potash, into acetic acid, whereby the presence of the methyl group is indicated;[1] and Kolbe found that when a solution of potassium acetate is electrolysed, carbon dioxide and ethane (methyl, as he thought it was), were liberated at the anode.[2] To take account of this behaviour of acetic acid, therefore, Gerhardt

wrote the formula, $\left.\begin{array}{c} CH_3.CO \\ H \end{array}\right\}O$; and the other acids were formu-

lated similarly. When, later, mixed types were introduced by Kekulé, the acids were formulated according to a mixed hydrogen-

water type, *e.g.*, $\left.\begin{array}{c} CH_3 \\ CO \\ H \end{array}\right\}O$

To obtain a solution of the problem of constitution, it was necessary to get behind the radicals, or atomic groups, to the atoms themselves, and to ascertain not only what radicals were supposed to be present in the molecule but what was the arrangement of the atoms in the radicals. The clue to the solution of this problem was found in the doctrine of *valency*, which was first definitely introduced into chemistry by Frankland.

It had, of course, long been known and quantitatively expressed in the law of multiple proportions that elements may combine in more than one proportion; and in 1834 Dumas had pointed out that while one atom of chlorine replaces one atom of hydrogen, one atom of oxygen replaces two atoms of hydrogen. Moroever, in the type formulae, it was clear that whereas one hydrogen atom or one chlorine atom could be linked with only one radical, one oxygen atom could be linked with two radicals, as in ether. Further, one nitrogen atom could be linked with three radicals, as in

[1] *Annalen*, 1848, **65**, 288.   [2] *Annalen*, 1849, **69**, 257.

triethylamine, or with five elementary or compound radicals, as in

tetramethylammonium iodide,[1]
$$\left.\begin{array}{l} CH_3 \\ CH_3 \\ CH_3 \\ CH_3 \end{array}\right\} NI$$

The conception of a combining capacity, however, and the idea that the combining capacity or the power of any given atom to combine with other atoms is limited, was first definitely expressed by Frankland[2] in a paper. *On a New Series of Organic Bodies containing Metals.* "When the formulae of inorganic chemical compounds are considered," wrote Frankland, "even a superficial observer is struck with the general symmetry of their construction; the compounds of nitrogen, phosphorus, antimony and arsenic especially exhibit the tendency of these elements to form compounds containing 3 or 5 equivalents of other elements, and it is in these proportions that their affinities are best satisfied; . . . Without offering any hypothesis regarding the cause of this symmetrical grouping of atoms, it is sufficiently evident, from the examples just given, that such a tendency or law prevails, and that, no matter what the character of the uniting atoms may be, the combining power of the attracting element, if I may be allowed the term, is always satisfied by the same number of these atoms." This hypothesis of a variable but limited combining power on the part of an element forms the basis of what came to be known as the doctrine of *atomicity*, *quantivalence*, *valency* or *valence*.[3] For a number of years the development of the doctrine of valency as a basis for the formulation of compounds and the representation of their constitution was retarded by the use of Gmelin's equivalents in place of Gerhardt's atomic weights or atomic weights based on the postulate of Avogadro; and it was not till 1858 that a satisfactory theory of molecular constitution was advanced, simultaneously and

[1] Hofmann, *Phil. Trans.*, 1851, p. 357.
[2] *Phil. Trans.*, 1852, p. 417; *Annalen*, 1853, **85**, 329. See also *Experimental Researches*, p. 153.
[3] The term "Valenz" or valence was introduced in 1868 by Carl Wilhelm Wichelhaus (1842–1927), Professor of Technological Chemistry, University of Berlin. At the present time both terms, valency and valence, are in use, the latter more especially in America; and the valency of an element is indicated by the adjectives univalent, bivalent, trivalent, quadrivalent, etc., which were suggested by Lothar Meyer (*Moderne Theorien*) in 1864. (The hybrids, mono-, di-, tri- and tetra-valent had been suggested by R. Erlenmeyer in 1860.) For a time the terms monad, dyad, triad, tetrad, proposed by Odling, were used in England, but never found general adoption.

independently, by two young chemists, Friedrich August Kekulé and Archibald Scott Couper.

The theory of molecular constitution put forward independently by Kekulé and by Couper[1] rested on two main postulates, the quadrivalency of carbon, already foreshadowed by the work of Kolbe and Frankland, and the capacity of the carbon atoms for mutual linking or combining together to form a carbon "chain." By this hypothesis of the mutual linking together of carbon atoms—which was later confirmed by experiment—it was possible to explain the formation of organic compounds containing a large number of carbon atoms. On the foundation of their two postulates, moreover, Kekulé and Couper showed how the molecular constitution or mutual linking together of the atoms of a compound could be represented diagrammatically and the relations between different compounds made readily intelligible.

In his classic paper, *On a New Chemical Theory*, Couper advanced beyond Kekulé, who was still fettered by the type theory, by representing the constitution of compounds by means of *graphic formulae* in which, as at the present day, the valencies of the atoms are represented by lines; and, except for the fact that Couper retained the value $O = 8$ and wrote a double oxygen atom, $O — O$, in place of a single oxygen atom, his formulae are similar to those at present in use, as the following examples show:

$$C\begin{cases}O-OH\\H_2\end{cases} \qquad C\begin{cases}O-OH\\H_2\end{cases} \qquad C\begin{cases}O-O\\H_2\,H_2\end{cases}C \qquad C\begin{cases}O-OH\\O_2\end{cases}$$
$$|\qquad\qquad\qquad |\qquad\qquad\qquad\quad |\qquad\quad |\qquad\qquad\quad |$$
$$C-H_3 \qquad\quad C-H_2 \qquad\qquad C-H_3H_3-C \qquad\quad C-H_3$$
$$\qquad\qquad\qquad |$$
$$\qquad\qquad\qquad C-H_3$$

Ethyl alcohol   Propyl alcohol   Ether   Acetic acid

In his *Textbook of Organic Chemistry* (1861), Kekulé made use of a graphic method of representation in which univalent atoms were represented by circles and multivalent atoms by figures formed by the coalescence of two, three or four circles, thus:

Ethyl alcohol

[1] Kekulé, *Annalen*, 1858, **106**, 129; Ostwald's *Klassiker*, No. 145; Couper, *Compt. rend.*, 1858, **46**, 1157; *Ann. Chim.*, 1858, **53**, 469; *Phil. Mag.*, 1858, **16**, 104; Ostwald's *Klassiker*, No. 183.

but neither this nor the symbolism suggested by Loschmidt in the same year[1] found favour with chemists.

In 1861, also, Alexander Crum Brown (1838–1922), Professor of Chemistry in the University of Edinburgh, unaware of Couper's work used a symbolism similar to his and practically identical with that employed at the present day.[2] Thus the molecule of the hydrocarbon propane, regarded as consisting of a "chain" of three carbon atoms to which eight hydrogen atoms are attached, can be represented by the graphic formula,

$$
\begin{array}{ccccccc}
 & H & & H & & H & \\
 & | & & | & & | & \\
H - & C & - & C & - & C & - H \\
 & | & & | & & | & \\
 & H & & H & & H &
\end{array}
$$

and acetic acid by the formula,

$$
\begin{array}{c}
H \qquad\quad O \\
| \qquad\;\; \nearrow\!\!\!\!\diagup \\
H - C - C \\
| \qquad\quad \diagdown \\
H \qquad\quad O - H
\end{array}
$$

Somewhat later the simpler *structural* or *constitutional formulae*, e.g., $CH_3 - CH_2 - CH_3$ or $CH_3 \cdot CH_2 \cdot CH_3$, were introduced, and are now most generally employed.

Although we do not know how the theory of molecular constitution arose in the mind of Couper, except that it seems to have developed from an investigation of the mutual affinities of the elements, we learn that in the artistic and imaginative mind of Kekulé the theory had its origin in the following circumstance.[3]

During a period of residence in London, Kekulé was returning from a visit paid to Hugo Müller at Islington to his lodgings at Clapham. "One fine summer evening," Kekulé relates, "I was returning by the last omnibus, 'outside' as usual, through the deserted streets of the metropolis, which are at other times so full of life. I fell into a reverie and lo! the atoms were gambolling

---

[1] *Chemische Studien* (1861). See Anschütz, *Ber.*, 1912, **35**, 539.

[2] *J. Chem. Soc.*, 1865, **18**, 232. In 1861, Crum Brown, taking the atomic weights of carbon and of oxygen as 6 and 8, used barred symbols (see p. 44). In 1864, however, he adopted the atomic weights 12 and 16 and used the symbols C and O. In 1864 Wurtz used graphic formulae similar to those of Crum Brown. The circles which Crum Brown placed round the symbols were later dropped. See R. M. Caven and J. A. Cranston, *Symbols and Formulae in Chemistry* (Blackie).

[3] Kekulé, *Ber.*, 1890, **23**, 1306; Japp. *J. Chem. Soc.*, 1898, **73**, 97.

before my eyes. Whenever, hitherto, these diminutive beings had appeared to me, they had always been in motion; but up to that time I had never been able to discern the nature of their motion. Now, however, I saw how, frequently, two smaller atoms united to form a pair; how a larger one embraced two smaller ones; how still larger ones kept hold of three or even four of the smaller; whilst the whole kept whirling in a giddy dance. I saw how the larger ones formed a chain. . . . I spent part of the night putting on paper at least sketches of these dream forms." From these sketches his graphic formulae were developed.

The theory of Kekulé and Couper gave to chemists the means of solving the problems of chemical constitution; and by means of the graphic or constitutional formulae it became possible to represent the molecular constitution of known compounds and to foresee the possible existence of isomeric compounds. Thus, it is clear that if one atom of hydrogen in propane, $CH_3 \cdot CH_2 \cdot CH_3$, is replaced by one atom, say, of chlorine, two and only two isomeric compounds can be obtained, represented by the formulae, $CH_3 \cdot CH_2 \cdot CH_2Cl$ and $CH_3 \cdot CHCl \cdot CH_3$.

While the molecules of a very large group of compounds, the *aliphatic*[1] compounds—so called because the natural fats belong to this group—could be represented as built up of chains of carbon atoms, it was not possible to formulate, in a similar manner, the *aromatic* compounds which were derived from various "essential oils"; compounds which are characterized by the fact that they contain a relatively high proportion of carbon and that they never contain less than six carbon atoms in the molecule. The simplest member of this group is the hydrocarbon benzene, which was discovered by Faraday in 1825.

In 1865 Kekulé, then Professor of Chemistry at Ghent, was engaged one evening in writing his textbook, but his thoughts were elsewhere. "I turned my chair to the fire and dozed," he relates. "Again the atoms were gambolling before my eyes. This time the smaller groups kept modestly in the background. My mental eye, rendered more acute by repeated visions of the kind, could now distinguish larger structures, of manifold conformation; long rows, sometimes more closely fitted together; all twining and twisting in snake-like motion. But look! What was that? One of the snakes had seized hold of its own tail, and the form whirled mockingly before my eyes. As if by a flash of lightning I awoke";

[1] From the Greek ἄλειφαρ (aleiphar), fat.

but the picture Kekulé had seen of the snake which had seized its own tail gave him the clue to the most puzzling of molecular structures, the structure of the benzene molecule, for which Kekulé suggested a closed ring of six carbon atoms, to each of which a hydrogen atom is attached.[1] Thus was evolved the structural formula of the benzene molecule, in which there exists a "ring" of six carbon atoms linked together in the form of a regular hexagon:

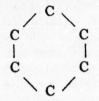

Kekulé's theory of the molecular constitution of benzene and of the aromatic compounds did not, as can readily be understood, meet with immediate and universal acceptance, but it was amply justified by the experimental investigations which it inspired and which, in the following decades, were carried out for the purpose of testing the theory and the predictions which it enabled one to make. To this work consideration will be given at a later point.

Although, as will be pointed out later, the simple theory of Couper and Kekulé has been found to be inadequate to account for all cases of isomerism met with, it has shown itself easily capable of the necessary expansion; and the doctrine of the linking of the atoms, of which the benzene theory of Kekulé forms the crowning achievement, and the method of structural representation developed therefrom, form the basis on which modern organic chemistry, defined by Gerhardt as the *chemistry of the carbon compounds*, has been built. Not only have these theories enabled chemists to interpret the relations between compounds already known but they have guided them also to the discovery and preparation of a multitude of new compounds. Without these theories of molecular constitution, organic chemistry would have remained such as it was described by Wöhler, "a monstrous and boundless thicket into which one may well dread to enter"; and its enormous development, which constitutes one of the most remarkable and important features of the chemistry of the second half of the nineteenth century, could not have taken place. More especially is this true of the benzene theory of Kekulé whose conception of closed

[1] *Bull. Soc. Chim.*, 1865, **3**, 98; *Annalen*, 1866, **137**, 129.

"rings" of carbon atoms has been extraordinarily fruitful and has shown itself capable of an almost limitless development so as to include compounds with "rings" of atoms varying in number and nature. As was pointed out at a later date,[1] "this theory found the chemistry of even the immediate derivatives of benzene an almost untilled field; it has transformed it into a fertile province, to which have been annexed regions the very existence of which was unknown." The truth of this statement will become clear in the sequel.

[1] Address presented by the Chemical Society to Kekulé in 1890 on the occasion of the twenty-fifth anniversary of the publication of the benzene theory.

CHAPTER THREE

THE DETERMINATION OF ATOMIC WEIGHTS AND THE
CLASSIFICATION OF THE ELEMENTS

THE theories of molecular constitution put forward by
Kekulé and by Couper, which came as the culmination of
long years of experimental investigation and of heated controversy, supplied chemists with, as it were, a map and compass
which made possible the survey and exploration of the ever-
widening domain of organic chemistry, and enabled them to understand and to visualize the linking together of the atoms within the
molecule. But the effective use of these theories was greatly
hampered by the persistent confusion of ideas regarding equivalents, atomic weights and molecular weights, which showed itself in
the number of different formulae used to represent the composition
of compounds. The necessity, therefore, for achieving agreement
regarding these matters became increasingly urgent.

In the opening chapter the difficulties which confronted chemists
in deciding what multiple of the equivalent or combining weight
should be taken as the atomic weight of an element were pointed
out. Berzelius, with wide knowledge and remarkable insight, was
guided in his choice of atomic weights by various considerations;
by analogies in chemical behaviour, by the densities of gases, by
the law of the constancy of atomic heat (specific heat multiplied
by the atomic weight) enunciated by Pierre Louis Dulong (1785–
1838) and Alexis Thérèse Petit (1791–1823) in 1819, and the law
of isomorphism which was enunciated in the same year by Eilhard
Mitscherlich (1794–1863). He was thereby enabled, in 1826, to
draw up a table of atomic weights of the elements which, with only
a few exceptions, were similar to those at present in use. The
foundations of his system of atomic weights, however, were in
many cases insecure, as he himself recognized;[1] and if the atomic
theory was to be accepted and general agreement regarding atomic
weights to be obtained, it was desirable to replace the arbitrary

[1] *Jahresbericht*, 1828.

method, so successfully used by Berzelius, "by something more fixed, more within the grasp of all kinds of intellect and less subject to the capricious modifications of each writer."[1] This sure basis for the determination of atomic weights was to be found, Dumas believed, in the hypothesis of Avogadro which, it will be recalled, postulated that equal volumes of gases, under the same conditions of temperature and pressure, contain the same number of constituent particles or *molecules*. It is clear, therefore, that according to this hypothesis the weights of equal volumes of gases, or the *densities* of the gases, must be proportional to the weights of the molecules; and the relative molecular weights of substances in the state of gas or vapour could therefore be ascertained from determinations of gas or vapour density,[2] as the following relations indicate:

$$\frac{\text{Density of A}}{\text{Density of B}} = \frac{\text{weight of 1 litre of A}}{\text{weight of 1 litre of B}} =$$

$$\frac{\text{weight of } n \text{ molecules of A}}{\text{weight of } n \text{ molecules of B}} = \frac{\text{weight of 1 molecule of A}}{\text{weight of 1 molecule of B}}$$

Such determinations were carried out by Dumas, using an apparatus devised by himself, in the case of substances which are volatile at not too high temperatures.

In order to explain Gay-Lussac's law of combination of gases by volume, however, it had to be assumed, as Avogadro showed, that even the molecules of elements must be regarded, in some cases at least, as composed of groups or clusters of atoms; but as there was no direct experimental method of determining the number of atoms in the molecule of an elementary gas, recourse was had to what may be called the principle of simplicity and adequacy. Thus, since one volume of hydrogen (containing n molecules) combines with one volume of chlorine (containing n molecules) to form two volumes of hydrogen chloride (containing $2n$ molecules) it follows that each of the molecules of hydrogen and of chlorine must contain at least two atoms. This conclusion is the simplest that can be drawn, and since it showed itself to be adequate to account for

[1] Dumas, *Ann. Chim.*, 1832, **50**, 170.
[2] It may be pointed out that even when the gas density is determined with great accuracy these values must be corrected for the deviation, shown by all gases and vapours, from the ideal Boyle's law, if accurate values of the molecular weight are to be obtained.

the facts, it was accepted as true. Similarly, in the case of oxygen, nitrogen, etc., the conclusion must also be drawn that the molecule consists of two atoms. If, however, the molecules of hydrogen, oxygen, chlorine, contain the same number (two) of atoms, then *in these cases*, not only the relative molecular weights, but also the *atomic weights* of the elements must be proportional to the densities of the gases. Unfortunately, Dumas wrongly assumed that the molecules of all elementary gases consist of two atoms and that, therefore, in all cases the atomic weights are proportional to the gas or vapour densities; and he assigned to sulphur (the molecule of which is hexatomic in the state of vapour at the temperature of investigation) the atomic weight of 94·4, to mercury (monatomic) the atomic weight 100·8 and to phosphorus (tetratomic) the atomic weight 68·5, in place of the Berzelius values, 32·24, 202·68 and 31·44. The values of atomic weight obtained by Dumas were not in harmony with the known chemical relations of the elements nor with the values indicated by other physical methods; and the result of Dumas' endeavour to place the values of the atomic weights on a sure foundation and to "replace by definite conceptions the arbitrary data on which nearly the whole of the atomic theory is based," was to discredit that theory still further and to drive chemists away from the uncertain atomic weights back to the experimentally determined combining weights—H = 1, C = 6, O = 8, S = 16, etc.—as advocated by Leopold Gmelin (1788–1853) in his widely used *Textbook*.

The confusion which reigned regarding equivalents, atoms and molecules was of course reflected in the varied formulae used to express the composition of a compound. Thus, using the atomic weights of Berzelius, the formula for acetic acid was written, $C_4H_6O_3,H_2O$ (with a molecular weight double that now assigned to the compound), whereas, when Gmelin's equivalents were used, the formula was written, $C_4H_3O_3,HO$. The confusion was rendered all the greater by the introduction, by Berzelius, of barred symbols in order to represent double atoms, *e.g.*, $\bar{H}=H_2$, $\bar{Cl}=Cl_2$, etc., and the formula of acetic acid was written,[1] $C_4\bar{H}_3O_3,\bar{H}O$, which gave a false impression of identity with the formula of Gmelin.

Although another attempt was made by Gerhardt and Laurent

[1] So many and varied were the formulae used by different chemists that nearly a whole page of Kekulé's *Lehrbuch* was filled with the various formulae proposed for acetic acid.

(p. 31) to introduce order, more especially into organic chemistry, by the use of formulae based on Avogadro's hypothesis, the confusion of formulae and atomic weights still persisted. In 1860, therefore, in the hope that some agreement on these matters might be reached through a personal exchange of views, a conference was arranged to be held at Karlsruhe.[1] Over a hundred of the leading chemists of the world were present but after three days the conference broke up with no other agreement reached apparently than that scientific questions cannot be settled by a vote and that full freedom must be left to the individual investigator! As each member left the conference, however, he received a copy of a *Sketch of a Course of Chemical Philosophy held in the Royal University of Genoa*,[2] drawn up in 1858 by the Italian chemist, Stanislao Cannizzaro, who was himself a member of the conference. In this *Course of Chemical Philosophy* Cannizzaro made clear to chemists the full importance of the hypothesis of Avogadro and the relation between the molecular and atomic theories, removed the misunderstandings which had prevented the general acceptance of these theories and, supporting his arguments by the results of experiments, showed how, on the basis of Avogadro's hypothesis, the values of the atomic weights could be decided. The reading of Cannizzaro's but little-known pamphlet soon brought conviction to most of those who had attended the conference, most quickly, perhaps, to the German chemist, Lothar Meyer, who after studying the pamphlet wrote: "The scales fell from my eyes, doubts vanished and were replaced by a feeling of the most peaceful assurance."

It has been pointed out that on the basis of Avogadro's hypothesis the molecular weight of a substance in the state of gas or vapour can be calculated from determinations of the gas or vapour density; and since one is dealing here not with absolute but with relative weights, it is convenient to calculate not the absolute density or mass per unit volume, but the density relatively to that of some other gas which is selected as standard. As standard gas, Cannizzaro selected hydrogen, as being the lightest gas; but as the hydrogen molecule consists of two atoms, the densities of the other gases were referred to that of hydrogen equal to 2. In this way the densities express the weights of the molecules relatively to the

---

[1] See note by Lothar Meyer in Ostwald's *Klassiker*, No. 30.
[2] *Il nuovo Cimento*, 1858, 7, 321; Ostwald's *Klassiker*, No. 30; *Alembic Club Reprints*, No. 18.

weight of the hydrogen atom taken as unity.[1] The determination of relative densities also has the advantage that as all gases and vapours increase or decrease in volume to approximately the same extent with change of pressure and temperature, the relative densities are almost independent of pressure and temperature. It is possible, therefore, to obtain the relative vapour density and molecular weight of substances which are solid or liquid under ordinary conditions and exist in the vapour state only at higher temperatures.

From determinations of molecular weights of compounds, atomic weights of elements can be decided. Thus, Cannizzaro pointed out, if the molecular weights of a series of compounds containing a given element be determined, the quantities of that element contained in the molecular weights of the different compounds are always whole multiples of the smallest quantity; and this quantity, therefore, represents the atomic weight of the given element, for an atom is the smallest amount of an element which can take part in the formation of a molecule. For example, from the analyses and determinations of the molecular weights of a number of compounds of hydrogen, the following table can be drawn up:

Substance	Molecular weight referred to the atomic weight of hydrogen equal 1	Weights of constituents in one molecular weight referred to H = 1
Hydrogen	2	2 Hydrogen
Hydrogen chloride	36·5	1 Hydrogen, 35·5 Chlorine
Hydrogen bromide	81	1 Hydrogen, 80 Bromine
Phosphine	34	3 Hydrogen, 31 Phosphorus
Water vapour	18	2 Hydrogen, 16 Oxygen
Ethylene	28	4 Hydrogen, 24 Carbon

One sees that in no case is the amount of hydrogen in the molecular weight of the above compounds less than one part by weight, and where the proportion is greater the higher proportions are whole multiples of the lowest. This holds, moreover, for all the compounds of hydrogen tested, and one can therefore say, with Cannizzaro, "The hydrogen atom is contained twice in the molecule of free hydrogen."

[1] Owing to the fact that many elements do not form compounds with hydrogen but do so with oxygen, it is now customary, by international agreement, to take oxygen as standard element with atomic weight equal to 16·00, the atomic weight of hydrogen on this scale being 1·0078. Since oxygen is bivalent, the equivalent of an element is taken as the weight of the element which combines with 8·0 parts by weight of oxygen.

The international unions of physics and of chemistry have adopted the C^{12} isotope of carbon as the basis of atomic weights.

Similarly, when the relative density and molecular weight of compounds of oxygen are determined, it is found that the smallest weight of oxygen in the molecular weight of any of its compounds is 16. This, then, is the atomic weight of oxygen. Further, since the molecular weight of free oxygen is found to be 32, the molecule of oxygen must be diatomic.

In like manner, the atomic weights of a number of other elements, carbon, nitrogen, sulphur,[1] etc., could be determined; and by taking the density as the measure of molecular weight, the complexity of the molecules of volatile elements could also be ascertained. Thus, it was shown that the molecules of phosphorus and of arsenic are tetratomic (represented by P_4 and As_4), and that the molecule of mercury is monatomic—the atom, therefore, being capable of existing in the free state.

The above deductions, one must remember, rest on the hypothesis that the molecule of hydrogen is diatomic, an hypothesis which was put forward in order to explain Gay-Lussac's law of combination of gases by volume. It may, however, be pointed out that just as the kinetic theory of gases (p. 73) gives support to the hypothesis of Avogadro, so also the deductions made from it regarding the specific heats of gases are in harmony with the values of molecular complexity calculated from the gas density.

The acceptance of Avogadro's hypothesis and the realization of its value for the determination of molecular weights and for the checking of atomic weights mark the beginning of a new era in chemistry; and the simplicity and consistency of the atomic and molecular weights and of the formulae of compounds based on these, more especially in the case of the increasingly numerous compounds of organic chemistry, won the adhesion, not immediate but gradual, of chemists. On the basis of Avogadro's hypothesis, definite information became available regarding the number of atoms in the molecule of a compound, and a firm foundation was thereby given for the building up of constitutional formulae and for the general formulation of compounds on the basis of the atomic theory. It was the molecular theory of Avogadro that made it possible for the atomic theory of Dalton to survive.

---

[1] The values of the molecular weight deduced from determinations of the vapour density are, in general, owing to the deviation of gases from Boyle's law, only approximate, and the atomic weights derived therefrom are, therefore, also only approximate (p. 72). They serve, however, to give a clear indication of the multiple of the equivalent, determined accurately by gravimetric analysis, which must be taken as the atomic weight.

The vapour densities of certain substances, *e.g.*, ammonium chloride, phosphorus pentachloride, etc., were for a time thought to be abnormal and to be exceptions to Avogadro's hypothesis; and this fact delayed the full acceptance of the hypothesis by Gerhardt. The anomalies, however, disappeared when Henri Sainte-Claire Deville (1818–81), in 1857, showed that at high temperatures ammonia and hydrogen chloride can exist side by side uncombined, and when Leopold von Pebal[1] (1826–87), Professor of Chemistry in the University of Lemberg, and Anton Karl von Than[2] (1834–1908), Professor of Chemistry in the University of Budapest, proved that in the vapour of ammonium chloride, ammonia and hydrogen chloride exist together in the uncombined state. Such cases, therefore, ceased to be regarded as exceptions to Avogadro's theorem, and from the value of the vapour density the extent to which the molecules of ammonium chloride are *dissociated* into ammonia and hydrogen chloride can be calculated.

The determinations of equivalents, carried out with considerable accuracy by Berzelius, were extended and improved by Dumas, Turner,[3] Marignac[4] and, more especially by Stas,[5] whose work will for ever remain a model of care and accuracy; and although many chemists, down to the present day, have devoted themselves to redetermining the values of equivalents and atomic weights with increasing precision and by the most refined methods, the atomic weights of most of the then known elements had been determined with considerable accuracy by the sixties of last century. These values of the atomic weights, checked by the method of Cannizzaro and by the law of Dulong and Petit, served not only as the foundation of quantitative chemistry but also as the basis of a classification of the elements which has exercised a very powerful influence on the development of chemical science.

With the introduction of the atomic theory and the assigning of definite atomic weights to the different elements, the existence of certain relationships between the chemical nature of the elements and the numerical values of the atomic weights soon came to be recognized. Leaving aside, for the present, the views put forward

[1] *Annalen*, 1862, **123**, 199.   [2] *Annalen*, 1864, **131**, 129.

[3] Edward Turner (1796–1837), Professor of Chemistry, University College, London.

[4] Jean Charles Galissard de Marignac (1817–94), Professor of Chemistry, University of Geneva.

[5] Jean Servais Stas (1813–91), pupil of Dumas and, later, Professor of Chemistry, University of Brussels.

in 1815 by Dr. William Prout, to which consideration will be given later, one finds that in 1817 attention was drawn[1] by Johann Wolfgang Doebereiner (1780–1849), Professor of Chemistry at Jena, to the fact that a curious numerical relationship exists between the atomic [molecular] weights of the oxides of the three chemically related elements, calcium, strontium and barium, known as the alkaline earth metals. Accepting the values of the atomic weights determined by Berzelius, with $O = 100$, Doebereiner pointed out that the atomic [molecular] weight of strontium oxide is nearly equal to the mean of the atomic [molecular] weights of the oxides of calcium and barium. Thus, $CaO = 356 \cdot 019$ and $BaO = 956 \cdot 880$, and $\dfrac{356 \cdot 019 + 956 \cdot 880}{2} = 656 \cdot 449$, which is approximately equal to the atomic [molecular] weight of strontium oxide, $647 \cdot 285$. Doebereiner, further, predicted that the atomic weight of bromine would perhaps be equal to the mean of the atomic weights of chlorine and iodine; and the approximate truth of this forecast was made clear when Berzelius determined the atomic weight of that element.

Doebereiner, at a later time,[2] developed his ideas and showed that it is possible to arrange a number of the elements into "triads" or groups of three elements related to one another in such a way that the atomic weight of one is approximately equal to the mean of the atomic weights of the other two. Such triads are formed by lithium, sodium, potassium; by chlorine, bromine, iodine; by sulphur, selenium, tellurium; and others. Doebereiner's ideas were extended in 1857 by Ernst Lenssen (1837–[?]) who arranged all the then known elements, except niobium, in groups of three;[3] and other arithmetical relationships between the atomic weights of related groups of elements were pointed out by a number of chemists, not only in Europe but also in America.[4] The regularities which were thus brought to light aroused, it is true, considerable interest among chemists and even moved Faraday[5] to express the view that

---

[1] *Ann. Physik*, 1817 (1), **26**, 332.   [2] *Ann. Physik*, 1829, **15**, 301.
[3] *Annalen*, 1857, **103**, 121.
[4] Among others, by Max Joseph von Pettenkofer (1818–1901) in 1850 (*Annalen*, 1858, **105**, 187; Ostwald's *Klassiker*, No. 66); by Dumas in 1851 and 1857 (*Athenaeum*, July 12, 1851; *Ann. Chim.*, 1859 [3], **55**, 129; *Annalen*, 1858, **105**, 74); by John Hall Gladstone (1827–1902) in 1853 (*Phil. Mag.*, 1853 [4], **5**, 313); by Josiah P. Cooke (1827–94) in 1854 (*Amer. J. Sci.*, 1854, **17**, 387); and by William Odling in 1857 (*Phil. Mag.*, 1857 [4], **13**, 423, 480).
[5] *A Course of Six Lectures on the Non-metallic Elements*, 1852.

"we seem here to have the dawning of a new light indicative of the mutual convertibility of certain groups of elements although under conditions which as yet are hidden from our scrutiny"; but they could not, on account of the confusion which then existed regarding equivalents and atomic weights, be other than approximations or be capable of more than a restricted application. Only when the atomic weights had been placed on a secure basis through the exertions of Cannizzaro was it possible for a satisfactory classification of the elements to be built up on the foundations of these revised values.

In 1865 John Alexander Reina Newlands (1837–98), a consulting chemist of Scottish and Italian parentage and a man with a "genial and enthusiastic nature," pointed out[1] that if the elements are arranged in the order of their atomic weights (as had first been done by Gladstone in 1853), with a few slight transpositions, as in the accompanying table,[2] elements belonging to the same group or family usually appear on the same horizontal line.

NEWLANDS'S TABLE OF THE ELEMENTS

H	1	F	8	Cl	15	Co, Ni	22	Br	29	Pd	36	I	42	Pt, Ir	50
Li	2	Na	9	K	16	Cu	23	Rb	30	Ag	37	Cs	44	Tl	53
G	3	Mg	10	Ca	17	Zn	25	Sr	31	Cd	38	Ba,V	45	Pb	54
Bo	4	Al	11	Cr	19	Y	24	Ce, La	33	U	40	Ta	46	Th	56
C	5	Si	12	Ti	18	In	26	Zr	32	Sn	39	W	47	Hg	52
N	6	P	13	Mn	20	As	27	Di,Mo	34	Sb	41	Nb	48	Bi	55
O	7	S	14	Fe	21	Se	28	Ro,Ru	35	Te	43	Au	49	Os	51

From this table it is seen that "the numbers of analogous elements generally differ either by 7 or by some multiple of 7; in other words, members of the same group stand to each other in the same relation as the extremities of one or more octaves in music. Thus, in the nitrogen group, between nitrogen and phosphorus there are 7 elements; between phosphorus and arsenic, 14; and lastly, between antimony and bismuth, 14 also."

In this relationship, which Newlands called the *law of octaves*, there is contained the important idea of a periodic recurrence of properties, but the significance of this fact was not realized. The classification, moreover, was obviously imperfect as it made no allowance for the existence of elements still to be discovered, and

---

[1] *Chem. News*, 1865, **12**, 83; Newlands, *On the Discovery of the Periodic Law*, 1884. An earlier classification of the elements had been given by Newlands in 1864 (*Chem. News*, 1864, **10**, 59).
[2] The numbers in the table are the serial numbers of the elements when arranged according to the value of the atomic weights.

elements which had no apparent connection with one another were found occupying corresponding positions in different octaves. The law, therefore, found no acceptance—met rather with ridicule —at the time,[1] and the truth which lay concealed in it was revealed only at a later date in the light of the greater and more clearly discerned generalization known as the *periodic law*.[2] This law, which was discovered independently and almost at the same time, in 1868-9, by the German chemist, Lothar Meyer, and by the Russian chemist, Mendeléeff, may be stated in the words: *The properties of the elements are a periodic function of (or vary in a periodic manner with) the atomic weight*.[3]

On arranging the elements in the order of their atomic weights, as then known, and keeping in view the possibility of undiscovered elements, a remarkable recurrence of properties was observed. This periodicity was rendered very clear by Lothar Meyer in the case of physical properties, by plotting the values of the physical property, *e.g.*, the atomic volume (volume of 1 gram multiplied by the atomic weight), against the atomic weights of the elements. On joining together the different points, the curve shown in Fig. 1 was obtained. The periodic variation of the atomic volume is thus represented very clearly; and it will be observed that elements of similar chemical properties, *e.g.*, the alkali metals, occur at corresponding points on the different parts of the curve.

[1] In 1887 the Royal Society awarded Newlands the Davy Medal.

[2] A first foreshadowing of the periodic law is found in two memoirs presented to the French Academy of Sciences in 1862 and 1863 by the geologist, Alexandre Émile Béguyer de Chancourtois (1819–86), Professor at the École des Mines, Paris, who arranged the elements spirally on a cylinder in the order of their atomic weights. In spite of a few uncertainties, due to the inaccuracies of the atomic weights, analogous elements, *e.g.*, the halogens, were found to lie on a vertical line and so to occupy similar positions on the spiral or *telluric helix*, as Béguyer de Chancourtois called it. The principle of periodicity of properties although not explicitly stated was, of course, implicit in this arrangement. The work of Béguyer de Chancourtois remained unnoticed until 1889. (See P. J. Hartog, *Nature*, 1889, **41**, 186.)

[3] L. Meyer, *Moderne Theorien*, 1864; *Annalen*, 1870, Supplement **7**, 354; Ostwald's *Klassiker*, No. 68; Mendeléeff, *J. Russ. Chem. Soc.*, 1869, **1**, 60; *Z. Chem.*, 1869, **5**, 405; *Annalen*, 1872, Supplement **8**, 133; Ostwald's *Klassiker*, No. 68. See also *J. Chem. Soc.*, 1889, **55**, 634; F. P. Venable, *The Development of the Periodic Law* (1896).

The question of priority in the discovery of the periodic law gave rise to some controversy. The first clear statement of the periodic law seems to have been *published* by Mendeléeff in 1869, but in 1868 a classification of the elements, similar to that published by Mendeléeff in 1872, had been drawn up, although not published, by Lothar Meyer who, even in 1864, had published tables of the elements which may be regarded as forerunners of the periodic classification. Honours may be regarded as even, and were recognized as such by the Royal Society which awarded the Davy Medal to both scientists in 1882. (See G. Rudorf, *The Periodic System*.)

Not only was periodicity observed in the case of physical properties but it was equally evident in respect of the general chemical properties and of the valency. Thus, when the elements were arranged in order of ascending atomic weights and in accordance with the valence, Lothar Meyer and Mendeléeff showed that they

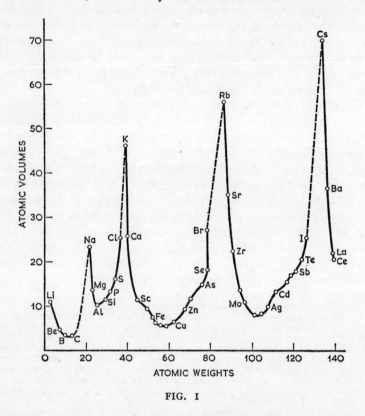

FIG. I

fell into a number of groups or families of related elements, or elements having analogous properties. In this way a natural classification of the elements, depending on the relation between the chemical properties of the elements and their atomic weights, was obtained. This is the so-called *periodic classification of the elements.*

In the table on p. 53, which is essentially the same as one given by Lothar Meyer, hydrogen occupies a unique position in a row by itself, and is followed by a *period* or *series* of seven elements beginning with lithium and ending with fluorine. This is followed by another *short period* beginning with sodium and ending with chlorine, and it will be observed that related elements fall into the same *group*. The elements of this period, it is seen, possess, member by member, properties which are analogous to those of the elements of the preceding period; that is, sodium is similar to lithium, magnesium to beryllium, and so on till we come to chlorine which has properties similar to fluorine. Owing to the periodicity of properties, elements of similar or analogous character fall into the same group or into corresponding places in the various periods.

MENDELÉEFF'S PERIODIC CLASSIFICATION OF THE ELEMENTS
(1872)

	Group I R_2O	Group II RO	Group III R_2O_3	Group IV RH_4 RO_2	Group V RH_3 R_2O_5	Group VI RH_2 RO_3	Group VII RH R_2O_7	Group VIII RO_4
1	$H=1$							
2	$Li=7$	$Be=9.4$	$B=11$	$C=12$	$N=14$	$O=16$	$F=19$	
3	$Na=23$	$Mg=24$	$Al=27.3$	$Si=28$	$P=31$	$S=32$	$Cl=35.5$	
4	$K=39$	$Ca=40$	$-=44$	$Ti=48$	$V=51$	$Cr=52$	$Mn=55$	$Fe=56, Co=59,$ $Ni=59$
5	$Cu=63$	$Zn=65$	$-=68$	$-=72$	$As=75$	$Se=78$	$Br=80$	
6	$Rb=85$	$Sr=87$	$?Yt=88$	$Zr=90$	$Nb=94$	$Mo=96$	$-=100$	$Ru=104, Rh=104,$ $Pd=106$
7	$Ag=108$	$Cd=112$	$In=113$	$Sn=118$	$Sb=122$	$Te=125$	$I=127$	
8	$Cs=133$	$Ba=137$	$?Di=138$	$?Ce=140$	—	—	—	— — —
9	—	—						
10	—	—	$?Er=178$	$?La=180$	$Ta=182$	$W=184$	—	$Os=195, Ir=197,$ $Pt=198$
11	$Au=199$	$Hg=200$	$Tl=204$	$Pb=207$	$Bi=208$	—	—	—
12	—	—	—	$Th=231$	—	$U=240$	—	—

In the Mendeléeff classification emphasis is laid mainly on the relationship between atomic weight and chemical properties, including valency, and this is clearly brought out in the table. Thus, in the case of the two periods referred to, each begins with a highly reactive, metallic or electropositive element (an alkali metal) having a valency of 1, and this is followed by elements having a maximum valency, in the case of their compounds with oxygen, increasing from 2, as shown by beryllium and magnesium, to a valency of 7, as shown by chlorine. In the case of the compounds with hydrogen, however, the valency rises to 4 in carbon and silicon, and falls again to 1 in fluorine and chlorine.

In the case of the short periods, 2 and 3, the seventh element, fluorine or chlorine, is strongly electronegative, and on passing to the element of next higher atomic weight, sodium or potassium, there is a sudden change of properties. When one comes to the fourth period, new types of elements occur and from potassium, which is analogous to sodium, to bromine, which is similar to chlorine, we have a series not of seven but of seventeen elements (including three unknown). Mendeléeff observed, however, that the later members of this *long period* resemble in certain respects (*e.g.*, valency), the members of the short periods; and he therefore divided this long period into two short periods, 4 and 5, joined together by three "transition elements" of only slightly differing atomic weights, iron, cobalt and nickel. Since these elements could form oxides of the type RO_4, and were therefore octavalent, they were placed in a new eighth group. Other long series or periods followed and these, likewise, were divided into even and odd series, joined together by transition elements. In any particular group the elements belonging to the even or to the odd series are more closely related than are the elements of the even to those of the odd series. The groups were therefore divided into two sub-groups, *e.g.*, potassium, rubidium, caesium and copper, silver, gold. This idea of sub-groups was unnecessarily extended by Mendeléeff, so that he placed sodium in the sub-group with copper, and fluorine with manganese. In order, moreover, to retain his succession of even and odd series, Mendeléeff inserted series 9, to be made up of elements still to be discovered. Later investigation has not justified this.

Not only does one find a general change in properties as one passes from left to right along a series, but within the groups and sub-groups one also finds a gradation of properties with increase of atomic weight. Thus, in the case of the alkali metals the melting point falls as the atomic weight increases; and in the case of the halogen elements the colour varies from pale yellow to greenish-yellow, reddish-brown and violet.

In order that related elements might fall into the same group, Lothar Meyer and Mendeléeff found it necessary to leave a number of gaps in their tables of the elements; and these gaps, Mendeléeff predicted, would one day be filled through the discovery of elements unknown at that time. Moreover, the changes of properties which are found as one passes from member to member of a series or of a group are fairly regular, and Mendeléeff therefore, with the confidence which comes from a recognition of the value and meaning of a scientific law, ventured to predict the properties which those hitherto unknown elements would be found to possess when they were discovered. Thus, in the case of the unknown element, referred to as "ekasilicon,"[1] for which a gap had been left in group 4, series 5, the properties which Mendeléeff foretold in 1871 were

[1] The prefix *eka* is the numeral 1 in Sanskrit.

found with an astounding degree of accuracy in the element germanium which was discovered in 1886 by the German chemist, Clemens Alexander Winkler (1838–1904). Here is the comparison between forecast and fact:

	"Ekasilicon"	Germanium
Atomic weight	72	72·6
Specific gravity	5·5	5·469
Specific gravity of oxide	about 4·7	4·703

In other cases, also, prediction was made with equal success.

Although certain anomalies are met with in the Mendeléeff classification, and although the classification does not perfectly reflect the properties of the elements, the periodic law has served to co-ordinate the different elements and to bring order and harmony into what would otherwise have been a disarray of isolated facts. As Mendeléeff said in his Faraday Lecture:[1] "Before the promulgation of the periodic law the chemical elements were mere fragmentary, incidental facts in Nature; there was no special reason to expect the discovery of new elements, and the new ones which were discovered from time to time appeared to be possessed of quite novel properties. The law of periodicity first enabled us to perceive undiscovered elements at a distance which formerly was inaccessible to chemical vision; and long ere they were discovered new elements appeared before our eyes possessed of a number of well-defined properties."

Not only were the gaps in the table of the elements a constant reminder that elements still remained to be discovered, but they guided the investigator towards their discovery. The Mendeléeff classification, also, could be used in order to check the values of atomic weight[2] and valency. Thus, in the case of beryllium, opinion was divided as to whether the oxide has the formula BeO or Be_2O_3. In the latter case the element would be trivalent and the atomic weight would be 14·1. The metal would then fall into the table beside nitrogen, but there was no room for it there. On

---

[1] *J. Chem. Soc.*, 1889, **55**, 634.
[2] Since tellurium must, on account of its chemical relations with selenium, be placed in group VI before iodine, Mendeléeff declared that the value of the atomic weight, 128, as determined by Berzelius must be wrong, and he reduced it to 125. The later and more accurate determinations of the atomic weights of tellurium and of iodine, however, have shown that Mendeléeff was wrong and that the atomic weight of tellurium is *higher* than that of iodine. This is one of the anomalies of the Mendeléeff classification of which the explanation will be given later.

the other hand, if the oxide has the formula BeO, the atomic weight of the metal would be 9·4 and its valency 2. The element therefore falls quite naturally into the second group between lithium and boron.

Although the great importance of the periodic law was not realized for some time, a number of chemists, among whom may be mentioned Thomas Carnelley (1852–90), Professor of Chemistry successively at Sheffield, Dundee and Aberdeen, later took up the study of the relations between the atomic weights and the properties, more especially the physical properties of the elements and of their compounds.[1] The applications of the law were thereby greatly extended, and it was shown that not only the properties of the elements but also the properties of the compounds are periodic functions of the atomic weights.

If the periodic law had important practical applications, the recognition of its essential validity also greatly stimulated speculation and investigation into the genesis of the elements, for as Lord Salisbury said:[2] "The discovery of co-ordinate families dimly points to some identical origin, without suggesting the method of their genesis or the nature of their common parentage." On these questions much light was thrown by the experimental investigations of the closing decades of the nineteenth century and the opening decades of the twentieth century, and out of these there developed the theory of the electronic constitution of matter by means of which the periodic law receives interpretation. The discussion of this, however, belongs to a later point.

[1] *Phil. Mag.*, 1879 [5], 7, 305.
[2] Presidential Address to the British Association, 1894.

CHAPTER FOUR

STEREOCHEMISTRY

THE doctrine of valency and the theory of molecular structure or constitution put forward by Couper and by Kekulé in 1858 afforded, it was thought, a satisfactory basis for the future development of organic chemistry. By that theory it was possible to account for the existence of isomeric compounds and one could represent the structure or mutual linking together of the atoms in the molecule by means of the formulae based on that theory and the doctrine of valency. It was not long, however, before the insufficiency of the Couper-Kekulé theory became apparent through the discovery that, in some cases, the number of isomeric compounds is greater than that theory allows. A new kind of isomerism was discovered with which that theory was incapable of dealing, an isomerism which manifested itself in the property known as *optical activity*.

Towards the close of the seventeenth century, the Dutch physicist, Christian Huygens (1629–95), observed that when light is passed through a crystal of Iceland spar, the waves of light "acquire a certain form and disposition" to which, in the early nineteenth century, the term *polarization* was applied. In a ray of polarized light, the ethereal vibrations which propagate the light and which ordinarily take place in all directions at right angles to the path of the ray are brought into one plane. In 1813 it was discovered by the French physicist, Jean Baptiste Biot (1774–1862)[1] that when polarized light is passed through a plate of quartz cut perpendicularly to the axis, the plane of polarization is rotated or twisted by an amount which is proportional to the thickness of the plate; and the rotation is, in some cases, to the right, in other cases to the left. Later, in 1815, Biot also found[2] that a rotation of the plane of polarization is produced when polarized light is passed through oil of turpentine, oil of laurel, oil of lemon, or through a

[1] *Mém. Inst.*, 1812, **13**, 1.
[2] *Bull. Soc. Philomath*, 1815, p. 190.

solution of camphor in alcohol. Substances which possess this property of rotating the plane of polarized light are now said to be *optically active*.

It is interesting to note that Biot observed that the amount of rotation depends also on the wavelength of the light used. Recently, some 150 years after the original discovery, C. Djerassi and others

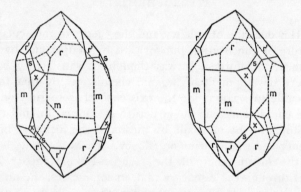

FIG. 2

QUARTZ CRYSTALS

have used this phenomenon to investigate a number of stereochemical problems. The method, known as optical-rotatory dispersion, has been very useful in determining the absolute configuration of a range of organic compounds, but especially in the field of steroid chemistry.[1]

While the rotatory power of quartz, which is shown only by the crystal and which disappears on fusion, could be explained as due to a helicoidal arrangement of the molecules in the crystal, as the French physicist Augustin Jean Fresnel (1788–1827) suggested,[2] the rotatory power of compounds in the liquid state, in which no fixed arrangement of the molecules is possible, must, as Biot concluded, be due to the molecular configuration or arrangement of the atoms in the molecule.

It had, moreover, been observed by the French mineralogist René Just Haüy (1743–1822), that some specimens of quartz crystals show hemihedral faces, and these hemihedral faces are disposed sometimes to the right and sometimes to the left (Fig. 2),

---

[1] *Endeavour*, 1961, **20**, 138.
[2] This spiral arrangement of the molecules has been confirmed by the X-ray examination of quartz crystals. See W. H. Bragg, *Proc. Roy. Soc.*, 1914, A, **89**, 575; Bragg and Gibbs, *ibid.*, 1925, A, **109**, 405; Gibbs, *ibid.*, 1926, A, **110**, 443.

the two forms being enantiomorphous and related to each other as object and mirror-image; and in 1820 (Sir) John F. W. Herschel (1792–1871) correlated the rotatory power of quartz crystals with the presence of hemidehral faces.[1]

The study of optical activity as a *molecular* property was inaugurated by the brilliant discoveries of Louis Pasteur in 1848 in connection with tartaric acid and the tartrates.[2] Two isomeric forms of tartaric acid were then known, the tartaric acid which was discovered in 1769 by Scheele and the optical activity of which in aqueous solution had been observed by Biot[3] in 1835, and paratartaric or racemic acid,[4] reference to which was first published in 1819 by Johann Friedrich John[5] under the name "Säure aus den Voghesen." This acid had been separated from argol by a wine manufacturer, Kestner by name, at Thann in the Vosges, and was shown by Gay-Lussac in 1826 and by Berzelius in 1830 to have the same composition as tartaric acid. It was later found to be optically inactive.

On carefully examining the crystals of tartaric acid and of the tartrates, Pasteur observed that all of them showed hemihedral faces, similarly disposed; but hemihedry was not exhibited by any of the racemates. In harmony with the view suggested by Herschel, the occurrence of hemihedral faces in the crystals of tartaric acid and the tartrates was regarded by Pasteur as the outward and visible sign of optical activity.[6]

Another discovery of great importance was made by Pasteur. In 1844 Mitscherlich had stated[7] that the crystals of sodium ammonium tartrate and sodium ammonium racemate were identical, although solutions of the former were found to be active but those of the latter were inactive. Since this statement was contrary to his views on the relation between crystalline form and optical activity, Pasteur examined these salts and found, it is true, that the crystals of the tartrate resembled the other tartrates which he had

---

[1] *Trans. Cambridge Phil. Soc.*, 1822, **1**, 43.

[2] *Compt. rend.*, 1848, **26**, 535; *Ann. Chim.*, 1848 [3], **24**, 442. See also *Œuvres*, Vol. I; *Alembic Club Reprints*, No. 14.

[3] *Compt. rend.*, 1835, p. 457.

[4] The name *racemic acid* (from Lat. *racemus*, a bunch of grapes) was introduced by Gay-Lussac in 1828, and *paratartaric acid* by Berzelius in 1830. (A. Findlay, *Nature*, 1937, **140**, 22.) For a full account of the history of racemic acid see M. Delépine, *Bull. Soc. Chim.*, 1941, **8**, 463.

[5] *Handwörterbuch der allgemeinen Chemie.*

[6] The view held by Pasteur that every optically active compound would, in the crystalline state, exhibit hemihedral faces has not been generally confirmed.

[7] *Compt. rend.*, 1844, **19**, 720.

examined in possessing hemihedral faces arranged in a similar manner. The crystals which were obtained from a solution of the inactive racemate, at the ordinary temperature, were also found, contrary to expectation, to have hemihedral faces; but these hemihedral faces instead of all being disposed in the same way, as in the case of the tartrates, were inclined, some towards the right and some towards the left (Fig. 3). On carefully separating the two sets of crystals and on examining their solutions, Pasteur observed,

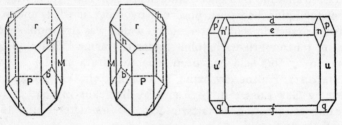

FIG. 3

HEMIHEDRAL CRYSTALS OF d- AND l-SODIUM AMMONIUM TARTRATE, AND HOLOHEDRAL CRYSTAL (RIGHT) OF SODIUM AMMONIUM RACEMATE

"with no less surprise than pleasure," that the crystals with right-handed hemihedry gave solutions which rotated the plane of polarized light to the right, whereas the crystals with left-handed hemihedry gave solutions which rotated the plane of polarization to the left.[1] When equal amounts of the two sets of crystals were dissolved, a solution was obtained which was quite inactive.

By this momentous discovery a new kind of isomerism was revealed.[2] The two salts into which the sodium ammonium race-

[1] The success of Pasteur's experiment may be regarded as due to a happy chance. It was shown at a later date that had the crystallization been allowed to take place at a temperature above 27° C., holohedral crystals of the racemate would have been deposited. (See A. Scacchi, *Rend. Accad. Sci. Fis. Mat.*, Napoli, 1865, **4**, 250; G. Wyrouboff, *Bull. Soc. Chim.*, 1884 [2], **41**, 210; *Compt. rend.*, 1886, **102**, 627; van't Hoff and C. van Deventer, *Z. physikal. Chem.*, 1887, **1**, 165; van't Hoff, H. Goldschmidt and W. P. Jorissen, *ibid.*, 1895, **17**, 49.)

[2] It is related that Pasteur, on making this discovery, rushed from his laboratory and, meeting the lecture assistant in physics, embraced him, exclaiming: "I have just made a great discovery! I have separated the sodium ammonium paratartrate into two salts of opposite action on the plane of polarization of light. The dextro-salt is in all respects identical with the dextro-tartrate. I am so happy and overcome by such nervous excitement that I am unable to place my eye again to the polarization apparatus." The importance attaching to the discovery is testified to by the fact, related by Pasteur, that when the separation of the racemate (paratartrate) into the two optically active acids had been demonstrated to Biot, the illustrious old man, very visibly affected, took him by the arm and said: "My dear child, I have so loved the sciences throughout my life that this makes my heart leap with joy."

mate had been separated by crystallization were found to be identical not only in their chemical but also in their physical properties, except in the arrangement of the hemihedral faces occurring on their crystals and in the property of rotating the plane of polarized light to an equal extent but in opposite directions. From these two salts Pasteur obtained two tartaric acids, one dextrorotatory and identical with the already known acid from grape juice, the other, hitherto unknown, having the power in solution of rotating the plane of polarized light to the left. On mixing in solution equal quantities of these optically active isomers, crystals of the inactive racemic acid were obtained, thereby showing that this acid is an equimolecular compound[1] of the two optically active acids.

Not only was Pasteur able to show, by his experiments with sodium ammonium racemate, that racemic acid could be resolved or separated into the two optically active tartaric acids, but he succeeded also in bringing about the racemization of dextrotartaric acid by the partial conversion of the dextro- into the laevo-acid. This conversion he effected by heating to 170° the salt of dextro-tartaric acid with the alkaloid cinchonine.[2] In this process, also, another isomeric tartaric acid, now known as *mesotartaric acid*, was formed, an acid which, like racemic acid, is optically inactive but which, unlike racemic acid, cannot be resolved into two oppositely active isomers.

The resolution of racemates into the optically active antipodes by crystallization, as in the case of sodium ammonium racemate, is applicable only in a few cases, but Pasteur introduced other methods of resolution which have played and continue to play an important part in the investigation of optically active compounds. One of these methods depends on the fact that although optically active isomers yield identical compounds, apart from hemihedrism and optical rotatory power, when combined with substances which are optically inactive, the compounds which are formed with optically active substances may show very marked differences in properties, *e.g.*, solubility. When, therefore, racemic acid was neutralized by an optically active alkaloid, and the solution allowed to

[1] Investigation has shown that racemic isomers may be compounds or mixtures (homogeneous or heterogeneous) of the active isomers in equimolecular proportions.
[2] *Compt. rend.*, 1853, 37, 162; *Œuvres*, I, 258. This racemization of an optically active compound is a general phenomemon, and takes place with varying degrees of ease—most readily in the case of compounds with only one asymmetric carbon atom. The process may be catalytically accelerated.

crystallize, a separation of the salts of the dextro- and laevo-acid was effected, owing to their different solubilities. This is the most generally useful method of resolving racemic compounds, and was greatly improved, in the case of racemic bases, by the introduction in 1899 by W. J. Pope (1870–1939), of the highly optically active camphor sulphonic acids,[1] and by the use of non-ionizing solvents in order to diminish the risk of racemization. Thereby such a vigorous fillip was given to the study of stereochemical problems that, in a few years, a number of different groups of optically active compounds were discovered (p. 265).

Moreover, Pasteur showed that resolution of the racemic acid can be brought about by the selective action of living organisms or of enzymes, which are the products of living cells. In the course of his investigations Pasteur had been impressed by the fact that whereas all artificial products of the laboratory and all mineral species are molecularly symmetric, most natural organic products— starches, sugars, tartaric acid, malic acid, morphine, quinine, etc. —the essential products of life, are molecularly asymmetric, and can exist as isomers which are not superposable on their mirror-images. Recognizing, therefore, the existence of asymmetric forces associated with the living organism, Pasteur brought about fermentation in a solution of ammonium racemate by means of yeast, and found that the naturally occurring dextro-rotatory tartaric acid was destroyed whereas the laevo-acid, an artificial product, was not attacked. The laevo-acid can in this way be obtained from the racemic acid.

The biochemical method is important but restricted in its application, and involves the destruction of half of the substance.[2]

With the discovery of the isomeric, optically active tartaric acids, there arose the problem of explaining their existence; and Pasteur put forward an hypothesis based on the crystallographic relations. All material things fall into two classes according as the image of the object, formed in a mirror, is or is not superposable on the object. In the case of a cube, for example, which is a symmetrical object, the image formed in a mirror is identical with the object and is superposable on it; but in the case of a screw or a hand, which have no plane of symmetry, the image is not superposable on the object. The mirror-image of a right hand represents a left hand,

[1] *J. Chem. Soc.*, 1899, **75**, 1105; Pope and Peachey, *ibid.*, p. 1066.
[2] Use has also been made of differences in the velocity of esterification and of hydrolysis of esters of a racemic acid with optically active alcohols (W. Marckwald and A. McKenzie, *Ber.*, 1899, **32**, 2130).

and a left hand cannot be superposed on a right hand. A hand, therefore, represents an asymmetric object which can exist in two distinct so-called *enantiomorphous* forms. Similarly, the crystals of the two optically active tartaric acids, with their hemihedral faces are found to be enantiomorphous; each represents the non-superposable mirror-image of the other. Pasteur, therefore, regarding the crystalline form as the visible manifestation of the internal structure, was led to the conclusion that the *molecular* structures of the two active tartaric acids are asymmetric and enantiomorphously related to each other as object and non-superposable mirror-image. "Are the atoms of the dextro-acid," he asked, "grouped according to the spirals of a dextrogyrate helix or placed at the summits of an irregular tetrahedron, or disposed according to some definite asymmetric grouping or other? We cannot answer these questions. But about one thing there cannot be any doubt, the atoms are arranged in some asymmetric grouping with a non-superposable mirror-image." Although Pasteur could not, in the existing state of knowledge, develop his views into a definite theory of chemical structure,[1] he introduced into chemistry a conception of extraordinary importance and fruitfulness, the conception of *molecular asymmetry*.[2] In this conception there is, of course, an implied *spatial* arrangement of the atoms, an idea which, although implicit in the atomic theory of Dalton, had received very little consideration by chemists, who were chiefly concerned with the *composition* of compounds. As early as 1808, it is true, William Hyde Wollaston (1766–1829) stated the opinion[3] that, as knowledge grew, it would be necessary to acquire a geometrical conception of the relative arrangement of the atoms in all the three dimensions of solid extension, and similar views were expressed by others from time to time; but, for the most part, chemists pursued the investigation of the constitution of molecules or the arrangement of the atoms in a plane, and did not trouble, did not in fact need to

---

[1] The helical or spiral arrangement of the atoms has been detected by Astbury (*Proc. Roy. Soc.*, 1923, A, **102**, 506; *Nature*, 1924, **114**, 122) in the structure of crystalline tartaric acid, and the tetrahedral arrangement of atoms or groups around a carbon atom is now generally accepted.

[2] Certain writers following F. G. Mann and W. J. Pope (*J. Soc. Chem. Ind.*, 1925, **44**, 833, 1225), draw a distinction between asymmetric figures, defined as possessing no axis, plane or centre of symmetry, and dissymmetric figures which may contain an axis of symmetry but not a centre of symmetry or plane of direct symmetry. In the latter case enantiomorphous isomerism, in which the isomers stand in the relation of object to non-superposable mirror image, may occur. This distinction has not been generally adopted.

[3] *Phil. Trans.*, 1808, p. 96.

trouble, about the *configuration* of molecules, or the arrangement of atoms in space. Their investigations culminated, as we have seen, in the structural theory and formulae of Couper and Kekulé.

For some time after the investigations of Pasteur on the tartaric acids, no other cases of "physical isomerism,"[1] necessitating a consideration of the spatial arrangement of the atoms in the molecule, became known; but in 1872 Johannes Wislicenus showed that there are two lactic acids which, from their chemical behaviour, must be represented by the structural or constitutional formula, $CH_3 \cdot CH(OH) \cdot COOH$.[2] One of these acids, ordinary or fermentation lactic acid, first isolated from sour milk by Scheele in 1780, is optically inactive, but the other isomer, paralactic or sarcolactic acid, discovered in 1808 by Berzelius in the juice of meat, is optically active. To explain this isomerism the structural formulae of Couper and of Kekulé were clearly inadequate, just as they were inadequate to explain the isomerism of the tartaric acids, and one had to assume, as Wislicenus pointed out,[3] that such cases of isomerism, for which he suggested the term *geometrical* in place of *physical* isomerism, are due to a difference in the arrangement of the atoms in space.

Stimulated by the experimental investigations of Wislicenus and

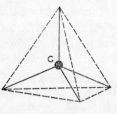

FIG. 4

the need of accounting for the existence of optically active isomers, J. H. van't Hoff and J. A. Le Bel, although guided by different considerations, developed, independently and almost at the same time, the idea of a geometrical spatial arrangement of the atoms and the conception of molecular asymmetry into a consistent theory of molecular structure, based on the so-called tetrahedral carbon atom.[4] Thereby were laid the theoretical foundations of a new chemistry, a "chemistry in space" or *stereochemistry*.[5]

[1] A term applied by L. Carius to cases in which isomerism is manifested only in physical, not in chemical, properties (*Annalen*, 1863, **126**, 217). Carius considered that "it is conceivable that in the formation of physical isomers, differences of condition may cause the production of substances with the same arrangement of atoms within the molecule, but with a different aggregation of their molecules; and that thereon depends the difference in their properties."

[2] *Annalen*, 1873, **167**, 302.   [3] *Ber.*, 1869, **2**, 620.

[4] Van't Hoff, *Voorstel tot uitbreiding der tegenwoordig in de scheikunde gebruikte structuurformules in de ruimte*; *Arch. néer.*, 1874, **9**, 445; *Bull. Soc. Chim.*, 1875, **23**, 295; *La Chimie dans l'Espace* (1875). Le Bel, *Bull. Soc. Chim.*, 1874, **22**, 337. Van't Hoff's Dutch pamphlet was dated September 5th, 1874; Le Bel's paper in the *Bulletin de la Société chimique de Paris* was published in November of the same year.

[5] From στερεός (stereos). solid. The term stereoisomerism was coined by Victor Meyer in 1888.

Since experiment had shown that the four valencies of the carbon atom are equivalent, it follows that they must be symmetrically arranged, either in a plane or in space. If, however, they are symmetrically arranged in a plane, the existence of two isomers would be possible in the case of a compound such as CH_2Cl_2,

represented by the formulae $H — \overset{\displaystyle Cl}{\underset{\displaystyle H}{|\ \ C\ \ |}} — Cl$ and $Cl — \overset{\displaystyle H}{\underset{\displaystyle H}{|\ \ C\ \ |}} — Cl$. No

such isomerism has been observed. To satisfy the chemical requirement of the equivalence of the four carbon valencies, van't Hoff adopted the view that these four valencies are directed towards the corners of a regular tetrahedron at the centre of which the carbon atom is supposed to lie[1] (Fig. 4). If, therefore, one imagines the four atoms or groups, with which the carbon atom can unite, situated at the corners of a tetrahedron, then so long as two, at

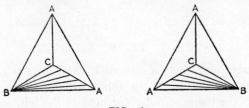

FIG. 5

[1] Although the general conclusions reached by van't Hoff and Le Bel were the same, van't Hoff, influenced by the work of Wislicenus, was guided by geometrical considerations and Kekulé's law of the quadrivalency of carbon, while Le Bel, following Pasteur, was guided by considerations of molecular asymmetry. In the mind of the former arose the idea of the tetrahedral carbon atom; in the mind of the latter, the idea of the asymmetric carbon atom. Although Le Bel accepted the idea of a tetrahedral "environment" for the carbon atom, a conception arrived at from a consideration of how four univalent radicals attached to a quadrivalent carbon atom should arrange themselves under conditions of equilibrium, he did not accept as necessary the view that the carbon valencies are definitely directed towards the corners of a regular tetrahedron. The tetrahedron might be deformed. By means of the tetrahedral carbon atom one can predict the number of possible isomers; the conception of the asymmetric carbon atom enables us to explain optical activity.

The idea of the tetrahedral grouping of radicals round a carbon atom was not entirely new. Wollaston, in 1808, had made such a suggestion and so also had Pasteur in 1860. In 1867 Kekulé (Z. Chem., 1867, 3, 217) pointed out that the carbon atom might be represented by a sphere furnished with four rods to represent the valencies, and that these rods should be placed in the direction of four hexahedral axes ending in the faces of a tetrahedron, and Emanuele Paternò (Giornale di Scienze nat. ed econom., 1869, 5, 117; Gazzetta, 1919, 49, 341), in 1869, made use of models in which the four atoms or groups attached to a carbon atom are arranged tetrahedrally.

least, of these atoms or groups are the same, the molecule, represented as a tetrahedron, will possess a plane of symmetry and its mirror-image will be superposable on and will be identical with the original; for it is clear that if the right-hand tetrahedron in Fig. 5 is turned through an angle of rather more than 90°, on the corner B as a pivot, it will become identical in disposition with the left-hand tetrahedron. If, however, the four atoms or groups attached to the carbon atom are all different, the molecule becomes asymmetric and gives a mirror-image which is no longer superposable on the original (Fig. 6). It is clear, therefore, that two isomeric forms are now possible, for on viewing these tetrahedra from a similar position, it is seen that the groups BCD are arranged from left to right in the one case, and from right to left in the other. If one of

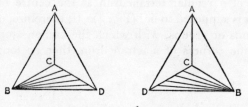

FIG. 6

these tetrahedra represents a molecule which rotates the plane of polarized light to the right, the other will represent a molecule which rotates the plane of polarized light to the left.

On the basis of his theory van't Hoff was able to calculate the number of possible isomers which could exist in the case of compounds containing one or more asymmetric carbon atoms. In the case of compounds in which only single bonds are present and in which all the atoms or groups attached to the asymmetric carbon atoms are different, the number of optically active isomers is given by 2^n, where n is the number of asymmetric carbon atoms. Besides the optically active isomers it will also be possible to have half their number of inactive racemic compounds. When the atoms or groups attached to the asymmetric carbon atoms are not all different, the number of possible optical isomers will be reduced. In the

case, for example, of tartaric acid COOH — C — C — COOH, in

with H, H on top and OH, OH on bottom of the two central carbon atoms.

which each of the asymmetric carbon atoms is attached to the same four groups, namely, ·H, ·OH, ·COOH and (·CHOH·COOH), the number of optical isomers is only two (not four). In such cases as tartaric acid, however, in which the two-dimensional structural formula is symmetrical, the existence of a so-called internally compensated inactive form (*e.g.*, mesotartaric acid) becomes possible. Thus, representing the configuration of mesotartaric acid by the model (Fig. 7), it will be observed that, when one examines the two halves of the molecule from a similar point of view, the groups H, OH, COOH are arranged in clockwise order in the one half, and in counter-clockwise order in the

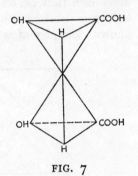

FIG. 7

other half; or, the one half of the molecule is the mirror-image of the other half. The rotation of the one half of the molecule, therefore, is balanced or compensated by the opposite rotation of the other half.

While the presence of an asymmetric carbon atom in a saturated compound, in which only single bonds are present, afforded an explanation of the occurrence of optical activity, it could not be concluded that all compounds are optically active in which an asymmetric carbon atom is present. As we have seen, optically active isomers may unite to form optically inactive racemic compounds; and compounds which are inactive by internal compensation may also exist.

That optical activity can be shown even by the simplest molecules containing an asymmetric carbon atom was proved by W. J. Pope and John Read[1] (1884–1963) who prepared (1914) optically active forms of the compound

The theory of the tetrahedral carbon atom was extended by van't Hoff to the case of compounds containing a double bond, or compounds of the type $RR'C = CRR'$. Such compounds can be

---

[1] Professor of Organic Chemistry, University of Sydney, 1916–23; Professor of Chemistry, University of St. Andrews, 1923–63.

represented by two tetrahedra united by a common edge (Fig. 8). Two isomers are possible. In one case the two groups R both lie on the same side of the plane containing the double bond, in the other case they lie on opposite sides. This kind of isomerism is known as *cis-trans* isomerism,[1] the earliest and one of the best-known examples of which is found in maleic and fumaric acids.

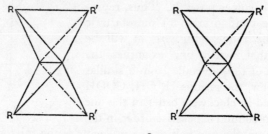

FIG. 8

Since these cis-trans isomers possess a plane of symmetry, neither is optically active.

Geometrical isomerism of the cis-trans type may be found not only in the case of compounds in which free rotation is prevented by the presence of a double bond between two carbon atoms, but also in the case of aldoximes and ketoximes.

$$\left(\begin{array}{ccc} R-C-R' & & R-C-R' \\ e.g. \quad \| & and & \| \\ N\cdot OH & & HO\cdot N \end{array}\right)$$

in which there is a double bond between nitrogen and carbon,[2] and of cyclic compounds, in which groups may lie on the one or the other side of the plane of the ring (*e.g.*, *cis*- and *trans*-hexahydrophthalic acids).[3]

Since molecular asymmetry is the factor essential for the occurrence of optical activity, this property, as van't Hoff pointed out, should be found in the case of compounds of which the molecule as a whole is asymmetric, although an asymmetric carbon atom may not be present. Thus, van't Hoff predicted that compounds of the allene type, $RR'C = C = CRR'$ (Fig. 9), the molecules of which, according to the theory of the tetrahedral carbon atom,

[1] A. von Baeyer, *Annalen*, 1890, **258**, 145.
[2] A. Hantzsch and A. Werner, *Ber.*, 1890, **23**, 11.
[3] A. von Baeyer, *loc. cit.*

possess no plane of symmetry, should be capable of existing in optically active forms, and this prediction was confirmed by Peter Maitland and W. H. Mills (1873–1959), in the laboratories of the

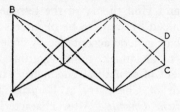

FIG. 9

University of Cambridge.[1] Analogous compounds which contain a six-membered ring in place of a double bond have also been resolved into optically active isomers. Thus, in the case of the compound 1-methylcyclohexylidene-4-acetic acid,

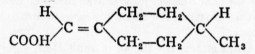

the molecule of which is represented by the model shown in Fig. 10, the groups H and CH_3 on the right lie in a different plane

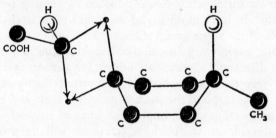

FIG. 10

from the groups H and COOH on the left. The molecule has, therefore, no plane of symmetry and its mirror-image, obtained by transposing the groups H and COOH on the left, is not superposable on, and therefore not identical with, the original molecule.

[1] *Nature*, 1935, **135**, 994; 1936, **137**, 542; *J. Chem. Soc.*, 1936, p. 987; Kohler, Walker and Tishler, *J. Amer. Chem. Soc.*, 1935, **57**, 1743.

William Hobson Mills was Head of the Chemistry Department, Northern Polytechnic Institute, London, from 1902 to 1912, and from 1912 to 1938, a member of the staff of the University Chemistry Department at Cambridge, receiving the title of Reader in Stereochemistry in 1931. He was Baker Lecturer, Cornell University, 1937, and President of Jesus College, Cambridge, 1940–48.

This compound has been obtained in two optically active forms.[1]

Other types of asymmetric molecules, giving rise to optically active isomers, have also been prepared.

Although the van't Hoff theory of the tetrahedral carbon atom was at once recognized by some to be of interest and value, it did not find immediate or general acceptance by chemists. It was even made the subject of an attack, more vigorous than effective, by that ardent opponent of structural chemistry, Hermann Kolbe, at that time Professor of Chemistry at Leipzig, and editor of the *Journal für praktische Chemie*. After characterizing van't Hoff's memoir as "teeming with freaks of the imagination," Kolbe wrote: "A certain Dr. J. H. van't Hoff, who holds a post at the veterinary school at Utrecht, has no taste, apparently, for accurate chemical investigations. He has thought it more convenient to mount Pegasus (evidently hired at the veterinary school) and to proclaim in his *La Chimie dans l'Espace* how, during his bold flight to the top of the chemical Parnassus, the atoms appeared to him to be arranged in cosmic space."

In spite of this attack, the theory of van't Hoff, surely if slowly, made for itself a place in the general body of chemical theory; and it achieved its first great and outstanding triumph, some thirteen years after its birth, when it was adopted as guiding principle by Emil Fischer in his laborious and magnificent researches into the compounds of the sugar group (Chapter IX). Since then the theory of molecular asymmetry has shown itself capable of extension in numerous directions to elements other than carbon and has satisfied, in a very remarkable manner, the requirements of chemical research. It has inspired the work of many chemists and has guided them in the erection of the imposing edifice of stereochemistry, to the later developments of which reference will be made in the sequel; and the essential truth of the underlying hypothesis, the tetrahedral arrangement of the four carbon valencies, has been proved by the recent and valuable methods of X-ray analysis. The rise and development of stereochemistry, which followed on the enunciation of the van't Hoff-Le Bel theory, may, in fact, be regarded as one of the most significant and characteristic features of the chemistry of the past hundred years, and as we shall learn in Chapter XII it has developed into and become merged in a wider stereochemistry which embraces all chemical compounds.

[1] Perkin, Pope and Wallach, *J. Chem. Soc.*, 1909, **95**, 1789.

CHAPTER FIVE

THE RISE AND DEVELOPMENT OF PHYSICAL CHEMISTRY
IN THE NINETEENTH CENTURY—I

ALTHOUGH certain of the physical properties of substances had no doubt been determined by chemists at an early time, chemistry remained, down to nearly the end of the eighteenth century, mainly a qualitative science. It accepted as its chief aim and motive, as expressed by Boyle in *The Sceptical Chymist*, "to find out nature, to see into what principles things might be resolved, and of what they were compounded," and was concerned, therefore, for the most part with a description of the composition, preparation and transformations of substances. With the introduction, however, of the balance as an indispensable instrument of chemical investigation and with the enunciation of the law of conservation of mass by Lavoisier, the foundation was laid for the development of chemistry as a quantitative and exact science; and the interpretation of the quantitative relations between substances became possible through the introduction of the atomic and molecular theories by Dalton and by Avogadro. Soon after these theories had been put forward, Dulong and Petit (p. 42) studied the relations between the specific heat of the solid elements and their atomic weights and enunciated the law of the constancy of atomic heats in 1819; and Faraday, in 1834, as a result of his experiments, was able to calculate electrochemical equivalents and to state the quantitative laws of electrolysis. Such investigations and conclusions, however, as well as those of Mitscherlich on isomorphism (1818–19), were isolated and unsystematic, and it was not till 1842 that the systematic investigation of the relations between physical properties and chemical composition and constitution was initiated by Hermann Kopp. Such investigations quickly increased in number, scope and variety, for the rapid development of organic chemistry after the first two or three decades of the nineteenth century furnished abundant material for the determination of physical properties; moreover, the discovery of homologous series and the evolution of the theories of chemical constitution made it

71

possible to bring the physical properties of substances into relation not only with chemical composition but also with molecular structure or constitution. The regularities which were thereby brought to light could, in their turn, be used to help in deciding the constitution of new compounds.

The scope and range of physical chemistry became greatly extended in the course of the century. With the development of the kinetic theory (1856–60), it became possible to interpret and coordinate the laws and properties of gases; through the enunciation of the law of mass action in 1867, chemists were enabled to obtain a greater insight into the processes of chemical change, and with the discovery of the law of conservation of energy (1842) and of the second law of thermodynamics (1867), chemical affinity acquired a definite meaning and became susceptible of exact measurement. The relations, moreover, between the energy of chemical reaction and other forms of energy—heat energy, light energy, electrical energy—could be studied on a basis of exact reasoning; and the recognition of the importance of energy and the application of the laws of thermodynamics to the study of chemical processes were probably the factors of greatest moment in the development of physical chemistry in the nineteenth century. The main lines and essential features of this development we shall now attempt to trace.

In 1811, as has already been mentioned, Amedeo Avogadro postulated that equal volumes of different gases contain the same number of molecules; and since, according to Boyle's law (1662) and Gay-Lussac's law (1802), the product of pressure and volume for a given mass of gas is proportional to the absolute temperature,[1] or $pv = r\mathrm{T}$, it follows that for equimolecular quantities of different gases under the same conditions of temperature and pressure, the expression pv/T must have the same value. For one gram-molecule of a gas (or the molecular weight expressed in grams), this constant value is represented by R, and one obtains the *general gas equation*, $pv = \mathrm{RT}$. It is clear, therefore, that under given conditions of temperature and pressure, the volume of a gas depends only on the number of molecules (or gram-molecules); or, to use a term introduced by Wilhelm Ostwald on the suggestion of the philosopher, W. Wundt, the volume of a gas is a *colligative* property.

[1] On the basis of the law of Gay-Lussac and of the experimental determinations of the coefficient of expansion of gases, $-273°$ C. becomes the zero point of what is called the *absolute* scale of temperature. This absolute zero of the gas thermometer is practically identical with the absolute zero deduced by Lord Kelvin on the basis of thermodynamics.

By determining the volume of a given mass of gas, or by determining the density of a gas, the molecular weight can, as Avogadro pointed out, be calculated.

In order to account for the laws of gases which were arrived at inductively, the hypothesis was adopted that matter in the gaseous state is composed of particles—the molecules of Avogadro—and that these molecules are in constant motion, moving with great velocity in straight lines, colliding ever and anon with one another and striking against the walls of the containing vessel. It is by virtue of this motion which, according to the hypothesis, is inherent in the molecules, that a gas can diffuse throughout all the space which may be offered to it; and the pressure of a gas is due to the incessant bombardment of the walls of the containing vessel by the molecules. The hypothesis that the particles of a gas are in constant motion had first been suggested in 1678 by Robert Hooke (1635–1703), and was adopted in 1738 by Daniel Bernoulli (1700–82), Professor of Experimental and Speculative Philosophy at Basle; and the mean velocity of motion of the molecules was first calculated, in 1851, by the English brewer and amateur of science, James Prescott Joule (1818–89). The hypothesis was developed and worked out, more especially by the German physicists Karl August Krönig (1822–79)[1] and Rudolf Clausius (1822–88)[2] in 1856 and 1857, and by the English physicist, James Clerk Maxwell (1831–79) in 1860.[3] Starting with this hypothesis it is possible, by applying the laws of mechanics, to deduce the various gas laws; and the hypothesis was thus developed mathematically into a *theory*, the kinetic theory, which co-ordinates not only qualitatively but also quantitatively the various gas laws and the theorem of Avogadro, and so enables one to "explain" the behaviour of a gas. On the basis of this theory also a definition of the temperature of a gas could be obtained, the temperature of a gas being measured by the total kinetic energy of the molecules.

The investigations carried out by various physicists and, more especially, by Émile Hilaire Amagat (1840–1915), Professor of Physics in the Catholic University, Lyons, between 1869 and 1893, showed that all gases deviate to a greater or less extent from the simple law of Boyle when the pressure is varied over a considerable range. At the ordinary temperature, in the case of all gases except

---

[1] *Ann. Physik*, 1856, **99**, 315.   [2] *Ann. Physik*, 1857, **100**, 353.
[3] J. J. Waterston, an English physicist, had already developed the hypothesis in a paper presented to the Royal Society in 1845, but not published till 1892, when it was discovered by Lord Rayleigh in the Society's archives.

hydrogen and helium, and at low temperatures in the case of these gases, the diminution of volume with increase of pressure is at first greater and, at higher pressures, less than corresponds with Boyle's law.[1] The general gas law, $pv = RT$, therefore, applies only to an *ideal* gas, and is valid in the case of actual gases only at very low pressures. These deviations are due to the mutual attraction of the molecules and to the volume occupied by the molecules themselves; and, taking these factors into account, Johannes Diderik van der Waals (1837–1923), in 1873, while a student at the University

of Leyden, deduced the expression $\left(p + \dfrac{a}{v^2}\right)\left(v - b\right) = RT$,

for the relation between the pressure and volume of a gas. In this expression a/v^2 represents the mutual attraction of the molecules and b is a quantity proportional to the actual volume of the molecules.[2] The factors a and b are constants which have different values for different gases. The van der Waals equation is applicable not only to highly compressed gases but also to liquids and represents, in a fairly accurate manner, the relation between pressure, temperature and volume. From it the continuity of the gaseous and liquid states can be deduced.

For the accurate calculation of molecular weights of gases from determinations of the density, account must be taken of the deviations from Boyle's law. This was done, more especially, by Daniel Berthelot (1865–1927),[3] Professor of Physics at the École de Pharmacie, Paris, and by Philippe Auguste Guye (1862–1922),[4] Professor of Chemistry in the University of Geneva. In recent years the methods introduced by them have found considerable application to the determination of exact molecular and atomic weights.

Deductions from the kinetic theory regarding the specific heats of gases have also proved of value in throwing light on the problem of the complexity of gas molecules. Thus, on the basis of the kinetic theory, the maximum value of the ratio of the specific heats of a gas at constant pressure and constant volume, in the case of ideal gases, should be $1 \cdot 67$ for monatomic gases, or gases the molecule

---

[1] Amagat, *Ann. Chim.*, 1881, **22**, 353; 1883, **28**, 456, 464; 1893, **29**, 68, 505; H. Kamerlingh-Onnes and F. M. Penning, *Arch. néerland*, 1923, **6**, 277.

[2] The value of b is, according to different assumptions, equal to 4 times or to $4\sqrt{2}$ times the actual volume of the molecules.

[3] *Sur les Thermomètres à Gaz*, 1903.

[4] *J. Chim. Phys.*, 1905, **3**, 321; 1908, **6**, 769.

of which consists of only one atom, and 1·40 for a diatomic gas. The greater the complexity of the molecule, the smaller, in general, is the ratio. For mercury vapour, experiment gave the value 1·67,[1] and for hydrogen, oxygen and nitrogen, the value 1·4. The conclusions, therefore, regarding the molecular complexity of these gases which were drawn by Avogadro on the grounds of simplicity and adequacy (p. 43), have thus been confirmed.

The kinetic theory enables one not only to obtain a clearer understanding of the behaviour of gases but also to calculate various molecular magnitudes, such as the mean velocity and the mean free path of a molecule; and the values obtained are in harmony with the *law of diffusion of gases* discovered experimentally by Thomas Graham in 1833. Moreover, although Avogadro postulated merely an equal, but unknown, number of molecules in equal volumes of different gases under the same conditions of temperature and pressure, it became possible, on the basis of the kinetic theory, to calculate the actual number of molecules in a given volume of a gas, as was first shown by the Viennese schoolmaster, Joseph Loschmidt (1821–95) in 1865. The number obtained, $2·691 \times 10^{19}$ in one cubic centimetre of a gas at 0° C. and under atmospheric pressure, is called the *Loschmidt number*. Since one gram-molecule of a substance in the state of gas occupies (ideally) the volume of 22,414 cubic centimetres (calculated on the basis of the atomic weight of oxygen equal to 16·00), the number of molecules in this volume, or in one gram-molecule, is $6·031 \times 10^{23}$. This is called the *Avogadro number*. This number is not only a deduction from the kinetic theory of gases but has, in more recent years, been confirmed experimentally in a number of different ways.[2] The hypothesis of Avogadro has become a law.

Further, if one divides the molecular weight of a substance by the Avogadro number, the weight of one molecule of the substance is obtained. Thus, the weight of one molecule of hydrogen is calculated to be $\dfrac{2·016}{6·031 \times 10^{23}} = 3·343 \times 10^{-24}$ gram, and the actual weight of the atom is $1·671 \times 10^{-24}$ gram.

Although men had long been familiar, from the practice of the art of distillation, with the condensation of vapours to liquids, it

[1] A. Kundt and E. Warburg, *Ann. Physik*, 1876, **157**, 353.
[2] See S. E. Virgo, *Science Progress*, 1933, **27**, 634; Millikan, *Ann. Physik*, 1938, **32**, 34, 520.

was not till towards the end of the eighteenth century that the view came to be generally held that even those substances which are gases under ordinary conditions may be liquefied by a lowering of temperature or an increase of pressure. As early as 1799 ammonia and sulphur dioxide had been liquefied by the French chemists, Fourcroy, Guyton de Morveau and others; and in 1805 chlorine and hydrogen chloride were liquefied by Thomas Northmore (1766–1851), classical scholar and amateur of science, in England. In 1823 a considerable number of gases were liquefied by Michael Faraday;[1] but a certain number, such as hydrogen, oxygen and nitrogen, proved refractory. As all attempts to obtain them in the liquid state, even under a pressure of 2,790 atmospheres, proved unsuccessful, these gases were regarded as being uncondensable and were spoken of as *permanent gases*. In 1869 the reason for the failure to liquefy the so-called permanent gases became clear through the work of Thomas Andrews, Professor of Chemistry in Belfast.[2] From a study of the behaviour of carbon dioxide and other gases, Andrews showed that there exists, in each case, a particular temperature, known as the *critical temperature*, such that below this point liquefaction can be brought about if the pressure is sufficiently high, but above this temperature liquefaction cannot occur, no matter how great the pressure may be.[3] Since the critical temperatures of hydrogen, oxygen and nitrogen are $-239 \cdot 0°$, $-118 \cdot 8°$ and $-147 \cdot 1°$ C. respectively, the failure of the early experimenters to achieve the liquefaction of these gases is intelligible.

After the conditions necessary for the liquefaction of gases had thus been established, Raoul Pierre Pictet (1842–1929),[4] in Geneva, and Louis Paul Cailletet (1832–1913),[5] ironmaster at Chatillon-sur-Seine, succeeded in liquefying on a small scale several of the "permanent" gases; and the problem of liquefying these gases on a relatively large scale was solved in 1883 by Sigmund Florenty von Wroblevski (1845–88), Professor of Physics, University of Cracow, and Karol Stanislav Olszevski (1846–1915),[6] Professor of

[1] *Phil. Trans.*, 1823, **113**, 160, 189; 1845, **135**, 155.

[2] *Phil. Trans.*, 1869, **159**, 575; 1876, **166**, 421.

[3] At the critical temperature the vapour pressure and the specific volume of the liquid will have definite values, known as the *critical pressure* and the *critical volume*. By applying van der Waals's equation to a liquid at its critical point a numerical relation is obtained between the values of the critical constants and the a and b of the equation.

[4] *Compt. rend.*, 1877, **85**, 1214, 1220; 1878, **86**, 37, 106.

[5] *Compt. rend.*, 1877, **85**, 1213, 1270.

[6] *Ann. Physik*, 1883, **20**, 243; 1885, **25**, 371; **26**, 134.

Chemistry, University of Cracow, and by (Sir) James Dewar (1842–1923), in London.[1]

By making use of the cooling which takes place when a compressed gas is allowed to expand below a certain temperature (*Joule-Thomson effect*), Dewar succeeded in 1898 in liquefying hydrogen, and Heikle Kamerlingh-Onnes (1853–1926), Professor of Physics in the University of Leyden, in 1908 liquefied the least readily condensable of all the gases, helium (boiling point − 268·7° C.).[2] The same principle was applied in 1895 in the construction of apparatus for the industrial liquefaction of gases by Karl von Linde (1842–1934) in Germany and by W. Hampson in England; and many millions of gallons of liquid air are now produced daily for use in industry and for purposes of refrigeration. The utilization of liquid air in scientific research and for other purposes was made possible, or at least was greatly facilitated, by the introduction by Dewar in 1893 of vacuum-jacketed vessels, the principle of which is now made use of in the well-known *Thermos flasks*.

When one considers the advances made in the more restricted domain of physical chemistry concerned with the relations between physical properties and chemical composition,[3] it may be noted that the law of Dulong and Petit, which was based on determinations of the specific heat of only thirteen elements, sometimes of doubtful purity, was found by later investigators, more especially by the French physicist, Henri Victor Regnault (1810–78)[4] and by Hermann Kopp[5] to be only approximately true; and in the case of certain elements of low atomic weight, *e.g.*, boron, carbon (diamond), and silicon, the values of the atomic heat were found to be so low (2·6, 1·5 and 4·8 respectively) that these elements were regarded as anomalies. An explanation of the anomalous behaviour is to be found in the fact that, as shown by the physicist, Heinrich Friedrich Weber (1843–1912),[6] and confirmed by many workers in more recent times, the atomic heat increases at different rates with rise of temperature. Whereas, in most cases, the atomic heat increases so rapidly that at the arbitrarily chosen room-temperature

[1] *Phil. Mag.*, 1884, **18**, 210; 1892, **34**, 205, 326; 1893, **36**, 328.
[2] Argon was first liquefied in 1895 by Olszevski (*Z. physikal. Chem.*, 1895, **16**, 380), and neon, krypton and xenon were liquefied by Ramsay and Travers in 1901 (*Phil. Trans.*, 1901, A, **197**, 47).
[3] See S. Smiles, *The Relations between Chemical Constitution and some Physical Properties*, 1910.
[4] *Ann. Chim.*, 1840, **73**, 5.   [5] *Annalen*, 1864, Supplement **3**, 1, 289.
[6] *Ber.*, 1872, **5**, 308; *Jahresbericht*, 1874, 64.

a value in the neighbourhood of 6 calories per degree is reached, in the case of carbon, boron, etc., the atomic heat approaches this value only at temperatures much above the ordinary. A theoretical basis for the empirical law was obtained in more recent times by P. Debye.[1]

Although the law of Dulong and Petit is only an approximate one, it proved, especially in the hands of Cannizzaro, to be of great practical value in helping to decide what multiple of the experimentally determined equivalent of a metal, or other solid element, should be taken as the atomic weight.

The experiments of Dulong and Petit were extended to solid compounds by Franz Ernst Neumann (1798–1895)[2] in 1831 and by Kopp[3] in 1864; and as a result it was found that the molecular heat, or the product of specific heat and molecular weight, is approximately equal to the sum of the atomic heats of the constituent elements. This law, known as *Neumann's law*, indicates that the molecular heat of a solid compound is mainly an *additive* property; a property, that is, which depends only on the atomic composition of the molecule.

The discovery of the additive character of molecular heat raised the question whether other physical properties might not similarly be found to be additive; and during the latter half of the nineteenth century, more especially, many workers, led by Kopp, undertook the systematic investigation of different physical properties in order to correlate them with the composition and, as knowledge of molecular structure developed, with the constitution, mainly of organic compounds. It was with this idea of the additive nature of physical properties of compounds in his mind that Kopp commenced his investigations of the molecular volume of liquids, and which he continued during the years 1842 to 1855.[4]

Owing to the variation of the molecular volume with temperature, uncertainty arose as to the temperature at which the comparison of

---

[1] *Ann. Physik*, 1912, **39**, 789; Schrödinger, *Physikal. Z.*, 1919, **20**, 420.
Petrus Josephus Wilhelmus Debye, born in Maastricht in 1884, was successively Professor of Physics in the Universities of Utrecht and Göttingen; in the Technical High School, Zurich; and in the University of Leipzig. In 1935 he was appointed Director of the Kaiser Wilhelm Institute of Physics, Berlin. He was awarded the Nobel Prize for Chemistry in 1936 and was Professor of Chemistry, Cornell University, 1940–50.
[2] *Ann. Physik*, 1831, **23**, 32.
[3] *Annalen*, 1864, Supplement 3, 1, 289.
[4] *Annalen*, 1842, **41**, 79; 1855, **96**, 1, 153, 303. See G. Le Bas, *The Molecular Volumes of Liquid Chemical Compounds*, 1915; van Haaren, *Chem. Weekbl.*, 1942, **39**, 474; Egloff, *Physical Constants of Hydrocarbons*.

the molecular volume of different liquids should be made. Kopp was led to adopt not one temperature, the same for all liquids, but the boiling point of the particular liquid under atmospheric pressure; and, as later investigation showed, this choice was a very fortunate one.[1]

From the values obtained for members of homologous series, Kopp was led to the conclusion that the molecular volume is an additive property and that isomeric compounds have the same molecular volume. Hence, values of atomic volumes of different elements, more especially carbon, hydrogen and oxygen, could be calculated. Numerous later investigations,[2] however, showed that the conclusion drawn by Kopp is not strictly correct, for oxygen in the hydroxyl group has a different atomic volume from the doubly-linked oxygen present in aldehydes and ketones. Branched-chain compounds, also, have a lower molecular volume than their straight-chain isomers, and other variations in the constitution exercise an influence on the molecular volume. The molecular volume of liquids, therefore, is not only an additive but also a *constitutive* property; and when one passes from a consideration of the broad regularities to a more searching investigation of the relations, it is found that these gain greatly in complexity owing to widespread constitutional and structural influences. The usefulness of molecular volume determinations in solving problems of molecular structure is, in consequence, considerably reduced. Use is no longer made of this property.

As in the case of the molecular volume so also in the case of many other physical properties which have been investigated, it is found that not only additive relations exist but constitutive influences also play a part. Of such properties one of the most important is the *molecular refractive power*.

The optical properties of liquids in relation to chemical composition were, at an early period, the subject of investigation. As early as 1805 Pierre Simon de Laplace (1749–1827), who became famous as a mathematician and astronomer, deduced, on the basis of the corpuscular theory of light, that the expression $(n^2 - 1)/d$, where n is the index of refraction and d is the density, should, for light

[1] In 1890 C. M. Guldberg (*Z. physikal. Chem.*, 1890, 5, 374) showed that the boiling points of different liquids under atmospheric pressure are approximately equal fractions of their critical temperatures. Liquids at their boiling points are, therefore, approximately in corresponding states.

[2] See, for example, H. L. Buff, *Annalen*, Supplement 4, 129; F. D. Brown, *Proc. Roy. Soc.*, 1878, 26, 247; T. E. Thorpe, *J. Chem. Soc.*, 1880, 37, 141, 327; R. Schiff, *Ber.*, 1882, 15, 1270; W. Lossen, *Annalen*, 1882, 214, 81.

of a definitive wave length, be constant for a given liquid independently of the temperature, and should have a value characteristic of the particular substance. This deduction did not agree well with experiment, but in 1863 the simpler expression $(n - 1)/d =$ constant, was established empirically[1] by John Hall Gladstone and (Rev.) Thomas Pelham Dale (1821–92), Rector of St. Vedast's, London.

In 1880, on the basis of the electromagnetic theory of light, Hendrik Anton Lorentz (1853–1928), Professor of Theoretical Physics, University of Leyden, and Ludwig Valentin Lorenz (1829–91), Professor of Physics at the Military High School, Copenhagen, deduced independently that the expression

$$\frac{n^2 - 1}{n^2 + 2} \cdot \frac{1}{d}$$

should be constant for a given substance;[2] and this expression for what was called the *specific refractive power* showed itself to be superior to that of Gladstone and Dale in that it is nearly independent even of change of state.

To obtain values which shall be comparable for different substances, it is usual, following Landolt, to consider the property known as the *molecular refractive power*, which is equal to the specific refractive power multiplied by the molecular weight of the substance and is represented by $M(n - 1)/d$ or by $\dfrac{n^2 - 1}{n^2 + 2} \cdot \dfrac{M}{d}$.

The systematic investigation of the specific and molecular refractive power of liquids, which was initiated by Gladstone and Dale[3] and by Hans Landolt,[4] was extended by Julius Wilhelm Brühl[5] (1850–1911), Professor of Applied Chemistry, Technical College, Lemberg, Eugen Conrady[6] and many others. From these determinations it became clear that the molecular refractive power is essentially additive in character, and from a comparison of the molecular refractivities it was possible to calculate a series of

[1] A theoretical basis for this expression was later obtained by H. Dufet (*Bull. Soc. Min.*, 1885, **6**, 261) and by W. Sutherland (*Phil. Mag.*, 1889, **27**, 141).

[2] Lorentz, *Ann. Physik*, 1880, **9**, 641; Lorenz, *ibid.*, **11**, 70.

[3] *Phil. Trans.*, 1863, **153**, 317; Gladstone, *Proc. Roy. Soc.*, 1868, **16**, 439; 1869, **18**, 49; 1881, **31**, 327.

[4] *Ann. Physik*, 1862, **117**, 353; 1864, **122**, 545; **123**, 595.

[5] *Annalen*, 1880, **200**, 139; **203**, 1; *Z. physikal. Chem.*, 1891, **7**, 140; *Ber.*, 1891, **24**, 657, etc.

[6] *Z. physikal. Chem.*, 1889, **3**, 210.

constants representing the atomic refractivities of different elements. By means of these constants the molecular refractivity may be calculated as the sum of the constituent atomic refractivities.

Investigation showed, however, most clearly that the molecular refractivity is influenced by constitution, by the arrangement of the atoms in the molecule and by the presence of double or triple bonds, etc.; and for this reason determinations of molecular refractivity have been widely used in solving problems of chemical constitution.

To the power of liquids of rotating the plane of polarized light reference has already been made, and the connection between this property and molecular asymmetry has been discussed.

Other properties of liquids, *e.g.*, the power, when in a magnetic field, of rotating the plane of polarized light; the selective absorption of light, both visible light (giving rise to colour) and of invisible ultra-violet light; the dielectric constant; the parachor, etc., have been studied in order to establish the relations between physical and chemical properties, but these cannot be discussed in detail here. Physical properties, moreover, have been applied for the purpose of throwing light not on molecular constitution but on molecular complexity, and of these one of the most important is the *molecular surface energy*.

If one imagines a gram-molecule of a liquid in the form of a sphere, the surface of the sphere may be called the molecular surface, and this molecular surface may be shown to be proportional to the two-thirds power of the molecular volume. The product of the surface tension of a liquid into the molecular surface, or $\gamma(Mv)^{\frac{2}{3}}$, where γ is the surface tension and v the specific volume of the liquid, is known as the molecular surface energy.

In 1886 it was deduced theoretically by Baron Roland von Eötvös (1848–1919), Professor of Physics in the University of Budapest,[1] and was later shown experimentally by (Sir) William Ramsay and John Shields (1869–1920), that the molecular surface energy decreases linearly with the temperature, and that the temperature coefficient of molecular surface energy is a *colligative property*. Determinations of the temperature coefficient of molecular surface energy, therefore, were largely employed for the purpose of calculating the molecular weights, and therefrom the molecular complexity, of substances in the liquid state.[2] Further,

[1] *Ann. Physik*, 1886, **27**, 452.
[2] *Phil. Trans.*, 1893, A, **189**, 647; *J. Chem. Soc.*, 1893, **63**, 1089.

it was shown in 1884 by Frederick Thomas Trouton (1863–1922), student of Trinity College, Dublin, and later Professor of Physics at University College, London, that in the case of substances having the same molecular weight in the gaseous and liquid state, the molecular heat of vaporization of the liquid divided by the absolute temperature of the boiling point has approximately the same value, whereas the value of this quotient is greater than normal in the case of substances which are associated in the liquid state.[1] From the determinations of the *viscosity* of liquids, or of the *fluidity*, which is the reciprocal of the viscosity, conclusions have also been drawn concerning the molecular complexity of liquids.[2]

That heat is very generally produced as a result of chemical change had long been recognized, and isolated determinations of the heat of combustion were carried out by Lavoisier and Laplace,[3] by whom also the first general law of *thermochemistry* was stated, namely: "If a chemical combination or any change of state whatever is accompanied by a decrease of the free heat, this heat will be completely taken up again when the substances return to their original state, and, conversely, if a chemical combination or change of state is accompanied by an increase of the free heat, this new heat will be evolved when the substances return to their original state." For this law, which was stated as a general principle which might be accepted as axiomatic, no experimental support or proof was offered.

The first to undertake a systematic experimental investigation of the heat effects accompanying chemical change was Germain Henri Hess (1802–50), Professor of Chemistry, University of St. Petersburg (Leningrad), who gave a more definite and practically useful statement of the principle put forward by Lavoisier and Laplace.[4] The *law of constant heat summation* enunciated by Hess and established by him on a basis of experiment, states that *the total heat of a reaction is constant no matter whether the reaction is allowed to take place directly or in stages*; or, *the heat of reaction depends only on the initial and final systems*. This law may be employed, and has in fact been widely employed, in order to calculate heats of reaction when, as in the case of many reactions in organic chemistry, direct measurement is impossible. This empirical law

[1] The Trouton law was modified by J. H. Hildebrand of the University of California (*J. Amer. Chem. Soc.*, 1915, **37**, 970). See also A. Byk, *Z. physikal. Chem.*, 1924, **110**, 291.
[2] See E. C. Bingham, *Fluidity and Plasticity*; Bingham and Miss J. P. Harrison, *Z. physikal Chem.*, 1909, **66**, 1.
[3] *Œuvres de Lavoisier*, II, 287.   [4] *Ann. Physik*, 1840, **50**, 385; 1842, **52**, 79.

can be deduced from the *law of conservation of energy*[1] which was discovered two years later (1842) by the German physicist, Julius Robert Mayer (1814–78), and which was more accurately defined and more fully illustrated by Hermann Ludwig Ferdinand von Helmholtz (1821–94) in 1847. On this law rests the whole edifice of thermochemistry.

Although a number of workers, and more especially Pierre Antoine Favre (1813–80), and Johann Theobald Silbermann[2] (1806–65), made important contributions to our knowledge of the heat effects accompanying chemical reactions, it is with the names of the Danish chemist Julius Thomsen and of the French chemist Marcelin Berthelot, whose experimental determinations cover almost the whole range of chemical reactions, that the development of thermochemistry is mainly associated.[3]

Thomsen was the first to apply the law of conservation of energy to the discussion of thermochemical phenomena and to point out that Hess's law of constant heat summation follows as a consequence of that law; and the heat of reaction, he recognized, was a measure of the difference of energy content of a system before and after the reaction. Thomsen found that the heat of combustion of an organic compound is, in large measure, an additive property and can be calculated from values assigned to the atoms and structural elements (*e.g.*, single, double and triple bonds), but constitutive influences are also present. Determinations of heats of combustion, however, find but uncertain application to the study of molecular constitution.

Although Thomsen had clearly interpreted the heat effect accompanying a chemical reaction as representing the change of energy, he later came to regard the heat of reaction also as a measure of chemical force or chemical affinity; and he laid down the general principle that every purely chemical reaction is accompanied by evolution of heat. This principle was restated by Berthelot[4] in a

---

[1] This law states that in an isolated system the sum total of energy remains unchanged no matter what changes may take place in the system. Energy can be neither created nor destroyed.

[2] *Ann. Chim.*, 1852 [3], **34**, 357; **36**, 1; 1853, **37**, 406. Favre was, in 1851, appointed Director of the analytical chemistry laboratory of the École Centrale des Arts et Manufactures, and, in 1854, became Professor of Chemistry at Marseilles. Silbermann was a physicist at the Conservatoire des Arts et Métiers.

[3] Thomsen, *Ann. Physik*, 1853, **88**, 349; **90**, 261; 1854, **91**, 83; **92**, 34; *Thermochemistry*, trans. by K. A. Burke. Berthelot, *Ann. Chim.*, 1865 [4], **6**, 290, 329, 442; 1873 [4], **29**, 94, 289, 433.

[4] *Essai de Mécanique Chimique fondée sur la Thermochimie*, 879; *Ann. Chim.*, 1875 [5], **4**, 5.

more definite and emphatic form in which it was laid down that only those reactions could take place spontaneously, *i.e.*, without the intervention of external energy, which are accompanied by evolution of heat;[1] and that of all possible reactions occurring spontaneously, that one takes place which is accompanied by the greatest evolution of heat. This *principle of maximum work*, as Berthelot rather unhappily called it, aroused much controversy, and although later investigation has shown that it holds good at and in the neighbourhood of the absolute zero of temperature, it has no general validity. This, indeed, is shown by the fact that not a few reactions are known which take place spontaneously with absorption of heat.[2] It may also be pointed out that the principle of maximum work is based on the assumption that the heat of reaction is equal to the decrease of total energy of a system, an assumption which is valid only when the reaction takes place without the performance of external work.[3]

The phenomenon of *chemiluminescence*, or the emission of light as an accompaniment of chemical reaction, has long been known; and the conversion of light energy into chemical energy, as in the darkening of silver chloride by light, was known as far back as the eighteenth century. Knowledge of this reaction led to the development of photography by Daguerre and by Talbot in 1839. The study of all such reactions and of the relations between the energy of light radiation and chemical energy constitutes the branch of chemistry known as *photochemistry*.

As a result of the earlier investigations of the influence of light on chemical reactions, it was recognized by Theodor von Grotthuss (1785–1822)[4] in 1818, and by John William Draper[5] in 1841, that only such rays as are absorbed are effective in producing chemical change, but all absorbed rays are not equally effective photochemically. This had, in fact, been recognized in some measure even by Scheele.

The pioneering work of Draper was succeeded by the quantitative investigations of Bunsen and (Sir) Henry Roscoe,[6] who, from

---

[1] Called by Berthelot "exothermic reactions."

[2] Called by Berthelot "endothermic reactions."

[3] In accordance with the law of conservation of energy, also known as the first law of thermodynamics, the decrease of the total or internal energy is equal to the heat evolved plus the external work done.

[4] Christian Johann Dietrich von Grotthuss assumed the name of Theodor.

[5] *Phil. Mag.*, 1841 [3], **19**, 195; 1842, **21**, 453; 1843, **23**, 401; 1845, **27**, 327.

[6] *Phil. Trans.*, 1857, **147**, 355; 1863, **153**, 139; *Ann. Physik*, 1857, **100**, 43; 1862, **117**, 529.

their study of the combination of hydrogen and chlorine, were able to give the proof of what had already been made probable by Draper, that the amount of photochemical change is proportional to the amount of light energy absorbed, or proportional to the intensity of the absorbed radiation multiplied by the time during which it acts. On the basis of this law, *actinometers* (*tithonometers*,[1] as Draper had called them), or instruments for measuring the intensity of light, were constructed. The phenomenon of *photochemical inductance*, discovered by Draper, was studied more fully by Bunsen and Roscoe in the case of the reaction between hydrogen and chlorine, but it was not till the early years of the twentieth century that the explanation of the phenomenon was found in the presence of impurities which react with the chlorine.[2]

Although other photochemical reactions became known and were investigated, no great advance was made till 1912 when the subject entered on a new and important stage with the deduction by Albert Einstein of the law of photochemical equivalence. To this, consideration will be given later.

The laws of *electrochemistry* and the relations between electrical energy and chemical energy, which could receive a comprehensive treatment only after the introduction of the theory of electrolytic dissociation, will be discussed in the following chapter.

Whereas at first, as has been pointed out, physical chemistry occupied itself mainly with the study of physical properties in relation to chemical composition and molecular constitution, a new interest began to develop soon after the middle of the century. Behind this development lay a desire to obtain a deeper insight into the process of chemical change and a fuller understanding of what had long been spoken of as chemical affinity. Thomsen and Berthelot, as we have seen, had thought to find in the heat of reaction a measure of chemical affinity, but the falsity of the view that the heat effect of a reaction is a criterion of the possibility of chemical change was pointed out by Lord Rayleigh in 1875. Understanding of the process of chemical change was greatly furthered by the experiments of Wilhelmy (1850) and others on the rate of reaction, and the state of equilibrium in a homogeneous system was brought into relation with the "active mass" of the reacting substances by

---

[1] From Tithonus, Prince of Troy, who, according to legend, married Aurora, Goddess of Dawn.

[2] C. H. Burgess and D. L. Chapman, *J. Chem. Soc.*, 1906, **89**, 1399; D. L. Chapman and P. S. MacMahon, *ibid.*, 1909, **97**, 135, 959, 1717.

Guldberg and Waage in 1867. The laws of thermodynamics, more-over, were applied to chemical reactions by Horstmann, Willard Gibbs, Helmholtz and others and new vistas began somewhat dimly to open before the eyes of chemists, while, more or less un-noticed, experiments began to be made by the botanists de Vriess and Pfeffer in measuring the osmotic pressure of solutions. But although, by the middle of the eighties, a large body of facts had been accumulated and although the laws of chemical affinity had been developed to a considerable extent, the systematic cultivation of the domain of physical chemistry was still lacking, workers were comparatively few in number and their labours unco-ordinated. Physical chemistry had not yet developed an individuality and self-consciousness.

In the years 1885–7, however, the whole spirit and life of physi-cal chemistry was altered, in no small measure, through the publi-cation in 1885 of Wilhelm Ostwald's *Textbook of General Chemistry* (*Lehrbuch der allgemeinen Chemie*). In this work the attempt was first successfully made to bring together in a comprehensive and orderly manner the known facts and laws of physical chemistry—or *general chemistry*, as Ostwald called it—and thereby to make avail-able for further cultivation a large and difficult region of physical science. Through the publication of this work the attention of chemists was focused on problems of wide general interest, and the stimulus to investigation which was thereby given was powerfully reinforced by the publication of van't Hoff's theory of solutions (1885) and of Arrhenius's theory of electrolytic dissociation (1887), and by the application of the laws of thermodynamics to the treat-ment of chemical affinity by Helmholtz, van't Hoff and others. The founding by Ostwald and van't Hoff in 1887 of the *Zeitschrift für physikalische Chemie, Stöchiometrie und Verwandtschaftslehre* united the workers in this domain and gave them their own organ of publi-cation, and thereby a concentration of forces was effected and a more rapid and effective advance made possible. Of the extraordinary development of physical chemistry which took place during the last fifteen years of the nineteenth century some account will be given in the next chapter.

CHAPTER SIX

THE RISE AND DEVELOPMENT OF PHYSICAL CHEMISTRY
IN THE NINETEENTH CENTURY—II

WHEN, in studying a chemical reaction, one passes from a consideration of the products of change to a study of the process itself, it is recognized that one of the most important factors in the process is the speed with which it takes place. In the case of many reactions, more especially of those which take place between inorganic salts or between acids and alkalis, the velocity with which the end of the reaction is reached is so great that its measurement is virtually impossible; and, on the other hand, there are reactions the rate of which is so slow as to be inappreciable. Between these extremes there are many reactions which take place with very varying velocity, depending on the nature of the reacting substances, the nature of the medium in which the reaction takes place and on the temperature.

It had long been known that the progress of a chemical reaction is influenced by the amounts of the reacting substances; and as early as 1777 Karl Friedrich Wenzel (1740–93),[1] chemist to the Freiberg foundries in Saxony, observed that in the case of the solution of zinc or copper in acid, "if an acid dissolves 1 drachm of zinc or copper in one hour, an acid of half the concentration will require two hours." It was, however, only at a much later date, in the year 1850, that Ludwig Ferdinand Wilhelmy (1812–64), Lecturer in Physics in the University of Heidelberg, was able, in the case of the inversion of cane sugar in presence of acids, to deduce a mathematical expression for the rate of the reaction, or for its progress with time.[2] On the assumption that the amount of cane sugar inverted in unit of time is proportional to the amount of sugar present (the amount of acid and other conditions remaining unchanged), the rate of inversion is given by $-\dfrac{dZ}{dt} = k.Z$, where

---

[1] *Lehre von der chemischen Affinität der Körper*, 1777.
[2] *Ann. Physik*, 1850, **81**, 413; Ostwald's *Klassiker*, No. 29. The influence of the nature of the acid on this reaction was studied by J. Löwenthal and E. Lenssen (*J. prakt, Chem.*, 1862, **85**, 321).

dZ is the loss of cane sugar in the time interval dt, Z is the amount of sugar present, and k is a constant. When this expression is integrated there results, $\log_e Z_o - \log_e Z = kt$; and this expression, Wilhelmy was able to show, is in harmony with experiment.

Important as the work of Wilhelmy is now recognized to be, it attracted but little attention at the time, for the energies of chemists were absorbed in the production and classification of organic compounds and in the discussion of problems of molecular constitution. In 1862, however, the study of reaction velocities was resumed by Berthelot and his pupil Péan de St. Gilles (1832–63), who investigated the rate of ester formation under various conditions;[1] and in 1866 a more general treatment of the problem of reaction velocity was undertaken at the University of Oxford by A. Vernon Harcourt and William Esson (1838–1916).[2] These chemists not only showed that the logarithmic equation obtained by Wilhelmy is valid for other reactions in which one substance undergoes change, but they also derived an expression for the velocity of a reaction in which two substances undergo change. Almost at the same time, in 1867, the full significance and generality of the problem were recognized by two Norwegian scientists, Guldberg and Waage, who were the first to state fully and clearly and to apply in many different directions the law that the velocity of a reaction at constant temperature is proportional to the product of the active masses of the reacting substances, the active mass being synonymous with the concentration expressed in gram-molecules per litre.[3] This law, generally called the *law of mass action*, is the fundamental law of chemical kinetics, and is applicable, in the first instance, to reactions which take place in homogeneous systems. At a later date it was applied by van't Hoff also to heterogeneous systems.

Although the law of the velocity of chemical reactions had thus been clearly stated by Guldberg and Waage, who, however, applied it mainly to the discussion of chemical equilibria, their work was practically unknown; and the first real stimulus to the systematic investigation of the rate of chemical change was given by the publication of van't Hoff's *Études de dynamique chimique* in 1884.

---

[1] *Ann. Chim.*, 1862, **65**, 385; **66**, 5; 1863, **68**, 225; Ostwald's *Klassiker*, No. 173.
[2] *Phil. Trans.*, 1866, **156**, 193; 1867, **157**, 117.
[3] Guldberg and Waage, *Études sur les affinités chimiques*, 1867; *J. prakt. Chem.*, 1879 [2], **19**, 69; Ostwald's *Klassiker*, No. 104.
Cato Maximilian Guldberg (1836–1902), was Professor of Mathematics and Peter Waage (1833–1900), his brother-in-law, was Professor of Chemistry in the University of Christiania (Oslo).

Guldberg and Waage, in deriving the law of mass action, introduced the idea of "chemical force", but van't Hoff, who had, in 1877, in ignorance of the work of the Norwegian scientists, also enunciated the law of mass action, considered only the velocity of a reaction as determined by a *reaction coefficient* characteristic of the reaction under given conditions, and the concentration of the substances undergoing change.[1] Moreover, in the derivation of the expressions for the velocity of reaction, van't Hoff grouped the different reactions according to the number of molecules taking part in the reaction rather than according to the number of substances. In this way he was able to determine the "order" of a reaction, as defined by the number of reacting molecules, and thereby to obtain an insight into the "mechanism" of a reaction. By his clear and inspiring treatment of the subject of chemical kinetics and of the "disturbances" which are found to influence the rate of a reaction; by his development of experimental methods of investigating reaction velocities; by his discussion of the influence of temperature on the velocity or reaction coefficient and of the conditions of equilibrium in heterogeneous as well as in homogeneous systems, van't Hoff brought the whole subject of chemical dynamics vividly before the minds of chemists and aroused a widespread interest in a field of investigation which had, until then, been but scantily tilled and from whose almost virgin soil a valuable harvest might be gathered. Since then many chemists, scattered over all the countries of the world, have taken up the investigation of reaction velocities and of chemical equilibria, seeking thereby not only to put the deductions from the law of mass action more fully to the test but also to gain an understanding of the mechanism of particular reactions and of chemical change in general.

The fact that the rate of reaction is increased by rise of temperature was recognized at an early time, and the influence of temperature on the velocity of inversion of cane sugar was made the subject of quantitative investigation by Wilhelmy in 1850. By him, as well as by other investigators at a later time,[2] the attempt was made to express in mathematical form the relation between the velocity coefficient and the temperature, and various expressions of a somewhat similar character were obtained. It was, however, first realized by Arrhenius that the study of the temperature

[1] *Ber.*, 1877, **10**, 669.
[2] Berthelot, *Ann. Chim.*, 1862 [3], **66**, 110; van't Hoff, *Études de dynamique chimique*, 1884; Arrhenius, *Z. physikal. Chem.*, 1889, **4**, 226; Kooy, *ibid.*, 1893, **12**, 155; Harcourt and Esson, *Proc. Roy. Soc.*, 1895, **58**, 108.

coefficients of reaction velocity is of importance from the point of view of the general mechanism of chemical change.

In the case of polymolecular reactions, or reactions involving two or more molecules, chemical change may be regarded as resulting from collisions between the reacting molecules, but the necessity for such collisions is not apparent in the case of unimolecular reactions. Moreover, in the simple case of a bimolecular reaction in the gaseous state, in which case it is possible to apply the kinetic theory to the calculation of the number of molecular collisions in unit time, it is found that the rate of chemical reaction cannot be explained merely on the basis that every collision leads to reaction. Arrhenius, therefore, expressed the view that, in a reacting system, there is only a certain number of "active" molecules which can undergo reaction, and that it is to the increase in the number of such molecules rather than to the increase in the number of collisions that the temperature coefficient of chemical reaction is due.

If the view be accepted that only those molecules can enter into chemical reaction which become "activated" by acquiring a certain amount of energy in excess of the normal amount,[1] it is possible to deduce the expression $d \log_e k/dt = E/RT^2$, first suggested by Arrhenius, for the effect of temperature on the rate of reaction; or, assuming that E is independent of the temperature,

$$\log_e \frac{k_2}{k_1} = \frac{E}{R}\left(\frac{1}{T_1} - \frac{1}{T_2}\right).$$

The symbol E represents the energy of the "active" molecules in excess of the mean energy of all the molecules and is called the *critical increment* or the energy of activation.

After some years this subject again attracted a large amount of attention,[2] and the view was held that even in the case of unimolecular reactions the molecules receive their energy of activation from collisions.[3] The view put forward and supported mainly by the French physicist, Jean Perrin, and by W. C. M. Lewis, of the University of Liverpool, that activation takes place through absorption of the energy of radiation, has not withstood the test of experiment.

.   .   .   .   .

[1] This energy of activation can be derived from the kinetic energy of the impact of molecules having a kinetic energy greater than the average.

[2] See C. N. Hinshelwood, *The Kinetics of Chemical Change in Gaseous Systems*; E. A. Moelwyn-Hughes, *Chemical Reviews*, 1932.

[3] F. A. Lindemann, *Trans. Faraday Soc.*, 1922, **17**, 598.

In the early decades of last century a number of phenomena were observed which, although otherwise apparently disconnected, were all characterized by the fact that the presence of certain substances greatly increased the rate at which a chemical reaction took place, even although these added substances underwent no change in amount. The conversion of starch to sugar, for example, was accelerated by dilute acids,[1] and the decomposition of hydrogen peroxide in alkaline solution was greatly accelerated by the presence of platinum, gold, fibrin and other substances.[2] In the presence of finely divided platinum, spirit of wine (ethyl alcohol) is oxidized to acetic acid by the oxygen of the air,[3] and hydrogen readily combines with oxygen.[4] "It is then proved," wrote Berzelius[5] "that several simple and compound bodies . . . have the property of exercising on other bodies an action very different from chemical affinity. . . . I do not believe that it is a force quite independent of the electro-chemical affinities of matter; . . . but since we cannot see their connection and mutual dependence, it will be more convenient to designate the force by a separate name. I will therefore call this force the *catalytic force*, and I will call *catalysis* the decomposition of bodies by this force in the same way that one calls by the name analysis the decomposition of bodies by chemical affinity."

The term catalysis was introduced by Berzelius in order to group together phenomena which exhibited the common characteristic that a change in the velocity of reaction was produced by relatively small quantities of material; and although the number of such cases to become known rapidly increased during the nineteenth century, it was not until the laws and experimental methods of chemical kinetics had been developed that catalytic phenomena and the effects of catalysts on the velocity of a reaction could be quantitatively measured. By his definition of a catalyst as any substance which, without itself appearing in the final product of reaction, alters the velocity with which that reaction takes place, Ostwald brought the study of catalytic phenomena within the domain of chemical kinetics and so made them amenable to quantitative investigation;[6] and it was in Ostwald's laboratory at Leipzig that the phenomena of catalysis were first most systematically and vigorously investigated.

[1] Kirchhoff, *Schweigger's Journ.*, 1812, **4**, 108.
[2] Thenard, *Ann. Chim.*, 1818, **9**, 314.
[3] E. Davy, *Phil. Trans.*, 1820, 108.
[4] Doebereiner, *Schweigger's Journ.*, 1822, **34**, 91; 1823, **38**, 321.
[5] *Jahresbericht*, 1836, **15**, 237; *Ann. Chim.*, 1836, **61**, 146.
[6] *Z. physikal. Chem.*, 1894, **15**, 705.

The essential postulate in the Ostwald definition is that a catalyst does not alter the energy relations of a reaction and has no influence therefore on the equilibrium in a reversible reaction. The question whether it can initiate a reaction or merely alter the velocity of a reaction already taking place has been much debated; and although it is very difficult to prove or to disprove the suggestion experimentally, there is undoubtedly much experimental evidence in favour of accepting the view that a catalyst may, in some cases at least, be necessary for the initiation of a reaction.[1]

The great magnitude of the effects produced relatively to the amount of the catalyst employed, which may be regarded as an outstanding characteristic of catalytic phenomena, has been strikingly demonstrated by a very large number of investigations.[2]

The influence of traces of moisture on the velocity of many gaseous reactions had for long attracted the attention of chemists, and examples of this were given, as early as 1794, in an "Essay on Combustion" by Mrs. Fulhame; and during the nineteenth century not a few reactions were met with which appeared to take place, with appreciable velocity, only when moisture was present. The systematic study of this catalytic action of water vapour was first undertaken in 1880 by H. B. Dixon (1852–1922),[3] Professor of Chemistry, University of Manchester, and was continued by his pupil, H. B. Baker (1862–1935),[4] who later became Professor of Inorganic Chemistry in the Imperial College of Science and Technology, London. As a result of these investigations the conclusion was drawn that intensive drying by means of phosphorus pentoxide inhibits the combination of carbon monoxide and oxygen, of hydrogen and oxygen, of ammonia and hydrogen chloride, and a number of other gaseous reactions. The influence of moisture, however, was again studied at a much later time by a number of chemists, especially by Max Bodenstein[5] and collaborators, who reached the conclusion that in the reactions studied there is no positive catalysis by water vapour, but that, in the drying process, traces of impurities are introduced into the gas phase which break

---

[1] See Lowry, *J. Chem. Soc.*, 1899, **75**, 211; Meyer and Schoeller, *Ber.*, 1920, **53**, 1410; Norrish, *J. Chem. Soc.*, 1923, **124**, 3006.
[2] See E. K. Rideal and H. S. Taylor, *Catalysis in Theory and Practice*; K. G. Falk, *Catalytic Action*; K. C. Bailey, *The Retardation of Chemical Reactions*; P. H. Emmett, *Catalysis*; G. C. Bond, *Catalysis by Metals*.
[3] *British Association Reports* for 1880, p. 593; *Proc. Roy. Soc.*, 1884, **37**, 56.
[4] *J. Chem. Soc.*, 1894, **65**, 611; 1902, **81**, 400.
[5] F. Bernreuther and M. Bodenstein, *Sitz.-ber. Preuss. Akad. Wiss.*, 1933. VI; Bodenstein, *Z. physikal. Chem.*, 1933, B, **20**, 451; B, **21**, 469.

reaction chains, so that reaction does not occur to any considerable extent.[1]

While the positive catalytic action of small traces of water vapour appears now to be unproved, other reactions may be mentioned which illustrate the great effects which may be produced by small quantities of a catalyst. Thus, the oxidation of solutions of sodium sulphite by oxygen is catalytically accelerated by the presence of copper, and so powerful is the action of this catalyst that a noticeable acceleration of the process is observed even when the water used in preparing the solution of sulphite is allowed to remain for only forty-five seconds in contact with metallic copper.[2]

Reactions may be catalytically retarded (*negative catalysis*) as well as catalytically accelerated (*positive catalysis*), and catalysis may be brought about by one of the reacting substances or by one of the products of reaction (*autocatalysis*). The accelerating action of a catalyst, moreover, may be inhibited ("poisoned") to a greater or less extent by another substance, and the amount of "poison" required to bring about an appreciable diminution of catalytic activity is sometimes exceedingly minute.[3]

Catalysis may take place not only in homogeneous but also in heterogeneous systems. As early as 1817 it was discovered by (Sir) Humphry Davy that when a spiral of hot platinum wire is suspended in a vessel containing the vapour of alcohol and air, oxidation of the alcohol vapour takes place so rapidly at the surface of the platinum wire that the latter is raised to incandescence.[4] Unlimited amounts of alcohol vapour are thus caused to burn rapidly by the same piece of platinum. Such cases of *contact catalysis*, or catalysis at a solid surface, are numerous and their investigation has been vigorously pursued, owing to their importance in industry (see Chapter XIII).[5] Moreover, not only may the reactions between substances be accelerated by such catalysts, but the direction of the reaction and the nature of the products obtained may be altered by alteration of the catalyst.

The characteristics of catalytic action which have been discovered in the case of inorganic catalysts are met with also in the case of the *enzymes* produced in living animal and vegetable cells. The

[1] See Schwab, *loc. cit.*; J. W. Smith, *The Effects of Moisture on Physical and Chemical Changes* (Longmans).
[2] A. Titoff, *Z. physikal. Chem.*, 1903, **45**, 641.
[3] G. Bredig, *Anorganische Fermente* (Engelmann, 1901).
[4] *Phil. Trans.*, 1817, **97**, 45.
[5] R. H. Griffith and J. D. F. Marsh, *Contact Catalysis*.

similarity of such processes as the conversion of starch into sugar by diastase[1] and the hydrolytic decomposition of amygdalin by emulsin[2] to the processes of hydrolysis accelerated by acids was recognized; and the more recent catalytic decompositions of hydrogen peroxide brought about by colloidal platinum agree, in a remarkable manner, with and form an inorganic prototype of the enzyme reactions.[3]

Although a very wide range of catalytic phenomena has been subjected to very careful investigation, the mechanism of catalysis cannot yet be said to have been fully and adequately explained; and in view of the variety of the phenomena, a single explanation need not be looked for to cover all cases. Berzelius had pointed out that a catalyst appeared to "arouse the slumbering affinities" of the reacting substances, and Liebig put forward the view that this effect was brought about through the "molecular oscillations" of the catalyst being communicated to the molecules of the reacting substances. This explanation, however, is of no scientific value, for it cannot be put to the test of experiment and it offers no suggestion for experimental investigation.

In the case of so-called *transfer* or *cyclic catalysis*, the most plausible and acceptable explanation is to be found in the assumption that a slightly stable compound is formed between the catalyst and one of the reacting substances, and that this compound then undergoes reaction with the other reactant with formation of the reaction product and regeneration of the catalyst. Such an explanation was first suggested by F. Clément and J. B. Désormes in 1806 to explain the catalytic action of the oxides of nitrogen in the manufacture of sulphuric acid,[4] and a similar view was expressed by Williamson in explanation of the catalytic action of sulphuric acid in ether formation.[5] This theory of intermediate compound formation has received much support from the investigations of the Friedel-Crafts reaction by Steele,[6] of the action of ether as a catalyst in the preparation of the Grignard reagent,[7] and from the study of enzyme reactions.

[1] A. Payen and J. F. Persoz, *Ann. Chim.*, 1833, **53**, 73.

[2] Liebig and Wöhler, *Annalen*, 1839, **22**, 1.

[3] See G. Bredig and R. Müller von Berneck, *Z. physikal. Chem.*, 1899, **31**, 258, and later volumes; Bredig, *Biochem. Z.*, 1907, **6**, 2834; *The Enzymes*, edited by Boyer, Lardy and Myrbäck; S. G. Waley, *Mechanisms of Organic and Enzymic Reactions*.

[4] *Ann. Chim.*, 1806, **59**, 329.

[5] *British Association Reports*, 1850; *J. Chem. Soc.*, 1852, **4**, 106, 229, 350.

[6] *J. Chem. Soc.*, 1903, **83**, 1470. Bertram Dillon Steele (1871–1934), Professor of Chemistry in the University of Queensland, Australia.

[7] V. Grignard, *Compt. rend.*, 1900, **130**, 1322.

In all cases of transfer catalysis, in which a reaction takes place through the alternate formation and decomposition of an intermediate compound or of intermediate compounds more rapidly than by a direct process, the catalyst acts by providing for the reaction an alternative path which requires a lower energy of activation (p. 90).

Even in the case of heterogeneous catalysis or *contact catalysis*, the formation of unstable compounds has been suggested,[1] although, in such cases, it seems probable that catalytic acceleration is brought about by an *adsorption* or condensation of the reacting substances on the surface of the catalyst.[2]

The question why, under given conditions, a chemical reaction should take place between substances is one which has exercised the minds of men from a very early period; and when Albertus Magnus (1193–1282), Bishop of Regensburg, explained chemical combination as due to *affinity*, he was merely reflecting a view, derived from Hippocrates and current at the time, that chemical action is due to similarity or kinship between the reacting substances. This view was abandoned during the eighteenth century, and early in the nineteenth century was replaced by the view, put forward by Davy and by Berzelius, that chemical affinity is electrical in character. But whatever the nature of the force which, acting between different kinds of matter, brings about, under certain conditions, a chemical reaction between them, the further questions arose how this force, so important for chemical change, could be quantitatively measured and how its action might be modified by changes of concentration, temperature and pressure.

At the end of the seventeenth century it was thought by the German chemist, Georg Ernst Stahl (1660–1734), that relative affinities could be determined by the order in which substances expel one another from compounds; and in 1718 Étienne François Geoffroy (1672–1731), Professor of Chemistry at the Jardin du Roi (Jardin des Plantes) in Paris, drew up a table of substances arranged in this way. Similar tables were also drawn up in 1775 by the Swedish chemist and mineralogist, Torbern Olof Bergman

---

[1] Brodie, *Phil. Trans.*, 1862, **151**, 855; Sabatier, *La Catalyse en Chimie Organique*; Bredig, *Anorganische Fermente*.
[2] Faraday, *Phil. Trans.*, 1834, **114**, 55; Turner, *Edin. Phil. J.*, 1824, **11**, 99, 311; J. J. Thomson, *Applications of Dynamics to Physics and Chemistry*; I. Langmuir, *J. Amer. Chem. Soc.*, 1916, **38**, 2221; W. G. Palmer and F. H. Constable, *Proc. Roy. Soc.*, 1925, A, **107**, 255; R. H. Griffith, *The Mechanism of Contact Catalysis* (Oxford University Press).

(1735–84), who, however, recognized that the affinity, as determined by mutual displacements, might be altered by temperature, especially when a substance is volatile. Wenzel, in 1777, held the erroneous view that the affinity can be measured by the velocity with which a reaction takes place, and Claude Louis Berthollet was led, by his investigations of the formation of trona in the soda lakes of Egypt, to the important conclusion that chemical affinity is not a constant and unvarying force but may be influenced by the mass of the reacting substances. A reaction which, under certain conditions, takes place in one direction may, by a variation of the masses of the reacting substances, be caused to take place in the reverse direction. The occurrence of reversible reactions leading to an equilibrium was thereby recognized.

Unfortunately, at this time the attention of chemists was distracted by the contention of Berthollet that the mass of the reacting substances has an influence also on the *composition* of the products of reaction—a view which was in conflict, as we have seen, with the law of definite proportions—and recognition of the importance of the mass of the reacting substances on the course of a reaction was thereby greatly delayed.

The views of Berthollet regarding the influence of mass on the course of a reaction were put to the test by various workers,[1] and more especially by Marcelin Berthelot and Péan de St. Gilles,[2] who made a systematic investigation of the formation of esters by the action of acids on alcohols.[3] By their experimental investigation of this reaction, which became the classical example of all reversible reactions, Berthelot and Péan de St. Gilles obtained abundant confirmation of the influence of mass not only on the rate at which ester formation takes place but also on the *amount* of ester produced. When an alcohol and an acid react together, conversion to ester and water is never complete, but the reaction appears to stop when a certain amount of ester and of water has been formed. A state of *equilibrium* is produced in which all four substances, alcohol, acid, ester and water, are present together. By increasing the relative

---

[1] For example, by J. H. Gladstone (*Phil. Mag.*, 1855 [4], **9**, 535; *Phil. Trans.*, 1855, 179) and by J. H. Jellett (1817–88), Professor in the University of Dublin (*Trans. Roy. Irish Acad.*, 1875, **25**, 271; Ostwald's *Klassiker*, No. 163).

[2] *Ann. Chim.*, 1862, **65**, 385; **66**, 5; 1863, **68**, 225; Ostwald's *Klassiker*, No. 173.

[3] The work of Berthelot and Péan de St. Gilles was supplemented by Nicolai Alexandrovitsch Menschutkin (1842–1907), Professor of Chemistry in the University of St. Petersburg, who studied the reaction between many different acids and alcohols and so obtained information regarding the influence of structure on ester formation (*Annalen*, 1879, **195**, 334; **197**, 193; etc.).

amount of alcohol, a greater proportion of the acid is converted to ester; and the greater the proportion of acid, the more complete is the conversion of the alcohol to ester. Moreover, when ester and water are brought together, alcohol and acid are formed, and the greater the relative amount of water, the more completely is the ester hydrolysed.

The law of the influence of mass on an equilibrium in a homogeneous system, which makes it possible to correlate the various facts discovered by Berthelot and Péan de St. Gilles, was first deduced by Guldberg and Waage in 1864 in a paper published in the Norwegian language, and more fully developed three years later.[1] The "chemical force," it was pointed out, with which two substances react is equal to the product of their active masses multiplied by the affinity constants. In the case of a reversible reaction, therefore, equilibrium will be attained when the "chemical forces" of the direct and reverse reactions are equal. It was also recognized that the "force" with which substances react is measured by the velocity of the reaction, and so there was obtained the law of mass action already stated (p. 88).[2] Equilibrium, therefore, will be attained when the opposing "forces" become equal, or, as was more clearly stated by van't Hoff in 1877, when the *velocities* of the opposing reactions become equal. Since, in the case of the reversible process, $A + B \rightleftharpoons C + D$, the velocity of reaction between A and B at constant temperature is given by the expression, $v_1 = k_1 [A] \times [B]$, where [A] represents the concentration of A and [B] represents the concentration of B; and since the velocity of the opposing reaction is given by $v_2 = k_2 [C] \times [D]$, it follows that, at equilibrium (when $v_1 = v_2$),

$$k_1[A] \times [B] = k_2[C] \times [D] \text{ or } \frac{[C] \times [D]}{[A] \times [B]} = \frac{k_1}{k_2} = K.$$

K is known as the *equilibrium constant*. Chemical equilibrium is thus regarded not as a static but as a *dynamic* equilibrium, as L. Pfaundler had taught,[3] brought about by the velocity of two opposing reactions becoming equal.

Since equilibrium depends on the concentrations of the participating substances and therefore on the volume of the system, it will

[1] *Forhandlinger i Videnskabs-Selskabet i Christiania*, 1864; *Études sur les affinités chimiques*, 1867; Ostwald's *Klassiker*, No. 104.

[2] Van't Hoff, as we have seen, deduced the law of mass action, independently of Guldberg and Waage, on the basis of reaction velocities and without the introduction of any hypothetical "chemical force."　　　　[3] *Ann. Physik*, 1867, **131**, 55.

also depend on the external pressure, in accordance with the theorem enunciated by Henri Louis Le Chatelier (1850–1936), Professor of Chemistry in the University of Paris, namely: When the pressure on a system in equilibrium is increased, a reaction will take place (if possible) which is accompanied by a diminution of volume; and when the pressure is diminished, a reaction will take place which is accompanied by an increase of volume. If the reactions concerned take place without change of volume, change of external pressure will have no influence on the equilibrium.[1]

In principle, all reactions must be regarded as reversible, and only when one or more of the products of reaction are removed, by precipitation, vaporization, etc., from the sphere of action can the reaction take place completely in one direction.[2]

Not only was the above equation found, by Guldberg and Waage as well as by van't Hoff, to be in harmony with the results obtained by Berthelot and Péan de St. Gilles, but its correctness was also confirmed by many later workers who also showed that the equation may be generalized to cover all cases of chemical equilibria. This well-established equation, therefore, enables one to calculate quantitatively what will be the state of a system in equilibrium and how the equilibrium will be affected by variation of the concentrations of the reacting substances.[3] No matter how the individual concentrations may be altered, the ratio of concentrations deduced according to the law of mass action and embodied in the equation given above will, at constant temperature, remain unchanged; and it becomes possible, therefore, if the value of the equilibrium constant is known, to calculate the conditions for the most effective utilization of the reacting substances. The importance of the equation, whether in the scientific laboratory or for the control of industrial production, it would be difficult to overestimate.

Although attempts had been made as early as 1869 to throw light on the affinity relations between acids and alkalis, through a study of the equilibria produced by the addition of an acid to a mixture of two bases,[4] a satisfactory interpretation of the observed facts could be offered only at a later time on the basis of the electrolytic dissociation theory; and it was only through the application of

---

[1] *Compt. rend.*, 1884, **99**, 786.   [2] Williamson, *Annalen*, 1851, **77**, 37.
[3] At the present day concentrations are replaced by activities. See Chapter XII.
[4] Julius Thomsen, *Ann. Physik*, 1869, **138**, 65, 201, 497; J. H. Jellett, *Trans. Roy. Irish Acad.*, 1875, **25**, 271; Ostwald's *Klassiker*, No. 163; Ostwald, *J. prakt., Chem.*, 1877 [2], **16**, 385.

thermodynamics to chemical processes that the exact definition of chemical affinity became possible and a basis was given for its quantitative measurement. Through the application of thermodynamics, moreover, it was possible not only to derive the law of mass action, but to calculate the effect of change of external conditions, *e.g.*, temperature and pressure, on a system in equilibrium, and thereby to obtain a clear insight into its behaviour. It may, indeed, be said that the introduction of thermodynamics, and more especially of the second law of thermodynamics, into the treatment and discussion of chemical and physico-chemical processes altered the whole character and outlook of physical chemistry and marks the beginning of a very definite and very important advance in the study of chemical affinity and in the general treatment of chemical phenomena.

The second law of thermodynamics, which deals with the conditions under which transformation of heat energy into other forms of energy takes place, proclaims that the extent to which such transformation can be effected is limited. The extent to which heat can be converted into work was first calculated, in the case of a perfect gas, by Nicolas Léonard Sadi Carnot (1796–1832),[1] Officer of Engineers in the French Army, in 1824; and ten years later the argument employed by Carnot was put into an analytic form by Benoit Paul Émile Clapeyron (1799–1864), Professor at the École de Ponts et Chaussées, Paris.[2] On the basis of the conclusions reached by Carnot and Clapeyron and with application of the first law of thermodynamics discovered in 1842, Rudolf Clausius (1822–88),[3] Professor of Physics in the University of Zurich, and William Thomson (Lord Kelvin) (1824–1907),[4] Professor of Natural Philosophy in the University of Glasgow, deduced the theorem generally known as the second law of thermodynamics, the importance of which in connection with the efficiency of steam engines, at that time the chief motive power, was at once obvious.

It was not, however, till 1869 that the law was first successfully applied to chemical processes. In that year August Friedrich Horstmann (1842–1929), Privat-Dozent and, later, Professor of Theoretical Chemistry at Heidelberg, on studying the sublimation of ammonium chloride found that at each temperature a certain

---

[1] *Réflexions sur la Puissance Motrice du Feu et sur les Machines propres à développer cette Puissance*, 1824; Ostwald's *Klassiker*, No. 37.
[2] *J. de l'École Polytech.*, 1834, **14**, 170; *Ann. Physik*, 1843, **59**, 446, 566.
[3] *Ann. Physik*, 1850, **79**, 368, 530; 1854, **93**, 481.
[4] *Phil. Mag.*, 1852 [4], **4**, 13.

definite vapour pressure was established, as in the case of a liquid.[1] To this process of sublimation, therefore, there could be applied the

same equation, $\dfrac{dp}{dT} = \dfrac{Q}{T(v_2 - v_1)}$, as had already been derived by

Clapeyron and by Clausius for the vaporization of a liquid. The change in the sublimation or vapour pressure with temperature, therefore, was brought into relation with the heat of vaporization or sublimation (Q), and the change of volume $(v_2 - v_1)$ which accompanies the process. This law may also be applied to the dissociation of solid substances, such as calcium carbonate, salt hydrates, etc., and has been amply confirmed by experiment.[2]

Although a thorough and comprehensive thermodynamic treatment of chemical equilibria was carried out by Josiah Willard Gibbs (1839–1903),[3] Professor of Mathematical Physics, Yale University, this work, on account of the highly abstract and somewhat involved method of treatment, remained for long practically unknown; and it was only later that the many important generalizations contained in this monumental work became revealed. One finds, therefore, that some of the conclusions reached by Gibbs were also arrived at independently by others. Thus it was not till 1887 that one of the great generalizations established by Gibbs and now known as the *phase rule* was made generally known and its practical applicability to the study of heterogeneous chemical equilibria made apparent by Roozeboom. This generalization[4] defines the conditions of equilibrium as a relationship between the number of what are called the *components*[5] of a system and the number of coexisting phases, or physically distinct portions of a system, and may be summarized in the equation $F = C + 2 - P$. Here F represents the number of "degrees of freedom" or the *variability* of the system; that is, the number of independent variables to which arbitrary values must be given in order to define the system. Thus, the system formed by water in equilibrium with

[1] *Ber.*, 1869, **2**, 137; 1871, **4**, 635; *Annalen*, 1870, Suppl. **8**, 112; Ostwald's *Klassiker*, No. 137. Similar applications of thermodynamics were made by M. Peslin (*Ann. Chim.*, 1871, **24**, 208) and by J. Moutier (*Compt. rend.*, 1871, **72**, 759).

[2] F. Isambert, *Compt. rend.*, 1881, **92**, 919; 1868, **66**, 1259; P. C. F. Frowein, *Z. physikal. Chem.*, 1887, **1**, 5.

[3] *Trans. Connecticut Acad.*, 1874–8.

[4] See Roozeboom, *Die Heterogenen Gleichgewichte*; A. Findlay, *The Phase Rule and its Applications*.

[5] By "components" are meant only those constituents of a system in equilibrium, the concentration of which can undergo independent variation in the different phases.

vapour—a system of one component existing in two phases (liquid and vapour)—possesses one degree of freedom, or is *univariant*. That is, it is possible for water to coexist with vapour at different values of temperature and pressure, but if *one* of the variable factors, pressure, temperature or volume (in the case of a given mass of substance), is arbitrarily fixed (*one* degree of freedom), the state of the system will then be defined. At a selected fixed temperature the vapour pressure has a definite value. On the other hand, the system formed by ice, water and vapour coexisting in equilibrium (*i.e.*, one component in three phases), the variability of the system is zero. The system can exist only at one definite temperature and under one definite pressure; and it is not possible arbitrarily to fix some other temperature without causing one of the phases to disappear and so destroying the system.

Moreover, when calcium carbonate is heated it dissociates into calcium oxide and carbon dioxide. At equilibrium, therefore, there are three phases (two solid and one gaseous phase) and since there are two components (calcium oxide and carbon dioxide), the system has one degree of freedom. At any given temperature, therefore, the dissociation pressure has a definite value, just as with the system water and vapour which also is a *univariant* system. On the other hand, when ammonium hydrosulphide, NH_4HS, is vaporized dissociation into ammonia and hydrogen sulphide occurs, and an equilibrium is established represented by $NH_4HS \rightleftarrows NH_3 + H_2S$. In this case, therefore, there is an equilibrium between solid ammonium hydrosulphide and the gases ammonia and hydrogen sulphide. The system consists of two components coexisting in two phases, solid and gas; and unlike the previous system is *bivariant* or has two degrees of freedom. At any given temperature solid ammonium hydrosulphide can exist in equilibrium with a mixture of ammonia and hydrogen sulphide of varying composition and of varying total pressure. If, however, *two* variables are fixed, *e.g.*, temperature and pressure, the nature of the system (composition of the gas phase) is defined.

The phase rule has now found very extensive application. It not only forms a basis of classification of the different cases of heterogeneous equilibrium and so enables one to co-ordinate the large number of isolated equilibria, but it gives an insight into the relationships existing between the different systems and is a guide to the investigation of unknown systems.

The "law of the incompatibility of condensed systems"

enunciated by van't Hoff in 1884 coincides with the phase rule in the case of systems from which the gas phase is absent.[1]

Chemical equilibrium may be altered not only by variation of the concentration of the participating substances, but also in general, by the temperature; and the influence of temperature on equilibrium was shown by van't Hoff to be related to the heat of reaction in accordance with the equation, known as the van't Hoff isochore, $\dfrac{d \log_e K_v}{dT} = -\dfrac{Q_v}{RT^2}$, where Q_v is the heat evolved at constant volume. From this expression it can be predicted that elevation of temperature will favour an endothermic reaction, and lowering of temperature will favour an exothermic one—a generalization known as van't Hoff's *law of mobile equilibrium*.

The important contributions made by Hortsmann, by Willard Gibbs and by Helmholtz, who in 1882 derived an expression for the relation between the electrical energy of a voltaic cell and the chemical energy (heat of reaction) of the reaction occurring in the cell,[2] were, in so far as they were known, appreciated mainly by physicists; and it was van't Hoff who first brought home to chemists the importance of the application of thermodynamics to the study of chemical processes. Moreover, in 1884 van't Hoff, in his *Études de dynamique chimique*, pointed out that, on the basis of the second law of thermodynamics, a measure of *chemical affinity* is to be found in the *maximum external work* (the "work of chemical affinity") which can be obtained when a chemical reaction is carried out reversibly and isothermally.

A system, van't Hoff pointed out, may be regarded as possessing a certain capacity for doing work, but the absolute amount cannot be determined. The change in this capacity, however, when one system passes spontaneously into another, can be measured by the maximum amount of work which can be done. This decrease in the capacity for doing work, or the maximum external work done, is, therefore, a measure of the affinity of a reaction, when the volume remains constant.

When a reaction is carried out under atmospheric pressure and at constant temperature the total external work is the sum of the electrical or other energy available for use and the work done by expansion against the constant atmospheric pressure. This

[1] *Études de dynamique chimique*, 1884.
[2] *Berlin Acad., Math. und Naturwissensch. Abt.*, 1882, **1**, 7.

"maximum work" was called by Helmholtz the "free energy" of the change, but this term is now, on the suggestion of G. N. Lewis (1876–1946), of the University of California, restricted to mean the "work available for use." When, therefore, one system passes spontaneously into another, the maximum *useful* work which becomes available represents the decrease in the free energy of the system and may be taken as representing the "affinity" of the chemical process. Its value can be calculated from a knowledge of the equilibrium constant and is equal to $RT\log_e K$. In the case of reactions involving electrolytes, the electrical energy supplied by a suitable voltaic cell is equal to the decrease of free energy or the affinity of the chemical reaction occurring in the cell.

To the first and second laws of thermodynamics, Hermann Walther Nernst (1864–1941), Professor of Physical Chemistry in the University of Berlin, added a third theorem on the basis of which it is possible to calculate the free energy or the maximum work from purely thermal measurements.[1]

As far back as 1748 it had been observed by the French *Abbé*, Jean Antoine Nollet (1700–70), a Professor of Physics in Paris, that when a bottle, filled with spirit of wine and closed by a piece of pig's bladder, was immersed in water, water passed through the membrane and created a pressure within the bottle which sufficed to burst the animal membrane. This phenomenon of the diffusion of a liquid through a membrane, or *osmosis* as it is now called,[2] was later rediscovered by others who found that a pressure is produced whenever an aqueous solution is separated from pure water by an animal membrane or by parchment paper. It was not, however, till 1827 that quantitative measurements of the pressure were carried out by the French physiologist, René Joachim Henri Dutrochet (1776–1847), who was attracted to the problem mainly on account of its physiological interest;[3] and the conclusion drawn by Dutrochet that the pressure is proportional to the concentration of the solution was confirmed by Karl Vierordt (1818–84) at Karlsruhe in 1848.[4]

The pressures produced in the experiments referred to were

[1] *Nachr. Gesell. Wissensch. Göttingen: Math.-phys. Klasse*, 1906, 1.
[2] The terms *endosmosis* and *exosmosis*, applied to the oppositely directed diffusion currents, were first used by Dutrochet. The term *osmosis* is derived from the Greek ὡσμός (ōsmos), a push.
[3] *Ann. Chim.*, 1827, 35, 393; 1828, 37, 191; 1832, 49, 411; 51, 159. See A. Findlay, *Osmotic Pressure* (Longmans).
[4] *Ann. Physik*, 1848, 73, 519.

obtained with membranes permeable both for solute and for solvent, and were due to a difference in the rate of diffusion through the membrane. The pressure, therefore, was only a temporary one and depended on the nature of the membrane.

In 1867, however, Moritz Traube (1826–94), in Breslau, showed that membranes could be produced artificially which were permeable for water but not for certain dissolved substances. They were similar, therefore, in this respect to the membranes surrounding plant and animal cells. With such membranes, which van't Hoff later called *semipermeable* membranes, the pressure produced by osmosis (called *asmotic pressure*), is permanent and is a maximum pressure for the given solution. It is to such pressures, produced with the aid of semipermeable membranes, that the term "osmotic pressure of a solution" is applied.

Among the semipermeable membranes prepared by Traube was one of copper ferrocyanide, and such a membrane, formed in the walls of a porous earthenware pot, was employed in 1877 by the German botanist, Wilhelm Pfeffer (1845–1920), for the quantitative measurement of the osmotic pressure of solutions of cane sugar and other substances.[1] The results obtained by Pfeffer showed that the osmotic pressure is proportional to the concentration of the solution and also that it increases with rise of temperature.

The osmotic investigations of Pfeffer were carried out on account of their botanical importance, but their central point of interest came later to lie in the theory of dilute solutions which was propounded by van't Hoff in 1885 and which opened a new and important chapter in the history of physical chemistry. In an account of the genesis of this theory,[2] van't Hoff relates how, as a young student of chemistry, his interest was aroused in the question of the influence of chemical constitution on the reactivity of organic compounds,[3] and how he was thereby led to the study of reaction velocities and to the mathematical and experimental investigation of chemical equilibria in gaseous systems and in liquid solutions. While his mind was exercised with the problem of chemical affinity, van't Hoff was made acquainted, by his botanical colleague, Hugo de Vries, with the measurements of osmotic pressure which had been carried out by Pfeffer; and he recognized, as also had Moritz Traube at an earlier date, that such determinations of

---

[1] *Osmotische Untersuchungen*, 1877.
[2] *Ber.*, 1894, **27**, 6.
[3] *Ansichten über die organische Chemie*, 1881.

osmotic pressure afford a means of measuring the "water attraction" of the solute[1] for water, or the maximum work obtainable by the addition of pure solvent to a solution. Moreover, as van't Hoff wrote:[2] "In the course of an investigation which had as its chief object a knowledge of the laws of chemical equilibrium in solutions, it became gradually apparent that there is a profound analogy, indeed almost an identity, between them and gases, more especially also in physical properties, provided that for the ordinary pressure of gases one substitutes, in the case of solutions, the so-called osmotic pressure."

Although the analogy between solutions and gases had doubtless been imperfectly perceived even by Gay-Lussac and others, it is to van't Hoff that the credit belongs of recognizing that the conception of osmotic pressure and of semipermeable membranes makes it possible to apply the second law of thermodynamics, with ease and clearness, to the theoretical investigation of the *quantitative* relations between the properties of solutions and their concentration, in a manner similar to that employed in the case of gases. This constituted a distinct advance on the earlier attempts made by Gustav Robert Kirchhoff (1824–87), Professor of Physics in the University of Heidelberg, and by Helmholtz to apply thermodynamics to the study of solutions.

The experiments of Pfeffer showed, as we have seen, that the osmotic pressure of solutions of cane sugar is proportional to the concentration, or inversely proportional to the volume of solution in which a given amount of solute is contained. This law can be explained, on analogy with gas pressure, as due to the bombardment of the semipermeable membrane by the solute molecules. Even if the osmotic pressure is regarded as due to an attraction between solvent and solute molecules, its value will also be proportional to the number of attracting molecules in unit volume provided the solution is so dilute that the mutual attraction between the molecules is negligible. This law, then, is analogous to Boyle's law for gases.

The osmotic pressure depends also on the temperature, as Pfeffer had shown; and van't Hoff, by the method of Carnot's thermodynamic cycle, deduced the law, confirmed by Pfeffer's measurements,

[1] This term was introduced by Prof. N. Story-Maskelyne in 1894 (see W. J. Pope, *Nature*, 1936, **137**, 782).
[2] *Z. physikal. Chem.*, 1887, **1**, 481; *Phil. Mag.*, 1888, **26**, 81. See also *Arch. néer.*, 1886, **20**, 239; *Kgl. Svenska Vetenskaps-Akad. Handlingar*, 1886, **21**, 3; Ostwald's, *Klassiker*, No. 110.

that the osmotic pressure is proportional to the absolute temperature—a law analogous to Gay-Lussac's law for gases.

A further important advance was made by van't Hoff when, by the application of thermodynamics, he showed that in the case of a dissolved gas the osmotic pressure of the solution is numerically equal to the pressure which the solute would exercise in the gaseous state if it occupied a volume equal to the volume of the solution. From this deduction, which forms the basis of the van't Hoff *theory of solutions*, the possibility emerges that, since dilute solutions even of non-gaseous solutes follow laws analogous to Boyle's law and Gay-Lussac's law for gases, they might also obey a law analogous to Avogadro's law. The truth of this remarkable suggestion was confirmed by the measurements of Pfeffer. In the case, therefore, of solutions containing one gram-molecule of solute in V litres of solution, the product of volume and osmotic pressure (P) is equal to RT, and R is found to have the same value in the case of dilute solutions as in the case of gases.

This law was obviously of the greatest importance for it pointed the way to a method of determining the molecular weight of a substance in solution. Since substances in dilute solution obey laws analogous to the gas laws—not only qualitatively but also quantitatively—it follows that solutions of the same molecular concentration have the same osmotic pressure at a given temperature. Moreover, just as one gram-molecule of a gas at o° C. occupies a volume of 22·4 litres under a pressure of 1 atmosphere, so a solution containing 1 gram-molecule of solute in 22·4 litres of solution has, at o°, an osmotic pressure of 1 atmosphere; and the molecular weight of a solute in grams will be that weight in grams which when dissolved in 22·4 litres of solution gives, at o°, an osmotic pressure of 1 atmosphere. The osmotic pressure of a solution, like the pressure of gases, is a colligative property.

In view of the difficulty of obtaining suitable semipermeable membranes and of determining the osmotic pressure with accuracy, van't Hoff's theory of solutions would have been of comparatively little practical importance if it had not been possible to bring osmotic pressure into definite relationship with other properties of solutions.

The osmotic pressure of a solution is a measure of the difference of the free energy of the solvent in the pure state and in the solution; and the osmotic work, PV, represents the energy required to separate, isothermally and reversibly, a volume V of the solvent from a dilute solution of osmotic pressure, P by means of a semi-permeable membrane. The same volume of solvent can, however, be separated from the solution by evaporation, and van't Hoff was able to show that there is a definite relationship between the osmotic pressure of a solution and its vapour pressure.

That the vapour pressure of a liquid is lowered by dissolving a substance in it, and that the relative lowering of the vapour pressure is proportional to the concentration of the solution, had been discovered by Clemens Heinrich Lambert von Babo (1818–99) in 1847 and by Adolf Wüllner (1835–1908)[1] in 1856. The earliest investigations of the lowering of vapour pressure were made on aqueous solutions of salts, and the complications due, as will be pointed out presently, to the ionization of the salts prevented the discovery of any well-defined laws. In 1886, however, François Marie Raoult (1832–1901), Professor of Chemistry in the University of Grenoble, by investigating the vapour pressure of solutions of non-electrolytes and in solvents other than water, was able to establish general laws of a simple character.[2] Thus, it was found that the relative lowering of the vapour pressure is proportional to the molecular concentration of the solute and that it is also equal to the ratio of the number of gram-molecules of solute to the total number of gram-molecules in the solution.[3] The vapour pressure of a solution, therefore, is also a colligative property, by means of which it is possible to determine molecular weights.

Just as a solute lowers the vapour pressure so also it raises the boiling-point and depresses the freezing-point of a solvent; and the elevation of the boiling-point[4] and the depression of the freezing-point[5] are proportional to the molecular concentration of the solution. By means, therefore, of determinations of the lowering of the vapour pressure, the elevation of the boiling-point and the depression of the freezing-point, it is possible to ascertain the molecular weight of substances in dilute solution, and the experimental technique of such determinations was developed by a number of workers.[6] By such determinations the theory of van't Hoff was confirmed.

[1] *Ann. Physik*, 1858, **103**, 529; **105**, 85; 1860, **110**, 564.
[2] *Compt. rend.*, 1886, **103**, 1125.
[3] *Compt. rend.*, 1887, **104**, 1430; *Z. physikal. Chem.*, 1888, **2**, 353.
[4] Faraday, *Ann. Chim.*, 1822, **20**, 324; Griffiths, *Ann. Physik.*, 1824, **2**, 227; Beckmann, *Z. physikal. Chem.*, 1889, **4**, 550.
[5] Blagden, *Phil. Trans.*, 1788, **78**, 277; Raoult, *Compt. rend.*, 1882, **94**, 1517; 1883, **95**, 187, 1030; 1892, **114**, 268, 440; 1897, **125**, 751.
[6] J. Walker, *Z. physikal. Chem.*, 1888, **2**, 602; Will and Bredig, *Ber.*, 1889, **22**, 1084; Beckmann, *Z. physikal. Chem.*, 1889, **4**, 550; S. L. Bigelow, *Amer. Chem. J.*, 1899, **22**, 280; and many later publications.

The theory of solutions put forward by van't Hoff was explicitly stated to be valid only in the case of infinitely dilute solutions, for only in such cases are the assumptions permissible which were made in the thermodynamic deduction. Fortunately, in the earlier years of the theory, this restriction was not felt too seriously by reason of the fact that in the chief applications of the theory to the determination of molecular weights and to the investigation of physiological problems, the errors involved were not sufficiently great or the solutions employed were, in any case, sufficiently dilute. This fact, however, combined with the extreme simplicity of the theory, doubtless led, in some cases, to an unwarranted extension of the theory and to the drawing of many false deductions regarding the constitution of solutions. During the earlier years of the present century attention was turned to the experimental study of concentrated solutions, and this served both to define the limits of concentration within which van't Hoff's theory may be accepted and to give a basis for a wider and more general mathematical treatment. To these investigations consideration will be given later (Chapter XII).

Although the simple van't Hoff law, $PV = RT$, was found to apply to solutions of organic compounds, like cane sugar, it did not apply to solutions of salts, the osmotic pressure of which had been found by de Vries to be greater than that of solutions of cane sugar of equimolecular concentration.[1] In the case of salt solutions, therefore, one has to write $PV = iRT$, the coefficient i representing the number of times the osmotic pressure of the given salt solution is greater than that of an equimolecular solution of cane sugar. Its value varies not only with the nature of the salt but also with the dilution. This anomaly in the behaviour of solutions of salts, as well as of acids and alkalis, an anomaly which, as the experiments of Raoult showed, is also found in the depression of the freezing-point, seriously affected the range of applicability of van't Hoff's theory and threatened to prevent its acceptance. The anomalous behaviour of salts, acids and alkalis, could of course, be explained—in analogy with the explanation of the abnormal vapour density of ammonium chloride (p. 48)—by assuming that the molecules of the solute undergo some sort of dissociation. The occurrence of dissociation, however, seemed too improbable, and the difficulty of answering the question, What then are the products of the assumed dissociation? seemed insurmountable. Nevertheless,

[1] *Pringsheim's Jahrbücher*, 1884, **14**, 427.

in 1887 an answer to this question and an explanation of the abnormal behaviour of salt solutions was given by the Swedish physicist, Svante Arrhenius, who brought the osmotic pressure into relation with the electrical properties of solutions and pointed out that the substances which yield solutions which do not behave in accordance with the van't Hoff theory are just those substances which, in solution, conduct the electric current with decomposition—*electrolytes* as Faraday had termed them. He put forward the view, therefore, that electrolytes (salts, acids and alkalis), in solution, undergo a spontaneous dissociation into positively and negatively charged part-molecules or *ions* (cations and anions), which, according to Faraday, transport the electric current through a solution during electrolysis.[1] From his determinations of the electrical conductivity of solutions of salts, acids, and alkalis, moreover, Arrhenius was led to conclude that this so-called *electrolytic dissociation* increases with dilution; and he was thereby able also to explain the increase in the value of i with dilution which had been observed by Raoult and by de Vries.

It is true that Williamson[2] and Clausius[3] had expressed the view that a process of mutual reaction and exchange takes place between the molecules of electrolytes and that dissociation of the molecules and recombination of their parts are perpetually occurring; and Hittorf was of opinion[4] that "the ions of an electrolyte cannot be combined in a stable form to whole molecules and these cannot exist in a definite regular arrangement." But although the revolutionary idea of a spontaneous dissociation of electrolyte molecules thereby found expression, no opinion was offered regarding the concentration of the ions at any given moment.

In his electrolytic dissociation theory, however, which he first stated in a letter to van't Hoff dated March 30th, 1887,[5] Arrhenius not only assumed that the degree or extent of electrolytic dissociation might be large but he also pointed out how it might be determined.[6]

---

[1] *Bihang till Kongl. Vetenskapsakad. Handlingar*, 1884, **8**, Nos. 13 and 14; Ostwald's *Klassiker*, No. 160.

Almost at the same time as Arrhenius propounded his theory of electrolytic dissociation, Max Planck (*Z. physikal. Chem.*, 1887, **1**, 577) suggested the occurrence of a dissociation of salt molecules in solution.

[2] *Ann. Chim.*, 1851, **77**, 37; *J. Chem. Soc.*, 1852, **4**, 110.

[3] *Ann. Physik*, 1857, **101**, 338.

[4] *Ann. Physik*, 1858, **103**, 54.

[5] Walker, *J. Chem. Soc.*, **1928**, 1392; *Z. physikal. Chem.*, 1887, **1**, 631.

[6] *Z. physikal Chem..*, 1888, **2**, 491. See also Ostwald, *ibid.*, pp. 36, 270; Planck, *Ann. Physik*, 1888, **34**, 139.

According to Arrhenius the molecular conductivity of a solution was a measure of the concentration of the ions, and the degree of electrolytic dissociation, or the "activity coefficient," a, was given by the ratio of the molecular conductivity at the given dilution to the molecular conductivity at infinite dilution, when all the molecules of electrolyte were regarded as being dissociated into ions. Moreover, the coefficient i, which is given by the ratio of the observed depression of the freezing-point to the depression which would be given if no dissociation took place, is equal to $1 + (n-1)a$, where n is the number of ions into which a molecule of the electrolyte dissociates. The degree of dissociation, a, could therefore also be determined by cryoscopic measurements, and experiment showed that the values of a obtained by these two very different methods were in remarkable agreement.

The theory of electrolytic dissociation, as is not to be wondered at, met at first with considerable opposition, largely owing to the difficulty of accounting for the formation and stable existence of the ions. By some, the formation of ions was attributed to a process of hydration, an idea first suggested by Arrhenius, and the assumption was made that the affinity between the water and the ions is greater than that between the water and the salt molecules, or that the "residual affinity" of the solvent molecules deflected or weakened the Faraday tubes of force by which the atoms were supposed to be held together in the molecule.[1] Although more recent work has demonstrated the necessity of taking account of the solvent and of the part which it plays in the formation of ions, the simpler view that the solvent acts as an inert medium was, in the early days of the theory, found to be adequate. Fuller light has been thrown on the process of ion formation by the modern theories of valency and the electronic constitution of the atom.

From the first, the electrolytic dissociation theory was enthusiastically accepted and acclaimed by Wilhelm Ostwald, who in fact became its apostle;[2] and the power of his advocacy, the extraordinary fruitfulness of the theory and the means which it offered of co-ordinating *quantitatively* the properties of electrolytic solutions soon overcame the opposition of all but a very few chemists. Inspired by the theories of van't Hoff and Arrhenius, the investigation of the properties of solutions of non-electrolytes and of

[1] One of the earliest and strongest champions of the view that there is an interaction between the electrolyte molecule and the solvent was H. E. Armstrong (*Proc. Roy. Soc.*, 1886, **40**, 268).

[2] *Endeavour*, 1959, **18**, 59.

electrolytes was pursued by the younger chemists of both hemispheres with all the fervour of a new age. It was, in fact, a new era which had opened, and the theories of van't Hoff and Arrhenius exercised a revitalizing action not only on the physical but also on the biological sciences and led to a recognition of the importance of the physico-chemical investigation of problems of plant and animal physiology.

Not only did the electrolytic dissociation theory reconcile the osmotic behaviour of electrolytic solutions with the theory of van't Hoff,[1] but it afforded an explanation of a large and varied range of properties and phenomena—the "avidity" of acids and alkalis as determined thermochemically by Thomsen[2] and by density measurements by Wilhelm Ostwald;[3] the catalytic action of acids and of alkalis; the colour and other physical properties of dilute solutions of salts;[4] Hess's law of thermoneutrality; the heat of neutralization; the "strength" of acids and alkalis; equilibria in solutions of electrolytes; reactions in analytical chemistry, etc.

Large, however, as was the field of fruitful application of the electrolytic dissociation theory, the behaviour of "strong" electrolytes gave rise to difficulties of interpretation. Whereas, in the case of "weak" electrolytes, the law of mass action could be applied to the equilibrium between undissociated molecules and ions (Ostwald's dilution law, 1888), this law was inapplicable to strong electrolytes. This was, of course, a grave defect of the theory, and it is only in recent years, as will be discussed later, that this defect was remedied and an explanation offered of the behaviour of strong electrolytes.

While the electrolytic dissociation theory was of inestimable value in the interpretation of the more purely chemical behaviour of electrolytes, it also afforded a means of interpreting and co-ordinating the phenomena associated with electrolysis and the behaviour of voltaic cells in a more satisfactory and adequate manner than had hitherto been possible. Faraday's laws of electrolysis were made intelligible through the conception of ions carrying

---

[1] Arrhenius developed his theory of electrolytic dissociation, the germ of which was contained in his doctoral dissertation of 1884, only after becoming acquainted with the van't Hoff theory of solutions in 1886–7. See Arrhenius, *J. Chem. Soc.*, 1914, **105**, 1414.

[2] *Thermochemische Untersuchungen*, I, 63.

[3] *J. prakt. Chem.*, 1877 [2], **16**, 385.

[4] H. Landolt, *Ber.*, 1873, 6, 1073; N. Kajander, *J. Russ. Phys. Chem. Soc.*, 1881, **13**, 474; A. C. Oudemans, *Ann. Physik. Beibl.*, 1885, 9, 635; Raoult, *Ann. Chim.*, 1885, **4**, 427; Ostwald, *Z. physikal. Chem.*, 1892, **9**, 579.

one or more units of electricity, according to the valency of the ion, the amount of electricity associated with one gram equivalent of a substance in the ionic state being the same for all ions. Further, on the basis of the electrolytic dissociation theory, the potential on the electrodes during electrolysis was now understood to be not disruptive—as had been thought to be the case by Theodor von Grotthuss (1785–1822) in 1805—but only directive in action.[1] By means of that theory, also, one can explain the fact that the equivalent conductivity of an electrolyte at infinite dilution is equal to the sum of two factors, one depending on the cation and one on the anion—the so-called law of the independent migration of the ions, discovered by Friedrich Wilhelm Kohlrausch (1840–1910), Professor of Physics in the University of Würzburg.[2] The unequal changes of concentration which take place at the electrodes during electrolysis, and which had been investigated more especially by Johann Wilhelm Hittorf (1824–1914), Professor of Physics and Chemistry in the University of Münster,[3] showed that the ions move with different velocities. It was, moreover, on the basis of the theories of van't Hoff and Arrhenius that Nernst developed his explanation of electrode potentials, liquid junction potentials and electromotive force of voltaic cells.[4]

[1] It had been pointed out by Clausius that since electrolytic solutions obey Ohm's law, none of the electrical energy could be used up in disrupting or decomposing the molecules of the electrolyte.

[2] *Göttinger Nachrichten*, 1876, 213; *Ann. Physik*, 1879, **6**, 167.

[3] *Ann. Physik*, 1853, **89**, 177; 1856, **98**, 1; 1858, **103**, 1; 1859, **106**, 337; Ostwald's *Klassiker*, Nos. 21 and 23.

[4] *Z. physikal. Chem.*, 1888, **2**, 613; 1889, **4**, 129; *Ann. Physik*, 1892, **45**, 360.

CHAPTER SEVEN

ORGANIC CHEMISTRY IN THE SECOND HALF OF THE NINETEENTH CENTURY. THE SYNTHESIS OF ORGANIC COMPOUNDS, AND THE THEORIES OF CHEMICAL STRUCTURE

ORGANIC chemistry, during the first half of the nineteenth century, was essentially a chemistry of analysis, which accepted, as at that time the only possible principle of development, the view of Lavoisier[1] that "chemistry advances towards its goal and towards its perfection by dividing, subdividing and re-subdividing." Chemists, therefore, devoted themselves more especially to isolating the compounds occurring in plants and animals, to determining their composition and to investigating their properties. The rapid increase in the number of compounds which resulted from these investigations, however, and the growing recognition of the importance not only of composition but also of constitution, led to the schemes of classification and the theories of molecular structure which have been discussed in previous chapters. In the sixties of last century the theories of Couper and of Kekulé had given to chemists a means of representing the constitution and of interpreting the chemical behaviour of organic compounds on the basis of the arrangement of the atoms in the molecule; and the special cases of geometrical isomerism were rendered intelligible by the later theories of van't Hoff and Le Bel. Not until the theories of chemical structure had been sufficiently developed could the second task of organic chemistry, that, as Gerhardt put it, "of acquiring a knowledge of the methods of building up organic substances outside the living organism," be successfully undertaken. Through the stimulus and guidance afforded by the theories of molecular structure, we find that organic chemistry, defined as the chemistry of the carbon compounds, became, during the second half of the nineteenth century, essentially synthetic; and it is in the synthesizing and production in the scientific laboratory not only of nature's own products, but of countless other compounds of carbon that organic chemistry has

[1] *Traité élémentaire de Chimie*; *Œuvres*, I, 137.

achieved its crowning glory.[1] In establishing the constitution and chemical nature of a compound, moreover, synthesis is the complement of analysis, and the synthetic method confirms and makes clear the indications of the analytic.

Although the synthetic period of organic chemistry may, perhaps, be said to have been ushered in by the French chemist, Marcelin Berthelot, who, in 1860, published his *Chimie organique fondée sur la synthèse*, certain naturally occurring and other organic compounds had long before then been prepared in the laboratory by processes of decomposition or transformation. Thus, oxalic acid was obtained by Bergman and by Scheele in 1766 by the oxidation of sugar, and again, in 1829, by Gay-Lussac by the action of caustic potash on wood; and by the action of dilute acids on starch, glucose was produced.[2] Urea (carbamide), as we have seen, was obtained by Wöhler in 1828 by heating ammonium cyanate (p. 19), and acetone, long known and formed as one of the products of the distillation of wood, was prepared by Eugène Melchior Péligot (1811–90), Professor of Chemistry at the Conservatoire des Arts et Métiers, Paris, by heating calcium or barium acetate.[3]

Of more strictly synthetic reactions, one may mention the production by Hennell in 1826 of alcohol from ethylene by way of sulphovinic acid (p. 19), and the remarkable synthesis of acetic acid by Kolbe in 1845, the first complete synthesis of an organic compound to be carried out in the laboratory.[4] By the action of chlorine on carbon disulphide, a compound which can be formed by the direct combination of its elements, Kolbe obtained carbon tetrachloride, CCl_4; and when this compound was acted upon by sunlight in the presence of water and chlorine, trichloracetic acid, $CCl_3 \cdot COOH$, was formed. By the action of potassium amalgam the three chlorine atoms were removed and replaced by hydrogen,[5] yielding acetic acid, $CH_3 \cdot COOH$. This acid was identical in all respects with the acid occurring as one of the products of distillation of wood or formed by the acetous fermentation (oxidation) of alcohol and present in vinegar.

These and not a few other organic substances had thus been produced by the art of the chemist before the middle of the nine-

---

[1] It may be noted that the term synthesis is used not merely to signify a "building up" or "putting together" of more complex from less complex molecules, but also to denote the production, through decomposition or transformation, of organic compounds outside the living organism.

[2] Kirchhoff, *J. Phys. Chim.*, 1812, **74**, 199.　　　[3] *Annalen*, 1834, **12**, 39.

[4] *Annalen*, 1845, **54**, 181.　　　[5] Melsens, *J. Pharm.*, 1842, **1**, 157.

teenth century. Important, however, as they were, the number of such syntheses was comparatively small; and it may be said that until the publication of Berthelot's work, no systematic or consciously directed effort had been made to produce in the laboratory the varied range of natural compounds with which chemists had become familiar, or to build up other carbon compounds from their elements or from less complex compounds. The chief aim and purpose of Berthelot was, therefore, to bring together and to direct the attention of chemists to the general principles and methods of synthetic organic chemistry, to stimulate investigation in the field of organic synthesis, and so to prove beyond all question that compounds identical with those produced by plants and animals can be synthesized from inorganic or mineral matter. This, Berthelot considered, was the fundamental problem of organic chemistry.

After the middle of last century the number of workers in the domain of organic chemistry rapidly increased—largely owing to the development of the important dye industry, to which attention will be directed later—and as the investigation of the behaviour of different compounds was more and more actively pursued, a continuously and rapidly expanding knowledge of the chemical behaviour and constitution of different classes of compounds was gained. In this way the number of methods of general applicability, by means of which definite transformations could be effected, rapidly increased.

Since the hydrocarbons constitute the simplest series of compounds from which other, more complex, compounds can be derived, an important advance was made when Berthelot showed that the hydrocarbon, acetylene, C_2H_2, discovered by Edmund Davy in 1836, can be obtained by the direct combination of carbon and hydrogen at the temperature of the electric arc, as well as in other ways. By acting on copper acetylide with nascent hydrogen, ethylene, C_2H_4, is formed.

Another fact of very great importance established by Berthelot in 1870 is that when acetylene is heated to a dull red heat three molecules of acetylene combine together and form benzene, C_6H_6, and at still higher temperatures more complex hydrocarbons, such as naphthalene, $C_{10}H_8$, and anthracene, $C_{14}H_{10}$, are formed. It was thereby shown to be possible to synthesize the hydrocarbons not only of the aliphatic but also of the aromatic series; and in this way the foundation was laid for the complete synthesis of many of the most important naturally-occurring compounds.

Even at a much earlier time, it may be mentioned, hydrocarbons of the methane series had been prepared indirectly, in 1849, by E. Frankland, who was, in fact, the first to carry out such syntheses.[1] Thus by heating alkyl halides, e.g., C_2H_5I, with zinc, zinc iodide was formed and the two ethyl groups, incapable of independent existence, combined to form butane, $CH_3 \cdot CH_2 \cdot CH_2 \cdot CH_3$. By the action of zinc on alkyl iodides, moreover, compounds of the type $C_2H_5 \cdot Zn \cdot I$ and $Zn(C_2H_5)_2$ are formed, and these are also capable of extensive use in synthetical operations. Although these organo-zinc compounds are now no longer used for synthetic purposes, they are to be regarded as the forerunners of the very important magnesium alkyl halides (Grignard reagents)[2] by means of which it is possible to synthesize a large number of different classes of compounds—hydrocarbons; primary, secondary and tertiary alcohols; phenols; acids; ketones, etc. The use of this synthetic agent has, in more recent years, led to a rapid development of organic chemistry.

In the course of an investigation which had as its object the synthesis of compounds containing an "asymmetric silicon atom" and the resolution of such compounds into optically active isomers (in analogy with compounds containing an asymmetric carbon atom), Frederic Stanley Kipping, Professor of Chemistry in University College, Nottingham, discovered, in 1904, that silicon tetrachloride and organo-silicon chlorides react with the Grignard reagent in accordance with the equations:

$$R \cdot MgBr + SiCl_4 = R \cdot SiCl_3 + MgBrCl$$
$$R \cdot MgBr + R \cdot SiCl_3 = (R)_2 \cdot SiCl_2 + MgBrCl$$

where R is a hydrocarbon radical. In this way mixtures of compounds, $R \cdot SiCl_3$, R_2SiCl_2, R_3SiCl, were obtained. These halogen compounds are readily hydrolysed by water, with replacement of the halogen atoms by $\cdot OH$ and formation of compounds called *silicols* (now called silanols). These undergo condensation with elimination of water and formation of compounds which, on analogy with ketones, received the name *silicones*. During and since World War Two they have found important industrial applications (p.298).

About the middle of the nineteenth century Kolbe[3] showed

[1] *Annalen*, 1849, **71**, 171; **74**, 41; **77**, 221; *J. Chem. Soc.*, 1850, **2**, 263. At a somewhat later time Adolphe Wurtz found that sodium could be used in place of zinc (*Ann. Chim.*, 1855 [3], **44**, 175).
[2] Victor Grignard, *Compt. rend.*, 1900, **130**, 1322; *Ann. Chim.*, 1901 [7], **24**, 433.
[3] *Annalen*, 1849, **69**, 279.

that when a solution of potassium acetate, $CH_3 \cdot COOK$, is electrolysed, a mixture of carbon dioxide and ethane is obtained; and at a much later date the method of electrolysis was applied by Crum Brown and James Walker in the synthesis of the higher dibasic acids and their esters.[1] Since then electrolysis has played an important part in organic syntheses.

To illustrate the advances made in the field of organic chemistry during the second half of the nineteenth century, brief mention may be made of a few of the more important synthetic methods which have found a very widespread or general application to the synthesis of organic compounds.

In the case of the aromatic or benzenoid compounds, hydrocarbons can be synthesized by means of a reaction introduced by the French chemist, Charles Friedel (1832–98), Professor of Mineralogy at the Sorbonne, Paris and the American chemist, James Mason Crafts (1839–1917), Professor of Organic Chemistry, Massachusetts Institute of Technology, in 1877. Thus, by treating benzene or one of its homologues with an alkyl chloride in presence of anhydrous aluminium chloride, the hydrogen atoms of the benzene ring can be replaced by alkyl groups; and in this way a great variety of benzene hydrocarbons, e.g., $C_6H_5 \cdot CH_3$, $C_6H_4(CH_3)_2$, $C_6H_4(CH_3) \cdot C_2H_5$, etc., can be built up.[2] This method also is capable of considerable extension to the synthesis of ketones, aldehydes, amides, etc., so that the Friedel-Crafts reaction has proved to be of very great synthetic value.

Benzene hydrocarbons, similarly, can be synthesized by heating a halogen derivative of benzene, e.g., C_6H_5Br, with an alkyl halide in presence of sodium. The halogen atoms are removed as sodium salts and the two hydrocarbon residues combine,[3] e.g., $C_6H_5 \cdot C_2H_5$.

For the purpose of increasing the number of carbon atoms in a molecule and thus building up more complex structures, the cyanide or nitrile radical, CN, is of importance. This radical may be introduced into a molecule in various ways. Thus, when an alkyl bromide or iodide is heated with potassium cyanide, the halogen is replaced by the CN-group, and one obtains a nitrile, e.g., $C_2H_5 \cdot CN$. When this compound is boiled with an acid or alkali, the nitrile group forms the COOH-group and ammonia.[4] From ethyl bromide, therefore, one obtains propionic acid,

---

[1] *Annalen*, 1891, **261**, 107; 1893, **274**, 41.
[2] *Compt. rend.*, 1877, **84**, 1392, 1405.
[3] This is known as Fittig's reaction (*Annalen*, 1864, **131**, 303).
[4] E. Frankland and H. Kolbe, *Annalen*, 1848, **65**, 298.

$CH_3 \cdot CH_2 \cdot COOH$. This synthetic process was made use of in 1860 by Maxwell Simpson[1] who, by replacing the bromine atoms in ethylene dibromide, $BrCH_2 \cdot CH_2Br$, by CN, obtained ethylene dicyanide, $CNCH_2 \cdot CH_2CN$. On hydrolysis this compound yields succinic acid, $COOH \cdot CH_2 \cdot CH_2 \cdot COOH$. From this acid W. H. Perkin and Baldwin Francis Duppa (1828–73) had already prepared racemic acid,[2] the synthesis of which from ethylene was therefore now complete. Use was also made of the above reaction by Kolbe and by Hugo Müller for the purpose of synthesizing dibasic acids from the monohalogen-substituted monobasic acids,[3] e.g.,

$$CH_2Cl \cdot COOH \rightarrow CH_2CN \cdot COOH \rightarrow CH_2(COOH)_2.$$

Further, owing to the fact that aldehydes and ketones readily react with hydrogen cyanide to form additive compounds, which, on hydrolysis, yield acids, a means is given of synthesizing hydroxy-acids, e.g.,

$$CH_3 \cdot CHO + HCN \rightarrow CH_3 \cdot CH(OH) \cdot CN \rightarrow CH_3 \cdot CH(OH) \cdot COOH$$
(lactic acid).

The presence of carbonyl (CO) groups in a compound renders the hydrogen attached to a neighbouring carbon atom more reactive, and such compounds become valuable synthetic agents. Thus the two carbonyl groups in malonic acid, $CH_2\begin{cases} CO \cdot OH \\ CO \cdot OH \end{cases}$ render the hydrogen atoms of the methylene (CH_2) group reactive, so that when the ethyl ester of malonic acid is mixed with sodium ethoxide, C_2H_5ONa, an atom of hydrogen is replaced by an atom of sodium and one obtains $CHNa(COOC_2H_5)_2$. When this compound is brought in contact with an iodide, e.g., C_2H_5I, sodium iodide is eliminated and the ethyl group takes the place of the sodium atom, giving $C_2H_5 \cdot CH(COOC_2H_5)_2$. The second hydrogen atom can similarly be replaced by an alkyl group, and one obtains, for example, $C\begin{cases} C_2H_5 \\ C_2H_5 \end{cases}\begin{cases} COOC_2H_5 \\ COOC_2H_5 \end{cases}$. In this way one can build up the esters of various complex dibasic acids.[4] Even

---

[1] Proc. Roy. Soc., 1860, 10, 574; Annalen, 1861, 118, 375.

[2] Annalen, 1861, 117, 130.

[3] Kolbe, Annalen, 1864, 131, 348; Müller, ibid., p. 350.

[4] The reactivity of the methylene group in malonic ester was first observed by van't Hoff (Ber., 1874, 7, 1383), by whom also the importance of malonic ester as a synthetic agent was realized. Such syntheses were later studied more fully by Max Conrad and C. A. Bischoff (Annalen, 1880, 204, 121), and others.

"rings" of carbon atoms can be built up by allowing a dihalogen compound such as $CH_2\begin{smallmatrix}CH_2Br\\CH_2Br\end{smallmatrix}$ to act on the disodium malonic ester, whereby there is formed $CH_2\begin{smallmatrix}CH_2\\ \\CH_2\end{smallmatrix}C(COOC_2H_5)_2$; and the hypnotic, diethylbarbituric acid or *veronal*,

$$CO\begin{smallmatrix}NH-CO\\ \\NH-CO\end{smallmatrix}C(C_2H_5)_2,$$

one of the products of modern synthetic chemistry,[1] may be prepared from carbamide (urea) and diethylmalonic ester.

By hydrolysing the esters obtained in these synthetic processes, the free dibasic acids are formed; and when these dibasic acids are heated carbon dioxide is evolved and a monobasic acid results. The higher monobasic acids of the formic acid series can in this way also be synthesized.

The use of malonic ester as a synthetic agent is similar to that of acetoacetic ester ($CH_3 \cdot CO \cdot CH_2 \cdot COOC_2H_5$), which was first studied by Anton Geuther (1833–89),[2] Professor of Chemistry at Jena, in 1863, and by E. Frankland and B. F. Duppa[3] in 1865. In this case also, the hydrogen atoms of the methylene group can be replaced by hydrocarbon radicals, with the formation therefore of more complex compounds. These ketonic esters, as they are called, undergo hydrolytic decomposition with production of ketones or acids, according to the conditions under which the decompositions are brought about, and the possibility is therefore given of synthesizing various ketones and acids. In more recent times, acetoacetic ester has been used in the synthesis of a number of special compounds having a cyclic or "ring" structure.

By these and other synthetic processes, *e.g.*, various so-called condensation processes, chemists have succeeded, more especially during the past half or three-quarters of a century, in preparing synthetically hundreds of thousands of organic compounds, some

---

[1] M. Conrad and M. Guthzeit, *Ber.*, 1882, **15**, 2844; E. Fischer and von Mering, *Therapie der Gegenwart*, 1903.

[2] *Jahresbericht*, 1863, p. 323; 1865, p. 302.

[3] *Annalen*, 1865, **135**, 217; 1866, **138**, 204, 328. See also J. Wislicenus, *ibid.*, 1877, **186**, 161.

of which, it is true, are naturally occurring, but most of which are known only as artificial products of the laboratory.

Although, in the minds of Gerhardt and of Berthelot, the most important task of the organic chemist was the artificial production in the laboratory of the compounds formed in the living animal or vegetable organism, the promulgation of the theories of molecular structure by Couper and Kekulé (p. 37) extended the horizon of organic chemistry; and this soon found a wider aim and motive in the desire not only to test the theories of molecular structure but also to gain a deeper insight into the constitution of the important groups of naturally-occurring compounds. To this aim and motive, also, there was added the attraction of exploring the great world of molecular architecture, a vision of which was opened up by Couper and Kekulé, and of learning how to build up molecules of ever-increasing size and variety. It is clear, moreover, that before the theories of molecular structure could be accepted as sure guides in unravelling the problems of constitution and in the work of synthesis, ample experimental confirmation of their validity had to be obtained and the predictions which could be made by their means had to be put to the test.

In the case of the aliphatic compounds it was found that the formulae of Couper and Kekulé, which represented these compounds as built up of open chains of carbon atoms (p. 38), interpreted satisfactorily the properties of the compounds in terms of the mutual linking of the atoms in the molecule, and sufficed, when modified by the theories of van't Hoff and Le Bel, to represent all the cases of isomerism which were encountered. The assumption of the universal quadrivalency of the carbon atom, however, which was made by both Couper and Kekulé, had to be modified, for not only did carbon in the well-known compound, carbon monoxide, act with a valency of two, but an increasing number of other carbon compounds, such as the isonitriles—first prepared in 1866 by Armand Emile Justin Gautier (1837–1920),[1] Professor of Chemistry in the University of Paris—the fulminates and others, were met with in which carbon similarly acted as a bivalent element.[2] The force of cumulative evidence gradually became irresistible; the assumption of the constant quadrivalency of carbon had to be abandoned.

---

[1] *Annalen*, 1866, **151**, 239.
[2] See J. U. Nef, *Annalen*, 1892, **270**, 267; R. Scholl and F. Kacer, *Ber.*, 1900, **33**, 51.

In 1900 the untenability of the assumption of the universal quadrivalency of carbon was rendered very evident through the preparation by Moses Gomberg (1866–1947), Professor of Organic Chemistry in the University of Michigan, of a highly reactive compound to which he gave the name triphenylmethyl $(C_6H_5)_3C$. Although much discussion took place as to the existence of a trivalent carbon atom since, in solution, this compound appeared to have a molecular weight corresponding to $(C_6H_5)_3C \cdot C(C_6H_5)_3$, the preparation of analogous compounds[1] put it practically beyond all doubt that carbon can act, in some cases, as a trivalent element.[2]

The Couper-Kekulé theory also met with another difficulty which, however, was later swept away. According to this theory each organic compound possesses a certain definite structure, a certain grouping of the atoms within the molecule, which should interpret the different reactions of that compound; but early in the eighties it was observed, in isolated cases, that a compound might behave as if it had two different structures. The first such case was isatin, which in some reactions, behaved as if it had the

structure $C_6H_4\underset{\diagdown NH\diagup}{\overset{\diagup CO\diagdown}{}}CO$ (lactam form), and in other reactions

as if it had the structure $C_6H_4\underset{\diagdown N\diagup}{\overset{\diagup CO\diagdown}{}}C(OH)$ (lactim form),

although only one form of isatin was known in the free state.[3] The difference in behaviour was found to be due to the hydrogen atom outside the benzene ring; and in the succeeding years a number of other substances were found to have, like isatin, apparently a dual nature. For this phenomenon, Peter Conrad Laar,[4] at that time Privat-Dozent at the Technical High School, Hanover, suggested the name *tautomerism*,[5] and he sought to explain the phenomenon as due to a continual oscillation of a hydrogen atom between two positions in the molecule with an accompanying change in the

[1] More especially by W. Schlenk and his collaborators, *Annalen*, 1910, **372**, 1; *Ber.*, 1910, **43**, 1753.

[2] The idea that elements may act with a variable valency, about which much discussion took place in the early days of the doctrine of valency, is now well established both in the case of inorganic and of organic compounds.

[3] Baeyer and Oekonomides, *Ber.*, 1882, **15**, 2093; 1883, **16**, 2193.

[4] *Ber.*, 1885, **18**, 648; 1886, **19**, 730.

[5] From the Greek ταὐτός (tautos), the same, μέρος (meros), part or role. See J. W. Baker, *Tautomerism* (Routledge).

bonds between the atoms. Acetoacetic ester, therefore, was

represented by the formula, $\begin{array}{c} CH_3 \cdot C - CH \cdot COOC_2H_5. \\ | \quad \diagup H \\ O \end{array}$ When the

compound is caught, so to speak, with the hydrogen in one position, it will act as a hydroxy-compound; while with the hydrogen in the other position, it will act as a ketone. In view of this assumed change of linkage, Paul Heinrich Jacobson (1859–1923),[1] Professor of Chemistry in the University of Berlin, suggested the name *desmotropy*,[2] and this, as well as the term tautomerism, passed into general use.

The study of tautomeric or desmotropic compounds was vigorously pursued, both by chemical and by physical methods, and in 1896 a great step in advance towards an understanding of the phenomenon was made when attempts to isolate the two desmotropic forms of a compound were crowned with success.[3] Instead, therefore, of the view which had been advanced by Laar that tautomerism is due to an intramolecular change, Ludwig Knorr (1858–1921), Professor of Chemistry in the University of Jena, showed that the change is really intermolecular. In other words, tautomerism is really due to the existence of two isomeric forms, in which a hydrogen atom occupies two different positions in the molecule. In the crystalline state each isomer has a definite constitution, but in solution or in the liquid state mutual transformation of one form into the other takes place, more or less rapidly, leading to an equilibrium mixture.[4]

Among the tautomeric compounds most widely studied were those which, like acetoacetic ester, $CH_3 \cdot CO \cdot CH_2 \cdot COOC_2H_5$, contain the ketonic grouping $- CO \cdot CH_2 -$ and which through tautomeric change pass into the hydroxylic or enolic form $- C(OH){:}CH -$. Since, in some cases, the rate of change of one or other tautomeric form is very great, and since, further, the direction of this keto-enol transformation is markedly affected by solvents and other substances (*e.g.*, acids and alkalis), some of which favour the change to the ketonic and others to the enolic form, the presence of both forms can be detected chemically only in certain cases. The physical properties, however, *e.g.*, magnetic

[1] *Ber.*, 1887, **20**, 1732; 1888, **21**, 2628.
[2] From the Greek δεσμός (desmos), a bond and τρέπειν (trepein), to change.
[3] L. Claisen, *Annalen*, 1896, **291**, 25; W. Wislicenus, *ibid.*, p. 147.
[4] *Annalen*, 1896, **293**, 70; 1898, **303**, 133; 1899, **306**, 332.

rotation, refractivity, etc., of the two tautomeric forms are different[1] and so the change of one form into the other and the existence of mixtures of tautomeric forms in the liquid state could be confirmed.[2] The correctness of the view that in the case of so-called tautomeric compounds one is dealing with two isomeric forms which undergo, more or less readily, transformation one into the other, was further confirmed when Wilhelm Gustav Wislicenus[3] (1861–1922) was able to determine, at least with approximate accuracy, the relative proportions of the keto- and enol-forms of formyl-phenylacetic ester in the equilibrium mixture, and still more so when it was found possible, in certain cases, to isolate the individual tautomeric forms and to determine the rate of transformation of the one into the other. This change was found to take place in accordance with the law of mass action.[4]

Since spontaneous transformation of one isomeric form into another has also been met with in the case of structural isomerides (*e.g.*, oximes), stereoisomerides, etc., Thomas Martin Lowry (1874–1936), Professor of Physical Chemistry in the University of Cambridge, suggested the general term dynamic isomerism for the phenomenon of the mutual transformation of isomerides in the liquid state leading to an equilibrium.[5] The phenomenon of *mutarotation*, or change in optical activity, which is met with in the case of certain optically active substances (*e.g.*, α- and β-glucose), has also been explained as due to dynamic isomerism, and the velocity of such change has, in several cases, been determined.[6]

Although, in the cases referred to, tautomerism may be regarded as due to the presence of a mobile hydrogen atom in the molecule, other cases of tautomerism are due to a change in the distribution of valencies in a ring system—so-called *intra-annular tautomerism*.[7]

While, as we have seen, the organic compounds belonging to the

---

[1] W. H. Perkin, sen., *J. Chem. Soc.*, 1892, **61**, 800; Brühl, *Ber.*, 1905, **38**, 1868; A. Haller and P. T. Muller, *Compt. rend.*, 1904, **139**, 1180; P. Drude, *Z. physikal. Chem.*, 1897, **23**, 267.

[2] The equilibrium conditions have also been very fully studied from the point of view of the *Phase Rule*. See Bancroft, *J. Physical Chem.*, 1898, **2**, 143; Roozeboom, *Z. physikal. Chem.*, 1899, **28**, 288; Findlay, *The Phase Rule and its Applications*.

[3] *Annalen*, 1896, **291**, 147. Wilhelm Wislicenus was a son of Johannes Wislicenus (p. 323), and became, in 1902, Professor of Chemistry in the University of Tübingen.

[4] Kurt H. Meyer, *Annalen*, 1911, **380**, 212.

[5] *J. Chem. Soc.*, 1899, **75**, 211; 1904, **85**, 1541.

[6] See Lowry, *loc. cit.* Also Reports of the *Union Internat. de Chimie*, 1931.

[7] See Thorpe and Ingold, *Report on some New Aspects of Tautomerism* (*Union Internat. de Chimie pure et appliquée*), 1931.

aliphatic series can be represented as built up of chains of carbon atoms, it was early realized that it is not possible to formulate benzene and its derivatives, which never contain less than six carbon atoms in the molecule, in a similar manner. In the case of benzene, C_6H_6, for example, which contains eight hydrogen atoms fewer than the aliphatic hydrocarbon hexane, C_6H_{14}, one might assume that the molecule consists of a chain of six carbon atoms linked by double or triple bonds; but the properties of benzene do not admit of this assumption. Benzene does not behave like an unsaturated aliphatic compound; it does not readily form additive compounds with halogens, nor is it, like the unsaturated aliphatic compounds, readily oxidized by alkaline permanganate solution (Baeyer's reagent).

As early as 1858 Couper,[1] for the first time, represented an aromatic compound, salicylic acid, by a structural formula,

$$C\begin{cases} C....H_2 \\ C....H \end{cases}$$
$$\vdots$$
$$C\begin{cases} C....H \\ C..O..OH \end{cases} \quad (O = 8)$$
$$\vdots$$
$$C\begin{cases} O_2 \\ O..OH \end{cases}$$

in which there are shown two groups of three carbon atoms, to which other groups and atoms are attached; but the mode of linking of the carbon atoms in each group is not indicated. In 1861 Joseph Loschmidt in a pamphlet, *Chemical Studies*, suggested the idea of a central nucleus of six carbon atoms, arranged in two layers of three, and he represented the molecule of benzene by a large circle (denoting the nucleus of six carbon atoms) to which six small circles (hydrogen atoms) were attached. Neither of these formulae was found satisfactory. In 1865, however, Kekulé postulated, as we have already seen (p. 40), that in the molecule of benzene there is a closed ring of six carbon atoms, that the structure of benzene is symmetrical and that, therefore, the six carbon atoms may be regarded as being arranged in the form of a regular hexagon. This hexagon formula for benzene enabled one for the first time to interpret, in a satisfactory manner, the behaviour of benzene and its derivatives.

[1] *Compt. rend.*, 1858, **46**, 1157.

Although the assumption of a closed ring of carbon atoms seems to be an almost inescapable one, in order to take account of the differences in behaviour as compared with compounds of the aliphatic series, it was important to obtain experimental confirmation; and such confirmation was definitely obtained when William Henry Perkin, junr., synthesized hexahydrobenzene and some of its derivatives by methods which proved that they contain a closed ring,[1] and showed that the compounds so prepared were identical with compounds obtained by the reduction of the benzene molecule[2] by Adolf von Baeyer, whose work contributed greatly to the confirmation of the Kekulé theory of the constitution of benzene.

In putting forward the hexagon formula for benzene, Kekulé postulated that the six carbon atoms are linked together in a perfectly symmetrical fashion, and that the six hydrogen atoms are equivalent.[3] From this it follows that there can be only one monosubstitution derivative of benzene of the type C_6H_5X. This prediction from the Kekulé theory was confirmed by Albert Ladenburg,[4] Wilhelm Koerner[5] and others, and more especially by Domingo Emilio Noelting,[6] who replaced each of the six hydrogen atoms in turn by the NH_2-group, and found that the same compound (aniline) was obtained in each case.

While, according to the simple hexagon formula for benzene, only one monosubstitution derivative should exist, the existence of three isomeric disubstitution compounds, of the type $C_6H_4X_2$ or C_6H_4XY, should be possible, according as the two substituents are attached to adjacent carbon atoms or are separated by one or by two carbon atoms respectively, as indicated by the formulae:

[1] *J. Chem. Soc.*, 1891, **59**, 798.

[2] *Annalen*, 1888, **245**, 103.

[3] *Annalen*, 1866, **137**, 158. It may be said that the hexagonal arrangement of the carbon atoms in the benzene ring, which was put forward by Kekulé on purely chemical grounds, has been confirmed, in recent years, by the X-ray examination of crystals of carbon (diamond and graphite), and of benzene derivatives (p. 256).

[4] *Ber.*, 1869, **2**, 272; 1874, **7**, 1684; 1875, **8**, 1666. Ladenburg (1842–1911) was Professor of Organic Chemistry, University of Breslau.

[5] *Giornale di Scienze Nat. ed Econom. di Palermo*, 1869, V; Ostwald's *Klassiker*, No. 174. Koerner (1839–1925), a pupil of Kekulé, worked under Cannizzaro and was later Professor of Organic Chemistry at Milan (see *Ber.*, 1926, **59**, 75).

[6] *Ber.*, 1904, **37**, 1015. Noelting was born in St. Domingo in 1851 and was, from 1880 to 1922, Director of the Chemistry School at Mulhouse, Alsace. He died in 1922.

these three isomers being known as *ortho-*, *meta-* and *para-*derivatives respectively. As a matter of fact, three and only three isomeric disubstitution derivatives of benzene are known; and in this respect, therefore, Kekulé's formula proved satisfactory. The more difficult problem, however, now emerged of deciding the position of the two substituents in the three isomers; of deciding, that is to say, which of the isomers is the ortho-, which the meta- and which the para-disubstitution compound.

It is probable that Kekulé regarded his theory mainly as an elegant philosophical system into which all the known facts relating to the aromatic compounds could be neatly and satisfactorily grouped together; and the first to regard the theory as capable of experimental proof was Kekulé's pupil, Koerner. He it was who first entertained the idea of attacking the problem of the position of the substituents in the benzene ring, obviously one of the most far-reaching in the whole of organic chemistry; and he was the first to solve it successfully.

Koerner prepared the mononitro-derivatives of the three isomeric dibrombenzenes, and he found that from one of these, *two* mononitro-dibrombenzenes were produced, while from a second, *three* mononitro-derivatives were obtained. From the third dibrombenzene only *one* mononitro-dibrombenzene was formed. From the Kekulé formula Koerner then deduced that the first dibrombenzene had the ortho-, the second the meta- and the third the para-configuration.[1]

The successful determination of the positions of the two substituents in the three isomeric disubstitution derivatives of benzene must be regarded as one of the triumphs of chemical acumen and of the methods of synthetic chemistry. By methods analogous to those indicated above, the positions of the substituents in other multisubstituted derivatives of benzene have also been ascertained, and the investigation of these compounds has in all cases confirmed the validity and practical utility of the Kekulé formula for benzene.

From the very first, however, a certain difficulty was encountered in seeking to harmonize the hexagonal formula for benzene with the universal quadrivalency of carbon postulated by Kekulé, a postulate which necessitates the linking together of the six carbon

---

[1] *Gazzetta*, 1874, **4**, 305; Ostwald's *Klassiker*, No. 174. By the converse argument Peter Griess later determined the positions of the two amino-groups in the three isomeric diamino-benzenes (*Ber.*, 1872, **5**, 192; 1874, **7**, 1223).

atoms of the ring by alternate single and double bonds, and this
was the formula

adopted by Kekulé. But this formula indicates the possibility of
the existence of two ortho-derivatives, and Kekulé, therefore,
made the *ad hoc* suggestion that there is an oscillation of the bonds,[1]
giving rise to the alternative ring structures,

The two ortho-positions and, similarly, the two meta-positions
then become identical. This dynamic formula for benzene recognizes for the first time what later came to be called desmotropic
change (p. 122), and may, perhaps, have been suggested by the
vision which Kekulé had had in 1865 of long rows of atoms "all
twisting and twining in snake-like motion" (p. 39).

Other suggestions, also, were made for getting over the difficulty
of disposing of the fourth valency of the carbon atoms. Adolph
Carl Ludwig Claus[2] (1840–1900), Professor of Chemistry in the
University of Freiburg, sought to overcome the difficulty by
linking together the fourth valencies of opposite carbon atoms, so

obtaining the *diagonal formula*, ⬡ ; while H. E. Armstrong,[3]

in his *centric formula* ⬡ , a formula already anticipated by

Lothar Meyer[4] in 1865, simply left the fourth valencies free and
unattached. A reconciliation between the formula of Kekulé and

---

[1] *Annalen*, 1872, **162**, 77; Michaelis, *Ber.*, 1872, **5**, 463.
[2] *Theoretische Betrachtungen und deren Anwendung zur Systematik der organischen
Chemie*, 1867.
[3] *J. Chem. Soc.*, 1887, **51**, 264. This formula was later adopted also by A. von
Baeyer.
[4] *Die modernen Theorien der Chemie*.

the centric formula was attempted by John Norman Collie (1859–1943), at that time Professor of Chemistry to the Pharmaceutical Society, London,[1] who regarded the benzene molecule as a vibrating or oscillating system; and according to this view, the dynamic formulae of Kekulé represent the extreme positions of oscillation of the carbon atoms, while the centric formula represents the intermediate position, when the carbon atoms are all in one plane.[2] As we shall see later (Chapter XII), modern physical methods indicate that the centric formula is the best general representation of the molecule of benzene.

In 1869 Ladenburg, who was at that time working in Kekulé's laboratory, suggested discarding the hexagon formula altogether

and substituting therefor a prism formula,[3] ; but

although this formula could take account of the various isomerides of benzene, it excluded the possibility of the existence of the double bonds which must be present in dihydro- and tetrahydro-benzene. For a time the prism formula was a serious rival of the hexagon, but the weight of cumulative evidence furnished by many investigations finally led to the abandonment of the prism formula[4] and the definite and universal acceptance of the hexagon formula as representing most satisfactorily the behaviour, properties and reactions of benzene and its derivatives.

Although the evidence regarding the arrangement of the bonds in the benzene molecule which was obtained from a study of the physical properties[5]—molecular volume, refractivity, heat of combustion, etc.—was rather conflicting,[6] it did, on the whole, favour the assumption that three double bonds are present; but it was clear, not only from the physical but also, more especially, from the chemical evidence, that the three double bonds are different from the double bonds met with in aliphatic compounds. Of this fact Friedrich Karl Johannes Thiele (1865–1918),[7] at that time Associate Professor of Chemistry in the University of Munich,

---

[1] Professor of Organic Chemistry in University College, London, 1902–30.
[2] *J. Chem. Soc.*, 1897, **71**, 1013; 1916, **109**, 561.
[3] *Ber.*, 1869, **2**, 140, 272; *Theorie der Aromatischen Verbindungen*, 1876.
[4] See F. R. Japp, *J. Chem. Soc.*, 1897, **71**, 123.   [5] See Chapter V.
[6] See, for example, R. Schiff, *Annalen*, 1883, **220**, 278; J. W. Brühl, *Ber.*, 1907, **40**, 878; J. Thomsen, *Ber.*, 1880, **13**, 1806; F. Stohmann, *J. prakt. Chem.*, 1893, **48**, 447.
[7] *Annalen*, 1899, **306**, 87.

and later Professor in the University of Strasbourg, sought to give an explanation by pointing out that, according to his *theory of partial valencies*, benzene possesses no residual or partial valencies with which to form additive compounds, whereas such valencies are present in the case of the aliphatic compounds with double bonds. Interesting and important as all these various suggestions undoubtedly are, however, there is not any one single formula which represents completely the physical and chemical behaviour of benzene (see Chapter XII).

The introduction of a ring structure for benzene soon led to the recognition of ring structures in other compounds, with varying numbers of atoms in the rings. In some cases the rings contained only carbon atoms,[1] *e.g.*, polymethylenes, naphthalene, anthracene, etc., but in other cases nitrogen, sulphur or oxygen atoms might also be present, as in the compounds pyridine, furfurane, thiophene, etc., and very many compounds with a great variety of single and multiple ring structures have been synthesized since the sixties of last century. While some of these are found in compounds produced in the animal or vegetable organism, many others have been built up in the laboratory by the art of the chemist.

If the tetrahedral arrangement of the four valencies of a carbon atom is accepted, the angle between the valencies will be 109° 28′, and when one atom is joined to another carbon atom union must take place at this angle. When a number of carbon atoms combine together, therefore, they cannot really form a "straight chain" of carbon atoms, as Couper supposed, but must have a zigzag arrangement, as X-ray examination has shown to be the case in aliphatic compounds, or must tend to form a "ring." In the latter case, since the angles of a pentagon are 108° the end members of a series of five carbon atoms must come very near together, and so, one may imagine, a ring of five carbon atoms can readily be formed with very little straining of the "bonds." Since the angles of a hexagon are 120°, the formation of a ring of six carbon atoms, as in hexamethylene, will involve a greater degree of straining, or deflection of the tetrahedral angles; and for rings containing fewer than five and more than six carbon atoms the strain will be greater. On the basis of the hypothesis of the tetrahedral carbon atom, therefore, A. von Baeyer, in 1885, advanced his "strain theory," that the less the "strain" between the valencies of the carbon

[1] For an account of the early history of the synthesis of closed carbon chains, see W. H. Perkin, junr., *J. Chem. Soc.*, 1929, p. 1347.

atoms, the greater will be the ease of formation and the greater the stability of closed-ring compounds. This theory was in general accord with the facts then known and received valuable confirmation through the work of W. H. Perkin, junr., who prepared compounds containing rings of 3, 4, 5 and 6 carbon atoms. It was, however, later shown by the work, for example, of (Sir) Jocelyn Field Thorpe (1872–1940), of the Imperial College of Science, and (Sir) Christopher Kelk Ingold,[1] of University College, London, that the natural angle at which two valencies emerge from a carbon atom is notably affected by attached groups (*e.g.*, by two methyl groups), with the result that the strain inherent in 3-, 4- or 5-membered rings may be reduced, and the stability of such rings increased. Such rings are found in the structure of many natural products.[2]

Investigations, however, carried out more especially by Leopold Ruzicka[3] and his co-workers, showed that the Baeyer "strain theory" in its original form cannot be entirely correct. By heating salts, especially the thorium salt, of dicarboxylic acids, an unbroken series of monocyclic ketones, with rings of from 8 to 21 carbon atoms, and also with rings of 23 and 29 carbon atoms, has been prepared, as well as rings containing an even number of carbon atoms from C_{22} to C_{34}.[4] Two such many-membered rings are found in the naturally-occurring odoriferous compounds *civetone*,

$$\begin{array}{l} CH \cdot (CH_2)_7 \\ \| \quad\quad\quad\quad\quad >CO, \text{ and } \textit{muskone}, \\ CH \cdot (CH_2)_7 \end{array} \quad CH_3 \cdot CH \overset{\displaystyle CH_2}{\underset{\displaystyle (CH_2)_{12}}{<\quad\quad>}} CO.$$

Many-membered rings containing a triple bond, $(CH_2)_{13} \overset{C}{\underset{C}{\underset{\|\|}{<\ }}}$,

or incorporating a benzene ring, and also heterocyclic compounds have been prepared.

Since these many-membered rings are shown by determinations of heat of combustion to be "strainless," their stability can be

[1] Professor of Organic Chemistry, University of Leeds, 1924–30; Professor of Chemistry, University College, London, 1930–61; Visiting Lecturer Stanford University, California, 1932; Baker Lecturer, Cornell University, 1950.

[2] See Ingold, *J. Chem. Soc.*, 1921, **119**, 305, 951.

[3] Professor of Organic Chemistry, Technische Hochschule, Zurich. Awarded the Nobel Prize for Chemistry in 1939.

[4] *Helv. Chim. Acta*, 1926, **9**, 249, 499. See L. Ruzicka, *Chem. and Ind.*, **1935**, 2; K. Ziegler, H. Eberle and H. Ohlinger, *Annalen*, 1933, **504**, 94.

explained only on the assumption that the ring members are not all in one plane, or that the molecule is multiplanar. This structure is in accordance with the modified "strain theory" suggested in 1890 by H. Sachse,[1] and is confirmed by X-ray analysis.

If the debt which pure science owes to Kekulé for his elucidation of the constitution of the benzene molecule is almost incalculable, the debt of chemical industry is not less great, for it is in the benzene theory that the industry of coal-tar dyes, synthetic drugs, photographic chemicals, etc., has its roots. To a brief discussion of the origin and growth of this industry, which has exercised a very great influence on the social, economic and political history of the world during the past hundred years, we shall proceed in the next chapter.

[1] *Ber.*, 1890, **23**, 1363; *Z. physikal. Chem.*, 1892, **10**, 203.

CHAPTER EIGHT

THE RISE AND DEVELOPMENT OF THE FINE-CHEMICALS
INDUSTRY

ONE of the outstanding features of the history of chemistry during the second half of the nineteenth century was, as we have seen, the development of synthetic organic chemistry; and during this period many thousands of compounds, which possessed both theoretical interest and practical importance, were prepared. The raw materials for the production of these compounds were not, as a rule, the elements, but a relatively small number of natural products or compounds readily obtainable from them; and although, at the present time, use is increasingly made of the hydrocarbons occurring in petroleum as raw materials for chemical syntheses, much of the synthetic organic chemistry of the nineteenth and twentieth centuries was built up on the products of distillation of coal contained in coal tar. It was out of and on the basis of the chemical investigations of the constituents of coal tar that the industrial production of dyes, medicinals, photographic chemicals and explosives developed.

Although, in the first half of the nineteenth century, a number of British chemists—Davy, Faraday, Graham and others—were known as brilliant investigators and exercised, through their gifts of exposition, a great influence in diffusing a popular knowledge of chemistry in England, there were very few laboratory facilities for the practical training of men who wished to devote themselves to the study and advancement of chemistry as their life's work. A hundred years ago, therefore, one finds that, when organic chemistry was rapidly developing, students migrated to the laboratories in Germany and France, more especially, perhaps, to the laboratory of Wöhler, at Göttingen, or to that of Liebig, at Giessen, where a chemical laboratory for the training of chemists was first founded. A great impetus to the more widespread study of chemistry in England was given when, in 1840, Liebig published his work, *Chemistry of Agriculture and Physiology*; and the stimulus was reinforced in 1842, when Liebig visited England in person and by his private interviews and his public lectures extended a knowledge of and increased the interest in chemical science. The immediate effect of Liebig's tour was an awakened realization of the importance of chemistry, more especially of organic chemistry, and universities and colleges hastened to give increased attention to laboratory teaching. Much of this teaching, however, was given for purely professional purposes and as ancillary to the training of doctors, engineers, etc. Something more, something different was necessary, and, taking advantage of the wave of enthusiasm for chemistry created by Liebig, the Prince Consort and Queen Victoria's physician, Sir James Clark, succeeded, with great foresight, in getting founded the Royal College of Chemistry in London. In this institution, it was hoped, chemistry would be studied for its own sake and students might be trained to advance the boundaries of the science by research. The founding of this College had not merely a national significance but was an act of great importance in the history of chemistry, for it was to this College that A. W. Hofmann, a young German chemist trained under Liebig at Giessen, came, in 1845, as its first Director; and it was by a young student at this College that the first coal-tar dye was prepared.[1] Liebig also played a part in the founding of the Chemical Society in 1841, the oldest society of its kind in the world. The German Chemical Society was founded in 1866, the year after A. W. von Hofmann's return from England to Germany. The chemical societies of France, Russia and America were founded in 1857, 1868 and 1876 respectively.

The introduction of coal gas as a general illuminant, which took place about 1810–12, led to the production of a large amount of coal tar. At first this was to a large extent a disagreeable waste product, and the difficulty of its disposal was solved, only to some extent, by burning the tar as a fuel. At a later time, about 1815–20, the more volatile portions were removed by distillation, the "spirit" so obtained being used either as a substitute for turpentine in making varnishes, or as a solvent for rubber in the production of a waterproof material, known by the name of its original

---

[1] The wave of enthusiasm which led to the founding of the College of Chemistry subsided, and landowners who had thought to find in Liebig's book and in the work of the College an immediate cure for their agricultural difficulties were disappointed and withdrew their support. When, therefore, Hofmann succeeded Playfair as Professor of Chemistry at the School of Mines in 1858, the College was absorbed in the latter institution (*J. Chem. Soc.*, 1896, **69**, 575).

manufacturer, Charles Macintosh.[1] Although, in 1838, the demand for coal tar was stimulated by the introduction of the creosoting process for the preservation of timber, much of the tar had still to be burned as a fuel or for the production of lamp black.

Towards the middle of last century, however, chemists had begun to interest themselves in the nature and composition of coal tar, and Hofmann, while still a student in Liebig's laboratory at Giessen, succeeded in isolating from coal tar a basic substance which he identified with the compound first obtained in 1826 by Otto Unverdorben (1806–73), who, at an early period, abandoned chemistry for commerce. For this compound Hofmann retained the name *aniline*.[2] On coming to London in 1845, Hofmann engaged energetically, along with his students, in the work of investigating coal tar; and one of the earliest results to be obtained was the discovery of the presence of benzene in the tar.[3] In 1848 Charles Blackford Mansfield (1819–55), a pupil of Hofmann, succeeded in isolating considerable quantities of benzene from the "light oil" of coal tar by fractional distillation; and he also isolated *toluene*, a hydrocarbon which the French chemist, Pierre Josephe Pelletier (1788–1842) and the Polish chemist Philipp Walter (1810–47) had obtained in 1836 by the dry distillation of pine resin, and which Sainte-Claire Deville had also obtained, in

[1] The solubility of rubber in coal tar naphtha was discovered by J. Syme, afterwards Professor of Surgery in the University of Edinburgh (*Annals of Philosophy*, 1818, **12**, 112). For an account of Macintosh see N. L. and A. Clow, *Chem. and Ind.*, **1943**, 104.

[2] Aniline (from the Portuguese *anil*, applied to indigo, and derived from the Arabic *an-nil*, the blue substance), was first obtained by Unverdorben by heating indigo, and was given the name *crystalline*. In 1841 Carl Julius von Fritzsche (1808–71), an assistant to Mitscherlich and, later, a member of the Academy of Sciences, St. Petersburg, obtained the same compound from anthranilic acid, which was produced by the action of caustic alkalis on indigo, and called it *aniline*. In 1842 the same base was obtained by the Russian chemist, Nicolai Nicolaiewitsch Zinin (1812–1911), by the reduction of nitrobenzene with ammonium sulphide (*Bull. Sci. St. Petersburg*, 1842, **10**; *J. prakt. Chem.*, 1844, **33**, 29). This base he called *benzidam*. In 1843 Hofmann showed that the three substances, crystalline, aniline and benzidam, were identical with the base isolated from coal tar (*Annalen*, 1843, **45**, 250; **47**, 37).

[3] Hofmann, *Annalen*, 1845, **55**, 200.

The hydrocarbon benzene was discovered in 1825 by Faraday, who isolated it from a liquid which condensed in cylinders of an illuminating gas formed by the decomposition of fish oils. Since analysis showed its composition to be C_2H ($C = 6$), Faraday called it bicarburet of hydrogen. In 1834 Mitscherlich prepared the same substance by distilling benzoic acid, obtained from gum benzoin, with lime. On account of its connection with gum benzoin, Mitscherlich called the hydrocarbon benzin, but this name was changed to benzol by Liebig. The systematic ending ene was introduced by Laurent, who also proposed for benzene the name *phéne* (from the Greek φαίνω (phaino), to shine), an account of the connection of benzene with coal gas. Hence arose the terms phenol (carbolic acid), phenyl, etc.

1841, by the distillation of Tolu balsam. The method of fractional distillation introduced by Mansfield became of paramount importance in later years for the separation of various constituents present in coal tar, when, as a result of the discovery of the first coal-tar dye, a demand arose for the production of these constituents on a large scale.

The discovery that benzene is a fairly abundant constituent of coal tar was of the highest importance, for in 1834 Mitscherlich had shown that when benzene is acted on by concentrated nitric acid, *nitrobenzene*—a substance introduced in 1847 as a cheap scenting material, more especially for soap, under the name of *essence of mirbane*—is formed; and since nitrobenzene could be reduced to aniline by the action of ammonium sulphide (Zinin, 1842), or, better, by the action of acetic acid and finely divided iron (A. Béchamp, 1854), coal-tar benzene became the raw material for the production of aniline.[1]

At the time when Hofmann came to London, organic chemistry was still mainly in the classificatory stage, and the attention of chemists was occupied with the theory of compound radicals, the doctrine of substitution, etc. Practically nothing was known regarding the internal constitution of compounds. The views regarding organic syntheses were, therefore, rather crude and were based largely on the method of addition and subtraction. W. H. Perkin, therefore, conceived the idea that the alkaloid, quinine, might be synthesized from toluidine, $C_6H_4(CH_3) \cdot NH_2$, by substituting allyl for hydrogen and so forming a compound of the composition, $C_{10}H_{13}N$; then, by removing two atoms of hydrogen and adding two atoms of oxygen by oxidation, quinine might be formed, thus: $2(C_{10}H_{13}N) + 3O = C_{20}H_{24}N_2O_2$ (quinine) $+ H_2O$. When the experiment was tried, no quinine was formed, but only a dirty reddish-brown precipitate. The matter, however, did not end there. "Unpromising though this result was," wrote Perkin[2] at a later time, "I was interested in the action, and thought it desirable to treat a more simple base in the same manner. Aniline was selected, and its sulphate was treated with potassium dichromate; in this instance a black precipitate was obtained and, on examination, this precipitate was found to contain the colouring matter since so well known as aniline purple or mauve, and by a number of other names.[3] All these experiments were made during

---

[1] Instead of acetic acid, Perkin later introduced the use of hydrochloric acid.
[2] *Hofmann Memorial Lecture, J. Chem. Soc.*, 1896, **69**, 5.
[3] See *Proc. Roy. Soc.*, 1856, **35**, 717.

the Easter vacation of 1856 in my rough laboratory at home. Very
soon after the discovery of this colouring matter, I found that it
had the properties of a dye, and that it resisted the action of light
remarkably well." This dye Perkin proceeded to produce on an
industrial scale at Greenford Green, near Harrow, and so began,
in 1857, the industry of synthetic-dye manufacture which, directly
or indirectly, has produced social, political and economic changes
of a most far-reaching character.

The introduction of the first coal-tar dye, *mauve* or *mauveine*,
met with great success, and stimulated chemists to further investigations. These led, in 1859, to the discovery by Verguin of another
dye called *aniline red*, *fuchsine* or *magenta*, named after the Italian
town where Napoleon III had that summer defeated the Austrians.
The formation of this red dye had been observed by the Polish
chemist, Jacob Natanson (1832–84)[1] in 1856, and by Hofmann in
1858, but was first successfully manufactured in small quantity in
France, in 1859, by heating commercial aniline (containing ortho-
and para-toluidine) with anhydrous stannic chloride. This manufacture was improved, in 1860, by the English chemists, H. Med-
lock and Edward Chambers Nicholson, former pupils of Hofmann,
who employed arsenic acid as oxidizing agent.

The brilliancy of the new aniline dyes and the great success
which they achieved made a powerful appeal to the scientific
chemist as well as to the manufacturer; and the extent to which the
energies of chemists were directed towards the discovery of new
dyes threatened to prejudice unfavourably the general advance of
the science.

But if advance was checked at all, the check was only a temporary
one, for the rapid development of the dye industry exercised a
most important and beneficial influence on the general development of chemical science, not only by creating a demand for an
increasing number of trained organic chemists, but also by placing
at the disposal of the investigator new substances and products
which could scarcely have been obtained without its aid. The
second half of the nineteenth century, therefore, was a period of
extraordinarily rapid expansion in synthetic organic chemistry.

With the introduction of magenta, a new starting-point was given
for the preparation of a series of dyes, the number of which now
began rapidly to increase. *Aniline blue* or *Lyons blue* was prepared
in 1861 by the French chemist, Charles Girard (1837–1918), and

[1] *Annalen*, 1856, **98**, 297.

the industrialist, Georges de Laire (1837–1909), by heating magenta with aniline in presence of benzoic acid; and in 1862 Nicholson, by heating aniline blue with concentrated sulphuric acid, obtained the more valuable dyes, Nicholson's blue (a monosulphonic acid) and water blue (a disulphonic acid). These dyes possessed the advantage of being soluble in water and in solutions of alkalis; and the process of sulphonation became of much importance in the development of dye manufacture.

Although the preparation of new dyes, along the lines opened up by W. H. Perkin, was carried on with much vigour, it was also realized by chemists that a satisfactory basis for further development could be obtained only through a knowledge of the chemical composition and nature of the dyes. In the work of investigating these Hofmann took a leading part, and in 1862 he showed that the dye magenta is the salt of a base which he called *rosaniline*.[1] In 1864, also, he confirmed a discovery made by Nicholson that magenta cannot be obtained by the oxidation of pure aniline, but only of commercial aniline, in which the isomeric ortho- and para-toluidines are present as impurities. In 1863 Hofmann showed that the aniline blue of Girard and de Laire is a triphenyl derivative of rosaniline;[2] and as a result of this he was led to try the effect of alkyl iodides on rosaniline. By means of these reagents different alkyl groups could be introduced into the rosaniline molecule, and various dyes of a purple or violet colour—*Hofmann's violets*—were produced.[3] Some of these dyes, it is interesting to note, had already been prepared in 1861 by Emil Kopp (1817–75), who also remarked[4] that "the red shade disappears and is converted into a violet, becoming bluer and bluer as the hydrogen is displaced by the hydrocarbon." Kopp, however, did not pursue the matter further.

Although Hofmann succeeded in throwing some light on the relationship which existed between magenta and a number of other dyes, which were shown to be derivatives of rosaniline, the constitution of rosaniline itself was unknown; and although a number of chemists actively studied this problem, it was not till 1878 that the two German chemists, Emil and Otto Fischer, basing their work on the principles of chemical structure which had been laid down by Couper and by Kekulé, showed that the parent of rosaniline, magenta, etc., was a hydrocarbon, *triphenylmethane*; that is, methane in which three hydrogen atoms are replaced by

[1] *Compt. rend.*, 1862, **54**, 428.   [2] *Compt. rend.*, 1863, **56**, 945; **57**, 25.
[3] *Compt. rend.*, 1863, **57**, 25.   [4] *Compt. rend.*, 1861, **52**, 363.

three phenyl ($C_6H_5 -$) groups.[1] This hydrocarbon is represented, therefore, by the formula, $HC(C_6H_5)_3$. When a mixture of aniline, ortho-toluidine and para-toluidine is oxidized, rosaniline is formed; and to this base, therefore, can be assigned the constitution:

$$HO \cdot C \begin{array}{l} \diagup C_6H_3(CH_3) \cdot NH_2 \\ - C_6H_4 \cdot NH_2 \\ \diagdown C_6H_4 \cdot NH_2 \end{array}$$

With hydrochloric acid water is eliminated and rosaniline chloride,

$$C \begin{array}{l} \diagup C_6H_3(CH_3) \cdot NH_2 \\ - C_6H_4 \cdot NH_2 \\ \diagdown C_6H_4 = NH_2Cl \end{array}$$

is formed.

When a mixture of aniline and para-toluidine is oxidized, another base, *para-rosaniline*, $HO \cdot C(C_6H_4NH_2)_3$, is obtained, and this also gives rise to dyes similar to those derived from rosaniline.

The elucidation of the constitution of rosaniline and para-rosaniline not only made clear the structural relations between magenta, Hofmann's violets and a number of other dyes, but led to the introduction of new and improved processes of manufacture and to the discovery of a large number of other dyes—crystal violet, methyl violet, malachite green, Victoria blue, etc.—possessing a wide range of colour. These newer methods of synthesis, moreover, required for their realization supplies of various compounds, such as alkyl chlorides, methyl and ethyl alcohol, benzaldehyde, benzoic acid, etc., and chemists were therefore led to devise new methods for the preparation of an ever-increasing number of secondary products or "intermediates," which industry then had to learn to produce economically. By this process of action and reaction between science and industry synthetic organic fore, an chemistry rapidly developed.

The number of dyes belonging to the triphenylmethane group rapidly increased and is now very large. In this group are representatives of almost every possible shade of colour, and the compounds are suitable for almost every kind of dyeing. The dyes belonging to this group are specially characterized by brilliancy, and although some are rather fugitive, many are very fast.

Although the constitution of a large number of "aniline dyes" had been unravelled by E. and O. Fischer, the constitution of

[1] *Annalen*, 1878, **194**, 256.

mauveine, the first of these dyes to be discovered, long remained unknown; and it was not till 1888 that Otto Fischer and Eduard Hepp (1851–1917), who later became an industrial chemist, were able to show,[1] on the basis of the investigations more especially of Rudolf Hugo Nietzki (1847–1917), Otto Nikolaus Witt (1853–1915), and August Heinrich Bernthsen (1855–1931), that mauveine is a phenyl derivative of phenosafranine, a compound containing

the azine ring, 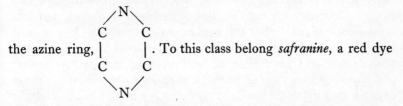 . To this class belong *safranine*, a red dye

formed as a by-product in the preparation of mauveine and first isolated by Perkin, and a large number of other dyes.

In 1858 Johann Peter Griess (1829–88), chemist in the brewery of Messrs. Allsopp and Sons, Burton-on-Trent, discovered that when nitrous acid acts on aniline, or any other derivative of benzene containing the amino-group (NH_2), an unstable compound—a so-called diazo-compound—is produced.[2] The diazo-compounds which were thus obtained are very reactive, and although not themselves dyes, possess the very important property of combining or "coupling" with aromatic amines to produce basic compounds, the salts of which are dyes. The formation of such a dye, known as *aniline yellow*, had been observed as early as 1861, and the dye was put on the market in 1863.[3] Although this was quickly followed by one or two other dyes of the same type, it was not till 1876, after the constitution of these compounds had been elucidated by Hofmann[4] and O. N. Witt,[5] and the molecule shown to contain the azo-group ($-N:N-$), that the production of azo-dyes began to undergo a rapid development. In the following years the azo-dyes came to occupy a position of the highest importance.

The diazo-compounds, it was found, could be "coupled" not

[1] *Ber.*, 1888, **21**, 2617.
[2] *Annalen*, 1858, **106**, 123; *Proc. Roy. Soc.*, 1859, **9**, 594.
[3] Mène, *Compt. rend.*, 1861, **52**, 311. The dye is para-aminoazobenzene hydrochloride, $C_6H_5.N = N.C_6H_4.NH_2$, HCl, the amino-group being in the para-position to the azo-group.
[4] *Ber*, 1877, **10**, 213.
[5] *Ber.*, 1877, **10**, 654.

only with compounds, containing the amino-group but also with compounds containing the hydroxyl-group[1]—phenols, cresols, salicylic acid, resorcinol, etc.—and in this way a new series of azo-dyes, first introduced by O. N. Witt under the name of *tropaeolines*,[2] was obtained.

The production of azo-dyes was greatly extended when derivatives not only of benzene but also of another coal-tar hydrocarbon, naphthalene, were brought into use. From naphthalene[3] two isomeric hydroxy-derivatives, alpha- and beta-naphthol, and two amino-derivatives, alpha- and beta-naphthylamine, are known; and by coupling the naphthols and naphthylamines with diazo-compounds derived from aniline, or by diazotizing the naphthylamines and coupling with various phenols, naphthols or amines, the range of azo-dyes can be and has been very greatly extended. Through the introduction into dye-chemistry of the naphthols and naphthylamines and of their sulphonic acid derivatives, naphthalene became one of the most important raw materials in dye manufacture.

In 1877 compounds containing two azo-groups were discovered,[4] and two years later, Nietzki produced the first dye belonging to this class, the dye *Biebrich scarlet*.[5] In 1884, moreover, the German industrial chemist, Paul Böttiger, discovered a new group of diazo-dyes which were able to dye cotton and linen directly without requiring a mordant; and of these direct cotton dyes the first to be put on the market was *Congo red*. In 1887 Arthur George Green (1864–1941)[6] introduced the *ingrain colours*. These are produced *in the fibre* by dyeing the fibre with a dye containing the amino-group (*e.g.* primuline), diazotizing this dye and immersing in a solution of β-naphthol, resorcinol, etc., which form a dye with the diazo-compound. A dye is thus developed in the fibre.

The great dye industry which arose from the discovery made by Perkin in 1856 not only placed a large and varied array of

[1] Kekulé and Hidegh, *Ber.*, 1870, **3**, 233.

[2] *J. Chem. Soc.*, 1879, **35**, 179.

[3] The hydrocarbon naphthalene was discovered in coal tar by Alexander Garden in 1819 (*Annals of Philosophy*, 1820, **15**, 74), and was isolated and studied in 1820 by John Kidd (1780–1851), Professor of Chemistry at Oxford (*Phil. Trans.*, 1821, p. 209). Its composition was determined by Faraday in 1826. Naphthalene occurs mainly in the fraction of coal tar which distils over between 170° and 230° C. and which is known as the "carbolic oils" or "middle oils."

[4] Caro and Schraube, *Ber.*, 1877, **10**, 2230.

[5] *Ber.*, 1880, **13**, 800, 1838.

[6] English industrial chemist and, from 1902 to 1915, Professor of Colour Chemistry, University of Leeds.

dyestuffs at the disposal of the dyer but soon began to have important social and economic repercussions. From a very early time, as its use in the dyeing of Egyptian mummy cloths bears witness, the dye *alizarin*, obtained from the root of the madder (*Rubia tinctoria*),[1] was employed for dyeing cotton of a bright red colour—the so-called Turkey red; and in the south of France, in Alsace and in Algiers great tracts of land were devoted to the cultivation of that plant. In 1868, however, Carl Graebe (1841–1927) and Carl Theodor Liebermann (1842–1914) showed that the dye alizarin is a dihydroxy-derivative of anthraquinone,[2] a compound formed by the oxidation of the hydrocarbon anthracene ($C_{14}H_{10}$),[3] a hydrocarbon discovered in coal tar by Dumas and Laurent in 1832. The way was thus opened for the industrial production of this important dyestuff. The decisive step in industrial production was made by Heinrich Caro (1834–1910), working for the Badische Anilin-und-Soda Fabrik. Caro lodged his patent in Britain the day before Perkin lodged his for the same process.[4] In the event, the Badische Fabrik gave Perkin licence to manufacture alizarin in Britain. For the first time the chemist had triumphed over nature in the production of a naturally-occurring colouring matter. The widespreading lands over which the madder once bloomed were soon compelled to bear crops of a different kind; and an industry of great monetary value passed from the field to the factory. The value of coal tar, moreover, was greatly enhanced, for the "last runnings" of the stills, in which the anthracene was present, and which previously had either been burned as

---

[1] In the madder root are contained certain compounds, known as glucosides, which, when allowed to ferment, undergo decomposition with production of glucose and various colouring matters. Of these the most important are *alizarin* (so called from the name *alizari*, given by the Arabs to the madder root), and *purpurin*, dyes which were first isolated in 1827 by the French chemists, Jean Jacques Colin (1784–1865), Professor of Chemistry at Dijon, and Pierre Jean Robiquet (*Ann. Chim.*, 1827, **34**, 225).

[2] *Ber.*, 1868, **1**, 49, 332.

[3] From ἄνθραξ (anthrax), coal. The hydrocarbon has a constitution represented

by three benzene rings fused together,

[4] Anthracene was converted into anthraquinone by heating with potassium dichromate and sulphuric acid, and the anthraquinone was then converted into its sulphonic acid derivative by heating with concentrated sulphuric acid. When this compound is fused with sodium hydroxide, in presence of sodium or potassium chlorate, alizarin is formed. In the early days of manufacture, owing to the difficulty of sulphonating anthraquinone, the synthesis was carried out by way of dichloranthracene, which could readily be sulphonated and oxidized to anthraquinone sulphonic acid.

a fuel or used as a lubricant, increased in value from pence to pounds.

Besides alizarin a large number of other derivatives of anthraquinone as well as other classes of dyes have been produced and play a part of great importance. In the present century, moreover, chemists have achieved, in the synthetic and industrial production of *indigotin*, perhaps the greatest or, at least, in view of the difficulties to be overcome and the great economic consequences, the most striking of their triumphs.

For thousands of years indigo, derived from various species of plants (such as *Isatis tinctoria*),[1] has been used as a blue dye, and during the Middle Ages commerce in indigo played a role of great social, economic and political importance. In the eighteenth century the ban against the introduction of Indian indigo into Europe was withdrawn, and in consequence the Indian indigo plantations came to control the markets of the world. But that is now changed. The Indian plantations have dwindled almost to vanishing point, and the production of indigo-blue has, like the production of alizarin, passed to the factory.

Although, throughout the first half of the nineteenth century, various chemists had obtained from indigo compounds which were later recognized as showing that indigo is related to benzene,[2] it was not possible to understand and interpret the transformations which indigo could be made to undergo until the theory of valency and the principles of structural chemistry had been established. It remained, therefore, for the chemists of the second half of the nineteenth century, more especially for Adolf von Baeyer,[3] through his researches begun in 1865, to elucidate the constitution of indigotin (indigo-blue) and to prepare the way for the artificial production of this dyestuff. By repeated reduction of isatin

---

[1] Indigo does not occur as such in the plant, but as a glucoside, *indican*, which on fermentation yields glucose and the so-called leuco-compound of indigo. When the solution is agitated with air, the leuco-compound is oxidized with production of indigo which separates out as an insoluble powder. Besides indigo-blue or indigotin, varying amounts of other compounds are also present in the natural indigo.

[2] In 1806 Fourcroy and Vauquelin, on oxidizing indigo, obtained an acid which R. F. Marchand (*J. prakt. Chem.*, 1842, **26**, 385) and Gerhardt (*Annalen*, 1843, **45**, 19) later identified as nitrosalicylic acid. In 1826 Unverdorben obtained aniline by distilling indigo with caustic potash, and Fritzsche, in 1841, obtained anthranilic acid by fusing indigo with caustic potash. In this year, also, Otto Linné Erdmann (*J. prakt. Chem.*, 1841, **24**, 1) and also Laurent (*Ann. Chim.*, 1841, **3**, 317) by the oxidation of indigo obtained isatin, which was found to have the composition $C_8H_5NO_2$.

[3] See papers published with various collaborators in the *Berichte der deutschen chemischen Gesellschaft* from 1868 to 1883.

Baeyer succeeded in obtaining dioxindole, oxindole and, lastly, indole itself,[1] a substance to which the constitution

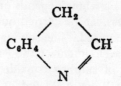

could be assigned. In 1878 Baeyer was able to synthesize isatin from phenylacetic acid and thereby to show that it has the formula,

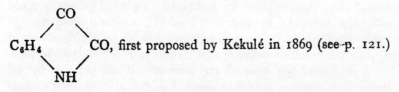

CO, first proposed by Kekulé in 1869 (see p. 121.)

With the synthesis of isatin accomplished, the synthesis of indigotin followed, by reduction of the chloride of isatin. The determination of the molecular weight of indigotin by Erwin von Sommaruga,[2] showed that in indigotin two indole nuclei are joined together,

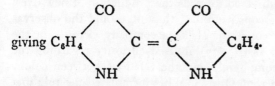

giving C₆H₄ ... C₆H₄.

The constitution of indigotin having been elucidated, its artificial production was advanced to a new stage in 1880 by Baeyer, who used as his starting material the hydrocarbon toluene. Successful as was this process in the laboratory, its commercial development failed owing to the cost of production being too great and the supply of toluene insufficient to make the complete displacement of natural indigo possible. Ten years later, however, Karl Heumann[3] discovered a new method of synthesizing indigotin from phenylglycine, and this method was subsequently developed along two different lines into commercially successful processes. Both methods involve a considerable number of separate reactions and

---

[1] In carrying out this reduction Baeyer passed the vapour of oxindole over heated zinc dust, an important reduction method not hitherto employed but later made use of by Graebe and Liebermann who were thereby enabled to show that alizarin, on reduction, yields anthracene.

[2] *Annalen*, 1879, **195**, 302. [3] *Ber.*, 1890, **23**, 3043.

require the use of a number of different substances—pyrosulphuric acid, ammonia, chlorine, acetic acid, etc.—and the economic success of the synthesis as a whole depends on the success with which each step of the process can be carried out, and on the cost of the different substances required. The solving of the various problems connected with the development of the "contact process" for the industrial production of sulphur trioxide and pyrosulphuric acid (technically known as "oleum"), and of the electrolytic process for the production of chlorine and caustic soda, etc., was achieved only after long years of intensive research;[1] and it was not till about 1897 that industrially produced synthetic indigotin could enter the market as the successful rival of the natural product.

During the present century research has led to the discovery of a large number and variety of new dyes (indanthrene dyes, acridine dyes, etc.), and this research has been stimulated not only by the desire to obtain new shades or faster dyes, but also by the introduction of new textile materials, such as the various kinds of artificial silk or rayon, for which, in some cases, special dyes are necessary.

Although, after the discovery of mauve, the attention of organic chemists was largely directed towards the production of new dyes, the elucidation of the constitution of these dyes and the observation that changes in colour or shade accompany changes in the composition of the dyes (p. 137) aroused curiosity regarding the origin of colour in compounds and the relations between colour and constitution.[2] In 1876 Otto N. Witt[3] formulated the rule that colour is associated with the presence of certain groups or arrangements of atoms in the molecule—these groups being known as *chromophores* or colour-bearing groups. Compounds containing a choromphoric group are known as *chromogens*, and these compounds may or may not be coloured. In order to produce a coloured compound, certain groups, known as *auxochromes*, may have to be introduced into the chromogen; and in order that the coloured compound may act as a dye, certain acid or basic groups (*e.g.*, OH, NH_2) must be present. Thus, the compound azobenzene,

---

[1] The successful development of the industrial production of indigotin is due mainly to Rudolf Knietsch and the chemists of the *Badische Anilin- und Soda-Fabrik*, an account of whose work is given in *Ber.*, 1901, **34**, 4069.

[2] See E. R. Watson, *Colour in Relation to Chemical Constitution* (Longmans); G. N. Lewis and M. Calvin, *The Colour of Organic Substances* (*Chem. Rev.*, 1939, **25**, 273); A. Maccoll, *Colour and Constitution* (Quarterly Reviews of the Chemical Society, Vol. I).

[3] *Ber.*, 1876, **9**, 522.

$C_6H_5 \cdot N = N \cdot C_6H_5$, contains the chromophore, $- N = N -$, and is coloured, but it is not a dye. If, however, the hydroxyl-group is introduced, a dye, $C_6H_5 \cdot N = N \cdot C_6H_4 \cdot OH$, is obtained.

While the potentiality of colour seems to depend on the presence of certain groups, the actual colour or shade may be considerably altered by the introduction of different atoms or groups into the molecule. Thus, as we have seen, Hofmann obtained violet dyes by introducing methyl or ethyl groups into the molecule of magenta (red); and by increasing the number of such groups the violet shade becomes bluer, as in the case of methyl violet and crystal violet, which contain five or six methyl groups. The phenyl, benzyl and other groups derived from benzene produce a still greater change of colour and, in 1879, Nietzki formulated the rule[1] that the colour of a dyestuff is deepened by the introduction of groups which increase the molecular weight, and the deepening of the colour increases with increase in the molecular weight. By deepening of the colour is meant a change of colour in the direction: yellow→ orange→ red→ violet→ blue→ green. Not only the molecular weight, however, but also the chemical nature, the number and the position of the auxochromes have been found to be of great importance;[2] and even a heightening of the colour may be produced by the introduction of the nitro-, the acetyl- and certain other groups.[3] Nevertheless, in spite of many exceptions, Nietzki's rule was found to be a valuable guide in the production of new dyes of a desired colour.

In 1888 H. E. Armstrong put forward the view, which was very widely accepted, that the origin of the colour of organic compounds belonging to the benzenoid series is to be found in the presence of a quinone structure,[4] $O = C \begin{array}{c} /CH = CH\backslash \\ \\ \backslash CH = CH/ \end{array} C = O$. Since,

however, it was found that some compounds having a quinonoid structure were colourless, while others, to which a quinonoid structure could not be assigned, were coloured, A. von Baeyer suggested that colour is due to an *oscillation* of the quinonoid state between two or more benzene nuclei;[5] and the same view was

---

[1] *Verhandl. des Vereins zur Beförderung des Gewerbefleisses*, 1879, **58**, 231.
[2] Liebermann and St. von Kostanecki, *Ber.*, 1885, **18**, 2142.
[3] M. Schütze, *Z. physikal. Chem.*, 1892, **9**, 109.
[4] *Proc. Chem. Soc.*, 1888, p. 27. [5] *Annalen*, 1907, **354**, 152.

expressed also by Richard Martin Willstätter (1872–1942),[1] Professor of Organic Chemistry, University of Munich, and Nobel Prizeman for Chemistry (1915). Again, however, coloured compounds were known to which this explanation was inapplicable; and other suggestions were similarly found to have only a restricted validity. It gradually became evident, therefore, that valuable as were the theories and rules put forward by Witt, Nietzki, Armstrong and others as guides in the practical operations of dye production—and in this respect they played a very important part— no satisfactory general relationship between colour and constitution could be found so long as one considered the selective absorption of light (to which colour is due) only in the visible part of the spectrum. The problem must be regarded from a much wider angle and one must take into account selective absorption in the ultra-violet and infra-red as well as in the visible region.

Investigations of the relations between chemical constitution and selective absorption of ultra-violet light had already been undertaken, at a comparatively early date, apart from any consideration of the question of the visible colour of dyes. Even as early as 1862 the English physicist (Sir) George Gabriel Stokes (1819–1903)[2] and William Allen Miller (1817–70),[3] Professor of Chemistry at King's College, London, had carried out investigations on the permeability of matter for rays of high refrangibility, and had found "no special connection between the chemical complexity of a substance and its diactinic power." It was not, however, till 1872 that the systematic investigation of the ultra-violet absorption spectra of organic compounds was begun by Walter Noel Hartley (1846–1915), at that time Demonstrator in Chemistry at King's College, London, who had at his disposal the apparatus which had been used by W. A. Miller. The pioneering investigations of Hartley,[4] in which he was at first associated with Alfred Kirby Huntington (1856–1920), who later became Professor of Metallurgy in King's College, London, were followed by a long series of investigations carried out by Hartley himself, by (Sir) James Johnston Dobbie, Edward Charles Cyril Baly (1871–1948),[5] and

[1] Willstätter and J. Piccard, *Ber.*, 1908, **41**, 1458, 3245.
[2] *Phil. Trans.*, 1862, **152**, 599.
[3] *Phil. Trans.*, 1862, **152**, 861; *J. Chem. Soc.*, 1864, **17**, 59.
[4] *Proc. Roy. Soc.*, 1879, **28**, 233; *Phil. Trans.*, 1879, **170**, 257.
From 1879 to 1911, Hartley was Professor of Chemistry in the Royal College of Science, Dublin. He was created a Knight in 1911.
[5] Lecturer at University College, London, and from 1910 to 1937 Professor of Inorganic Chemistry in the University of Liverpool.

many other chemists, mainly English, down to the present day. From the general point of view these investigations have shown little more than that substances with similar constitution have similar absorption spectra in the ultra-violet; but from the special point of view of colour it has been found that while aliphatic compounds, in general, show no absorption bands in the ultra-violet, aromatic compounds have the property of selective absorption, and are, therefore, strictly speaking, coloured, although the colour is not visible to the human eye. The position of these absorption bands, moreover, can be altered by changes in the structure or constitution, and the shift of the absorption bands may bring them into the region of the visible spectrum. The compound then becomes visibly coloured. The introduction of the quinonoid structure into a molecule has a powerful influence in shifting the absorption bands towards longer wave-lengths and so in producing visible colour. This is clearly shown by the shift of the absorption bands of phenolphthalein (colourless) into the visible region (red) when the quinonoid structure is brought about by alkali. Similarly the azo-group and certain other groups bring about a shift of the absorption bands, although the shift may not in all cases suffice to produce a visible colour.

Although aliphatic compounds do not show selective absorption in the ultra-violet, they may do so in the infra-red;[1] and the position of the absorption bands may also be shifted into the visible region of the spectrum by variation of molecular structure, more especially by the introduction of conjugate double bonds. As a general conclusion from the investigations so far carried out, one may state that visible colour is shown only by unsaturated carbon compounds. Saturated compounds possessing a visible colour are not known.

The success which attended, and continues to attend, the efforts of chemists to produce synthetically a wide range of varied colouring matters was paralleled, on a much less extensive scale, by their achievements in the domain of synthetic drugs and perfumes. Only some of the main lines of development can be sketched here.

Although ether had been known even to the medieval alchemists and chloroform had been prepared by Liebig in 1832 and its narcotic action had been discovered by (Sir) James Young Simpson in 1847; and although alkaloids and other physiologically active

---

[1] The investigation of infra-red absorption spectra was later energetically prosecuted from the point of view of molecular constitution (see Chapter XII).

principles had been extracted from plants and used medicinally from an early period, it was not till 1881 that the era of synthetic organic drugs may be said to have opened. After the discovery in 1880 by Zdenko Hanns Skraup (1850–1910) of a method of synthesizing the base quinoline[1]—discovered in coal tar by Friedlieb Ferdinand Runge (1794–1867), Professor of Chemistry, Breslau, in 1834 and obtained by Gerhardt in 1842 by distilling the alkaloid cinchonine with alkali—Otto Fischer, in 1881, prepared the first synthetic alkaloid, *kairine*, which is a derivative of hydroxyquinoline. This and certain other derivatives of quinoline, all of which have antipyretic properties, had but little success; but it was followed by the important febrifuge, *antipyrine* or *phenazone* (phenyl-dimethylpyrazolone),[2] discovered by Ludwig Knorr (1859–1921), in 1883. Although this and other synthetic antipyretics, such as *pyramidon* (dimethylaminoantipyrine), and *antifebrin* (acetanilide), the physiological properties of which were accidentally discovered in 1886, have proved of much value, they are drugs which combat the symptoms of disease (high temperature) and not the disease itself, and they do not possess the specific curative properties shown, for example, by quinine in relation to malaria.

The investigation of the physiological action of organic compounds which had been stimulated by the discovery of the behaviour of antifebrin, showed that physiological action is in many cases due to the presence of certain groupings of atoms and can be modified or even entirely altered by the introduction of different groups into the molecule. Consideration of these facts led to the introduction of the analgesic drug *phenacetin* (*p*-ethoxyacetanilide) by Oscar Hinsburg[3] in 1887. Further it was found that by acetylating salicylic acid a compound was obtained which undergoes decomposition only in the alkaline juices of the intestine and therefore does not produce the gastric disturbances caused by salicylic acid itself. Under the name *aspirin*, acetylsalicylic acid was introduced as an anti-rheumatic and anti-neuralgic drug in 1899 and has found a very widespread use.

The introduction in 1884 of the alkaloid cocaine (first isolated in a pure state in 1860 by Wöhler), for the production of anaesthesia

[1] *Ber. Wien. Akad.*, 1880, **81**, 593. Skraup was at that time an assistant at the *Handelsakademie*, Vienna. He later (1906) became Professor of Chemistry in the University.

[2] *Annalen*, 1887, **238**, 137.

[3] *Annalen*, 1899, **305**, 276.

in operations on the eye, aroused much interest among surgeons regarding the use of local anaesthetics; and the elucidation of the constitution of cocaine by Willstätter[1] in 1898 led chemists to study the physiological action of the decomposition products of this alkaloid. Although the relation between physiological or pharmacological action and constitution is much less definite than between colour and constitution, the systematic investigation of substances related to cocaine led to the synthesis of a number of compounds possessing a local anaesthetic action.

As early as 1890 Eduard Ritsert had observed the anaesthetic action of the ethyl ester of aminobenzoic acid,[2] but this compound was introduced as a local anaesthetic, under the name *anaesthesine*, only in 1902. Meanwhile, Alfred Einhorn observed that the esters of aminohydroxybenzoic acid have anaesthetic action and two such esters were introduced for use under the name *orthoform* and *new orthoform*.[3] These were followed in 1904 and 1905 by *stovaine* (E. Fourneau), *alypine* (F. Hofmann) and *novocaine* (A. Einhorn), of which the last mentioned has found widespread use, especially in conjunction with adrenalin, which, by its vasoconstrictive effect, reduces its absorption by the tissues. Investigation of a considerable number of other derivatives of aminobenzoic acid led at last to the discovery of *pantocaine* (the hydrochloride of the β-dimethylaminoethyl ester of *p*-butylaminobenzoic acid), which has proved to possess advantages over novocaine owing to increased penetration and prolongation of anaesthesia. It is stated to be a perfect substitute for cocaine for all purposes, has greater stability and is not habit-forming.[4]

In the field of general anaesthetics, an outstanding recent development is *fluothane* ($C_2HClBrF_3$).[5] This provides smooth and easily reversible anaesthesia for almost every surgical procedure in patients of all ages. It has a great advantage in being non-explosive.

Adrenalin, to which reference has just been made, is also now prepared synthetically. It was first isolated in 1901 by the Japanese chemist, Jokichi Takamine, from the suprarenal glands of sheep and oxen, but within a few years its chemical constitution had been elucidated and a method for its synthesis worked out in the

[1] *Ber.*, 1898, **31**, 1534, 2498, 2655; 1899, **32**, 1635.
[2] *J. Pharm.*, 1890, **22**, 21.
[3] A. Einhorn, *Annalen*, 1900, **311**, 26; A. Einhorn and E. Ruppert, *ibid.*, 1902, **325**, 305.
[4] *Medicine in its Chemical Aspects* ("Bayer").
[5] Ferguson, J. 25th Hurter Memorial Lecture, Chem. & Ind. 1964, 818.

laboratories of Meister, Lucius and Brüning in Germany. It is a derivative of *o*-dihydroxybenzene or *catechol*, and when injected subcutaneously it produces a violent contraction of the arteries and a rapid increase of blood pressure.

Just as certain colouring matters will dye wool but not cotton, so it was found, as the result of many investigations carried out towards the end of last century and the beginning of the present century, that various living tissues show selective absorption for different dyes. Thus, methylene blue is absorbed by and stains only the living nerve, and when this dye is injected into a living animal the nerve tissues but not the surrounding structures are stained.[1] Similarly, different bacteria can be distinguished by their selective absorption of dyes, many of which act also as bactericides. Guided by these observations and the principle that a drug acts only on organisms by which it is absorbed, the idea arose of making use of this property of selective absorption for the treatment of diseases due to protozoal parasites, by introducing into an organism substances which are poisons for the parasites but which are not absorbed by, and are therefore not harmful to, the cells of the organism itself. There thus began a close co-operation between chemist and physiologist, and there developed a science of chemo-therapy of which the founder was the German physiologist, Paul Ehrlich (1854-1915).[2]

Investigation of various dyes showed that trypan red, a dye of the type of Congo red, destroys the trypanosome of the South American horse disease, "mal de caderas" (Ehrlich), and Maurice Nicolle (1862-1932) and Felix Mesnil, of the Pasteur Institute, Paris, found that another azo-dye, trypan blue, destroys the trypanosome of the cattle disease "piroplasmosis."[3]

The principle of selective absorption was very successfully utilized in finding a cure for syphilis, which is due to a micro-organism known as *Spirochaeta pallida*. It had been known from the time of Paracelsus that mercury is a specific against this disease, but mercury is harmful also to the human organism; and the problem, therefore, which Ehrlich set himself was to prepare a compound—a 'magic bullet'—which would contain a toxic substance and which would be absorbed by the organism producing the disease, but not by the human organism. After many trials and many failures,

[1] Paul Ehrlich, *Biol. Zentralbl.*, 1887, **6**, 214.
[2] Jenkins, Hartung, Hamlin, junr., and Data, *The Chemistry of Organic Medicinal Products*; *Medicinal Chemistry*, edited by Alfred Burger.
[3] *Ann. Inst. Pasteur*, 1906, **20**, 417.

Ehrlich prepared the compound known as *salvarsan* (or *arsphenamine*),[1]

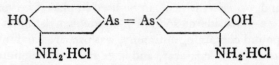

$$\text{HO} \diagdown \diagup \text{As} = \text{As} \diagdown \diagup \text{OH}$$
$$\text{NH}_2 \cdot \text{HCl} \qquad\qquad \text{NH}_2 \cdot \text{HCl}$$

(the hydrochloride of diaminodihydroxyarsenobenzene), a compound analogous to the azo-dyes, but containing the group − As = As − in place of − N = N −. A little later the more soluble derivatives *neoarsphenamine* and *sulpharsphenamine* were prepared, and the former product has been widely used in the treatment of syphilis, yaws and some other spirochaetal infections.

The trypanocidal properties of a number of derivatives of acridine, the parent substance of a number of valuable dyes, were also studied by Ehrlich, who prepared the compound known as *trypaflavine* (or *acriflavine*).[2] Although this substance (3 : 6-diamino-N-methylacridinium chloride) was found to have only very restricted trypanocidal properties, it was found by Dr. Carl Hamilton Browning, of the Bland-Sutton Institute of Pathology, London, that trypaflavine has the most valuable property of destroying the germs producing sepsis in wounds, without seriously interfering with the leucocytes or "warrior cells". It is, therefore an excellent antiseptic.[3]

Owing to the ravages produced by sleeping-sickness, much attention was given to the discovery of a specific against the parasite, *Trypanosoma gambiense*, which is the cause of this disease.

In 1920 a compound of undisclosed composition known as "Bayer 205" (now officially called *suramin*), was introduced in Germany and was found to be a valuable drug for the early treatment of sleeping-sickness. Its composition was ascertained by E. Fourneau,[4] of the Pasteur Institute in Paris, who showed it to be a very complex derivative of carbamide. Although this compound was effective in the early stages, it had little effect in the later stages of the disease, in which the central nervous system is involved. Fortunately, however, Walter Abraham Jacobs, of the Rockefeller Institute of Medical Research, New York, and Michael Heidelberger had, in 1919, prepared an arsenic-containing compound, *tryparsamide*, which was found to be effective in the later

[1] P. Ehrlich and A. Bertheim, *Ber.*, 1912, **45**, 756.
[2] L. Benda, *Ber.*, 1912, **45**, 1787. [3] *Brit. Med. J.*, 1917, I, 73.
[4] *Compt. rend.*, 1924, **178**, 675.

stages of sleeping-sickness, and could be successfully administered after treatment with suramin. In 1923 E. Fourneau also prepared a compound, *orsanine*, which can be used in place of tryparsamide.

In 1925 a glucoside of sodium *p*-aminophenylstibinate, a complex compound containing antimony, was prepared and introduced[1] for the treatment of kala-azar, and later aromatic diamidines, especially stilbamidine, were also widely and successfully used as curative agents for this disease.

Later years, moreover, saw some of the greatest achievements of chemists in the synthesis of drugs for the treatment of what is perhaps the most serious of all human diseases, malaria. For the treatment of this disease, affecting some 300 million people, one had, until 1926, been almost entirely dependent on the alkaloid, quinine, or on a mixture of the alkaloids present in cinchona bark. In that year, however, *plasmoquine* (now known as *pamaquin*), a derivative of quinoline, prepared in the laboratories of the German Dye Industry, and in 1930 the more effective compound *atebrin* (or *mepacrine hydrochloride*), a derivative of acridine, were introduced. These compounds are complex and difficult to produce, and the problem of synthesizing a simpler and more effective drug was attacked by a number of chemists in England and America. As it had been found that the compound, sulphapyrimidine (p. 153), was effective to some extent in controlling malaria, the pyrimidine nucleus was taken as the starting-point, and the results obtained led to an investigation of biguanides. In November 1944 F. H. S. Curd and F. L. Rose, of Imperial Chemical Industries, succeeded in synthesizing a compound which has been found to be more powerful and less toxic and to exert a wider influence on malaria than any other specific. This compound, N(1)-parachlorophenyl-N(5)-isopropylbiguanide,

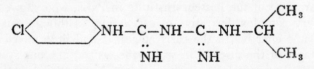

was called *paludrine*.[2] It combines the properties both of prevention and of cure.

[1] Burroughs Wellcome & Co., Eng. Pat., 1925, No. 234677.
[2] *Annals of Trop. Med. and Parasit.*, 1945, **39**, 139, 157, 208; *Chem. and Ind.*, 1946, 75.
The name paludrine is derived from the Latin *palus*, a marsh. Owing to the association of malaria with marshy land, the disease has also been called paludism.

The year 1935 witnessed another very notable advance in the science of chemotherapy. Hitherto, the treatment of diseases by means of drugs synthesized by the chemist was largely restricted to diseases of protozoal or spirochaetal origin, but in that year Gerhard Domagk,[1] working in the laboratories of the German Dye Industry, at Elberfeld, found that a red azo-dye, derived from sulphanilamide, $NH_2 \cdot C_6H_4 \cdot SO_2NH_2$ and m-phenylenediamine, had marked antibacterial properties. Introduced into medicine under the name *prontosil*, this compound was found to be effective against the streptococci which cause, for example, puerperal fever, scarlet fever, and erysipelas. It was then shown by French bacteriologists[2] that the antibacterial activity of prontosil is due to its decomposition in the body with production of sulphanilamide, a compound which had been synthesized in 1908 by Paul Gelmo at the Vienna Institute of Technology. Recognition of the antibacterial properties of sulphanilamide led chemists to prepare and investigate the antibacterial properties of derivatives of this compound. As a result of investigations by Ewins and Phillips of May and Baker, in England, it was found that if one of the hydrogen atoms of the $\cdot SO_2NH_2$ group is replaced by various molecular structures (pyridine, thiazole, guanidine, pyrimidine, etc.), the effectiveness and range of usefulness of the so-called sulpha-drugs can be greatly extended. Thus, sulphapyridine (M. & B. 693) and sulphathiazole (M. & B. 760) are effective in combating not only streptococci but also pneumococci and meningococci, the organisms responsible for pneumonia and meningitis.[3] In recent years sulphadimidine has been widely used, especially for children. Its relatively rapid absorption and slow excretion mean that doses need be less frequent; a further advantage is that it is rather more soluble than most sulphonamides and therefore there is less risk of damage from crystallization in the kidneys.

A new, and spectacularly successful, phase in chemotherapy began with the introduction of penicillin into general medical practice during and immediately after World War Two. The detailed history of this discovery is too long to be recounted here, but it is interesting to note that its beginnings can be traced back to the nineteenth century.

---

[1] *Deutsch. med. Wochenschr.*, 1935, **61**, 250.
[2] *Compt. rend. Soc. Biol.*, 1936, **120**, 756.
[3] The sulpha-drugs are bacteriostatic, not bactericidal, in action; that is, they prevent the multiplication and arrest the action of the bacteria, which are then destroyed by the leucocytes of the blood.

As much as eighty years ago it was well known that many species of bacteria and fungi produce substances toxic to other microbial species; if mixed cultures are grown one species may destroy the other. This antagonism between micro-organisms—or antibiosis as it is now called, whence antibiotic—seems first to have been turned to practical account by Pasteur and Joubert in 1887. They discovered that the anthrax bacillus is destroyed by a certain bacterial species. Prophetically, they recorded that "these facts perhaps justify the highest hopes for therapeutics." Thereafter, there is a continuous record of research in this field, though useful applications in medicine proved very elusive; the fundamental difficulty was that the toxic substances involved were not specific—they were poisonous not only to bacteria but to animal tissue. Before the end of the century the first crystalline antibiotic had been isolated; this was mycophenolic acid, obtained by Gosio from a mould. In 1913, Vaudremer claimed some success in treating tuberculosis in guinea-pigs with cultures of the mould *Aspergillus fumigatus*; much later work has shown that this mould produces at least four antibacterial substances, of which one (helvolic acid) is active against the tubercle bacillus. In 1925, two Russian workers, Zukerman and Minkewitsch, reported that guinea-pigs could be protected against diphtheria by means of the culture fluid on which the bacterium *B. mesentericus vulgatus* had been grown.

The way was thus well paved for Fleming's classic observation of 1929. In his laboratory at St. Mary's Hospital, Paddington, a culture of *Staphylococcus aureus* had accidentally become contaminated with an air-borne mould, subsequently identified as *Penicillium notatum*. He noticed that round the mould colony the bacteria had disappeared. Investigation showed that the mould produced a substance toxic not only to staphylococci but to a wide range of other pathogenic organisms. The culture on which the mould had grown appeared to be no more toxic to animal tissues than ordinary untreated culture medium. Nevertheless it was not clear that this was an example of antibiosis altogether different from the many others already recorded in the literature. In particular, penicillin is extraordinarily active against many bacteria; the most sensitive are affected by it at a dilution of one part in a hundred million. No less important, penicillin is virtually non-toxic to animal tissues.

Fleming's observation led to an attempt by H. Raistrick and others at the London School of Hygiene and Tropical Medicine to isolate penicillin from the mould cultures. For two practical reasons, the attempt failed. Firstly, penicillin is a very labile substance, easily destroyed by the normal processes of extraction. Secondly, it was present only in traces—about one milligram per litre—in the culture fluid, which contained also a complex mixture of other natural products. Undoubtedly, a contributing factor was that the extreme value of the prize to be won was not then apparent. In the event, Raistrick and his colleagues abandoned their attempt when they were close to success, and virtually nothing more was done about penicillin for several years.

Then, in 1938, H. W. Florey and E. Chain—at the Sir William Dunn School of Pathology, Oxford—resumed investigation of penicillin as part of a general study of the phenomenon of antibiosis. Extraction methods were developed, in which chromatography (p. 246) played a prominent part, and the first clinical trial was made early in 1941. Thereafter events moved rapidly. Industrial and academic aid was enlisted in both Britain and the United States and the investigation developed into one of the greatest of all wartime projects. Progress was such that by 1944 sufficient penicillin was available for all D-day casualties who needed it and within a few years the drug was universally and cheaply available. Fleming, Florey and Chain became Nobel Laureates and the two former were knighted for their achievement.

Intensive chemical research[1] established the formula of penicillin as

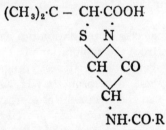

There proved to be several forms of penicillin, in which R may be C_5H_9-, $C_6H_5CH_2-$, or $HO\,C_6H_4\,CH_2-$; these differ significantly in their chemotherapeutic properties. A synthetic penicillin, identical with natural benzyl penicillin, was obtained in very small yield in 1944, but all the world's supply of the drug is still produced by the fermentation method.

The immense success of penicillin greatly stimulated the search for other new chemotherapeutic agents among the metabolic

[1] E. Chain, *Endeavour*, 1948, 7, 83, 152.

products of micro-organisms. Despite the investigation of tens of thousands of examples of antibiosis in laboratories throughout the world, surprisingly few new antibiotics have satisfied all the stringent tests necessary for their admittance to general medical practice and it is fair to say that none has the exceptional combination of qualities found in penicillin. Nevertheless they are of first-class medical importance, for growing clinical experience showed some limitations even in penicillin. Apart from the fact, noted at an early stage, that by no means all pathogenic bacteria respond to penicillin treatment there is the important complication that with wide— and possibly excessive—use of penicillin there have grown up large numbers of penicillin-resistant species which are often, however, sensitive to other antibiotics. Further, although the toxicity of penicillin is extremely low, a few people show an abnormal sensitivity to it. Unlike penicillin, some of these newer antibiotics— such as chloramphenicol—are available in synthetic form. Among important antibiotics in regular use are bacitracin, erythromycin, neomycin, streptomycin (particularly valuable in the treatment of tubercular infections), and the tetracyclines. Chemically, these substances differ widely among themselves; bacitracin, for example, contains a thiazoline ring. Another antibiotic, griseofulvin, has proved particularly useful in treating some troublesome fungal infections. The demand for antibiotics of all kinds is now so considerable that their manufacture is an important branch of the chemical industry.

The spectacular success of the antibiotics has perhaps distracted too much attention from the exploration of new classes of synthetic chemotherapeutic agents. Nevertheless, progress has been made. Certain sulphones, such as dapsone ($C_{12}H_{12}O_2N_2S$), have proved to be very effective in the treatment of leprosy. In the veterinary field, antrycide (quinapyramine) has proved valuable in treating the deadly cattle disease, carried by the tsetse fly, known as nagana.

Further development of the fine-chemicals industry has taken place in another important direction. For many years now it has been known that a number of protozoal and bacterial diseases are transmitted by insect carriers, and these diseases, therefore, may be combated not only by means of prophylactic drugs, natural or synthetic, but also by destroying the insect carriers. Until 1940 two of the most important insecticides used were the pyrethrins, present in the extract of pyrethrum flowers, and rotenone, the chief active principle of derris root.

On the entrance of Japan into the war of 1939–45, the supply of these insecticides, vitally important for troops fighting in tropical and jungle-covered countries, was greatly reduced, and research chemists and chemical industry were called upon to provide adequate supplies of effective substitutes.

The first such compound to come into extensive use was *p.p*-dichlorodiphenyl-trichloroethane,

$$\text{Cl·C}_6\text{H}_4 - \text{CH} - \text{C}_6\text{H}_4\text{·Cl}$$
$$\mid$$
$$\text{CCl}_3$$

a compound which was referred to officially and became known popularly as DDT. This compound had first been synthesized in 1874 by Othmar Zeidler, a young chemist working for a doctorate degree in the University of Strasbourg, and had been re-synthesized by Paul Müller,[1] a chemist in the dye-manufacturing firm of J. R. Geigy, in Basle, Switzerland, by whom also its marked insecticidal properties were discovered in 1936–7. It was first successfully used on a large scale in 1939 against the Colorado beetle.

After 1942 the industrial production of DDT was developed in Great Britain and in America, and was used with spectacular success in combating an epidemic of typhus in Naples in 1943, a disease of which lice are the carriers. It is a most powerful insecticide and can be used for the destruction of house flies (carriers of dysentery), mosquitoes (carriers of malaria), lice (carriers of typhus), fleas (carriers of bubonic plague), and many other insect pests, domestic and agricultural.

More recently the chemists of Imperial Chemical Industries discovered and produced a compound which is, in some cases, an even more powerful insecticide than DDT. This substance, to which the name *gammexane* was given, is one of the four isomeric forms, the so-called gamma-isomer, in which the compound hexachloro*cyclo*hexane, or benzenehexachloride, $C_6H_6Cl_6$, exists. This compound was first made by Faraday in 1825, and the existence of four isomers was established by Teunis van der Linden[2] in 1912. In 1943 the gamma-isomer was isolated and its powerful insecticidal properties demonstrated by the chemists of Imperial Chemical Industries.[3] Gammexane is particularly toxic to locusts.

The war against insects is complicated by the same phenomenon

---

[1] See Läuger, Martin and Müller, *Helv. Chim. Acta.*, 1944, **27**, 892.
[2] *Ber.*, 1912, **45**, 236.   [3] R. E. Slade, *Chem. and Ind.*, 1945, 314.

as is found in the war against bacterial infection, namely that after a time resistant strains of the organism in question begin to appear. The user of insecticides therefore needs a battery of chemical weapons just as the physician does. Neither DDT and its analogues, nor benzene hexachloride solve all the world's insecticide problems. Other classes of substances that have been successfully developed include *cyclo*pentadiene derivatives, nitroparaffins, and organophosphorus compounds. An example of the last class is parathion (*oo*-diethyl *o-p*-nitrophenyl phosphorothionate), particularly active against red-spider. The use of some of these substances is greatly complicated by their high toxicity to man, which demands extreme care in their application. Lately, some progress has been made in the development of systemic insecticides, that is insecticides that have a sufficiently low phytotoxicity to be absorbed into the sap of plants without damaging them. Such insecticides are of great potential importance in attacking such sap-sucking pests as aphids and the scale-insects that attack citrus fruits. The latter are particularly resistant to ordinary insecticide sprays because throughout the greater part of their life-cycle they are most effectively protected by their scale.

The economic consequences of the new synthetic insecticides have been enormous. But insects are not the only pests that the farmer and horticulturist must contend with; weeds, unless controlled, may smother growing crops. Traditional methods—ploughing, harrowing, hoeing, and so on—remain essential but chemical methods of weed control are of growing importance. For certain purposes, for example bringing waste land under cultivation, total destruction of all existing plant growth is desirable. For this purpose, sodium chlorate is commonly used. This has the disadvantage, however, that it persists in the ground for long periods, during which no fresh growth can be started. This disadvantage is not present in two new related herbicides, diquat and paraquat, recently introduced into agricultural practice. Diquat is 1, 1'-ethylene-2, 2'-dipyridylium dibromide.[1]

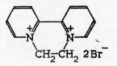

There herbicides have the advantage that unlike chlorate they are immediately inactivated in the soil, so that replanting can be undertaken forthwith. These substances promise to revolutionize certain branches of agricultural practice, for example the re-seeding of old grassland; the indications are that this can quite satisfactorily be done immediately after application of paraquat, without any other preparation whatsoever.

Very frequently, however, it is not total but selective destruction that is required—the elimination of charlock, for example, from growing corn. In recent years, a considerable number of very active, highly selective herbicides have been developed for agricultual and horticultural use. They are, indeed, now so widely used that they are to be found in one form or another in every household with a garden, where they are particularly useful in eliminating weeds from lawns. Much of this work derives from the isolation of plant growth substances (auxins) from human urine by F. Kögl in 1933.[1] Auxin-a has the formula

$$CH_3-CH_2-\underset{\underset{CH_2}{|}}{\overset{\overset{CH_3}{|}}{CH}}-CH-CH$$
$$CH_3-CH_2-\underset{\underset{CH_3}{|}}{CH}-CH-C-CH(OH)-CH_2-CH(OH)-CH(OH)-COOH$$

and auxin-b is very similar. A third auxin is indole-3-acetic acid.

During the course of an investigation which was carried out in the laboratory of Professor Ira Remsen in the Johns Hopkins University, Baltimore, Constantin Fahlberg (1850–1910), who later became a manufacturing chemist, accidentally discovered, in 1879, that one of the compounds he had prepared possessed a remarkably sweet taste. This compound had been synthesized from toluene and was found to have the constitution, $C_6H_4\diagup\overset{CO}{\underset{SO_2}{\diagdown}}\diagdown NH$.

It formed a white crystalline powder, and had a sweetness five hundred times as great as that of cane sugar. In 1887 this compound was placed on the market under the name *saccharin*, and although it was at first feared that its advent would spell the ruin of the cane and beet-root sugar plantations, the fear was not realized, for it was soon recognized that saccharin has no food value and that its uncontrolled and unlimited consumption is

[1] F. Kögl. *Z. Physiol. Chem.*, 1933, **214**, 241; 1934, **228**, 90.

harmful to the human organism. It was treated, therefore, as a medicament and its manufacture and sale put under licence. Saccharin proved very welcome during the prolonged sugar short-age of the last war and is also appreciated by diabetics whose intake of natural sugar must be strictly regulated. Another important artificial sweetening agent is sodium cyclamate ($C_6H_{11}NHSO_3Na$).

The extraordinary development which occurred in the produc-tion of synthetic dyes and medicinals during the latter half of the nineteenth century and which has continued down to the present time, has, by its widespread human appeal, made a very strong impression on the popular mind; and this impression has been deepened by the notable success which has been achieved by chemists in the artificial production of those perfumes and spices which have been so highly prized by all peoples in all ages. While, in some cases, the synthetic products merely imitate the naturally-occurring compounds, in other cases chemists have been able to synthesize in the laboratory compounds identical with those to which the odour of flowers is due.

Just as W. H. Perkin was the first chemist to synthesize a dye, so he was also the first chemist to prepare a naturally-occurring perfume, *coumarin*. This compound, which constitutes the fragrant principle of the Tonka bean and of the sweet woodruff (*Asperula odorata*), is used in the preparation of the perfumes known as Jockey Club and New Mown Hay. It was synthesized by Perkin in 1868 from salicylic aldehyde[1] which, in turn, is prepared from phenol (carbolic acid), one of the constituents of coal tar.[2] Since that time, vanillin, the active principle occurring in the vanilla bean; oil of wintergreen (methyl salicylate); oil of cinnamon (cinnamic aldehyde); the perfume of meadow sweet (salicylic aldehyde); the perfume of hawthorn blossom (anisic aldehyde), and

---

[1] J. Perkin, *J. Chem. Soc.*, 1868, **21**, 53.
Other methods of synthesizing coumarin, especially that by Léonce Bert (*Compt. rend.*, 1942, **214**, 230) have been devised. The coumarins have been the subject of extensive investigation, both chemical and physico-chemical, in recent years (*Chem. Rev.*, 1945, **36**, 1).

[2] By allowing salicylic aldehyde, $C_6H_4\diagup^{OH}_{\diagdown CHO}$ to react with acetic anhydride, two hydrogen atoms of a CH_3-group are eliminated along with the oxygen atom of the aldehydic group, with production of a double bond (Perkin's reaction), and forma-mation of acetyl coumaric acid. By heating this compound acetic acid is eliminated and coumarin, $C_6H_4\diagup^{O-CO}_{\diagdown CH=CH}$, is formed.

other naturally-occurring fragrant compounds, have been produced artificially by the chemist. Compounds having an odour similar to that of violets, musk, etc., are now also prepared synthetically.

The development of the art of photography, which at the present time plays a part of great importance in the life of all peoples, has been largely dependent on the work of the synthetic chemist. Synthetic organic chemistry has provided not only developers of the photographic image (pyrogallol, hydroquinone, amidol, etc.), but also dyes which render the photographic film sensitive to different rays of light, not only in the visible but also in the invisible (ultra-violet and infra-red) parts of the spectrum. Not only can one, by this means, secure a truer representation of different colour values, but by using photographic emulsions rendered sensitive to infra-red light it is possible to obtain, even at a distance of many miles, clear photographs of objects which, with ordinary photographic films, would be blotted out by a haze of scattered light. In addition, the use of dyes allied with physical processes has resulted in such progress in colour photography that this is now commonplace.

CHAPTER NINE

THE CONSTITUTION AND SYNTHESIS OF
NATURALLY-OCCURRING COMPOUNDS

ALTHOUGH, after the discovery of mauveine by Perkin in 1856, many chemists devoted themselves to the production of new dyestuffs and to the study of the relations between colour and constitution, not a few, and these some of the most eminent, turning back to nature and the traditional aim of organic chemistry, took up the investigation of the more important classes of compounds produced in plants and animals. Thereby they sought to gain a fuller understanding of the chemical nature of these compounds and of their mutual relationships, and to unravel, both by analysis and synthesis, the inner structure of their molecules; and in this work they have gained a great success. Before the closing decades of the nineteenth century the chemical structure of only a few plant and animal products was known; but today not only has the molecular structure of very many of these substances been ascertained, but methods also have been devised for their production synthetically in the laboratory.

It can therefore be claimed with complete assurance that what Berthelot considered to be the fundamental problem of organic chemistry has been solved, namely, to prove beyond all question that compounds identical with those produced by plants and animals can be synthesized from inorganic or mineral matter. It must, however, at the same time be pointed out that, wonderful as the success of chemists has been in building up the complex molecules of many plant and animal products, the methods of the chemist are not the methods of nature, and that the chemist knows very little about the mechanism of the processes carried out in the living organism and cannot reproduce them in the laboratory.

Among natural products, few, if any, are of greater importance, viewed from the chemical, biological and economic standpoint, than the compounds grouped under the general name of *carbohydrates*, a name under which are included the sugars, starches

162

and cellulose. With the existence of such substances, for example, in honey and the vegetable fibres used for weaving into cloth, man had, of course, long been familiar; but it was not till 1884 that the exact scientific study of these substances, a study surrounded by many experimental difficulties, was undertaken, with the object of elucidating their chemical constitution and stereochemical configuration. Guided by the structural theory of Couper and Kekulé and the stereochemical theory of van't Hoff and Le Bel—with the aid of which the successful investigation of the molecular structure of the carbohydrates was alone made possible—Emil Fischer succeeded, in great measure, in unravelling the inner structure and spatial arrangement of the atoms in the molecules not only of the naturally-occurring but also of the artificially synthesized hexose and pentose sugars. This brilliant series of researches,[1] remarkable no less for the skill with which they were carried through than for the acuteness of the reasoning based on the experimental results, was accomplished during the years 1884 to 1900, and constitutes the first great test and vindication of the importance and validity of the van't Hoff-Le Bel theory.

Before 1887 four naturally-occurring, isomeric hexose sugars having the formula $C_6H_{12}O_6$ were known, namely, glucose, fructose, galactose and sorbose; and in 1887 a fifth isomeric sugar, mannose, was isolated by Emil Fischer. In the case of glucose and fructose, two sugars which are present in honey and various sweet fruits, the chemical behaviour pointed to the presence in each of five hydroxyl groups, and of either an aldehydic group (glucose) or a ketonic group (fructose). The former sugar, therefore, was to be regarded as an aldehyde-alcohol (aldose) and the latter as a ketone-alcohol (ketose). Moreover, as Heinrich Kiliani (1855–1945) showed,[2] these sugars are normal, so-called straight chain compounds, the structure of which could be represented by the formula,[3]

$$CH_2OH \cdot CHOH \cdot CHOH \cdot CHOH \cdot CHOH \cdot CHO \text{ (glucose),}$$

or the formula

$$CH_2OH \cdot CHOH \cdot CHOH \cdot CHOH \cdot CO \cdot CH_2OH \text{ (fructose);}$$

[1] Published in the *Berichte der deutschen chemischen Gesellschaft* from 1887 onwards.

[2] *Ber.*, 1885, 18, 3066; 1886, 19, 221, 767, 1128. Kiliani was from 1897 to 1920, when he retired, Professor of Chemistry, University of Freiburg.

[3] The carbon atoms printed in heavier type are asymmetric, in the sense of the van't Hoff-Le Bel theory.

and galactose, also, was shown to have the same structure as glucose.

Considerable as was the advance which had thus been made, the more intimate relations between the simple hexose sugars still remained unravelled; and the experimental difficulties which faced the investigator appeared almost insuperable, owing to the resistance which the compounds offered to the ordinary methods of purification and owing to the apparent impossibility of separating and identifying the different members of a closely related family of compounds.

It was at this point that Fischer entered with a new weapon of attack, the compound phenylhydrazine, which he had first prepared in 1875. This compound, which reacts with aldehydes and ketones and has been widely used for the purpose of characterizing these, gives rise with glucose, fructose and the other aldoses and ketoses to compounds known as *osazones*. As the osazones formed by the different sugars are sparingly soluble, crystalline compounds, of characteristic appearance and of definite melting point, they could be employed for the purpose of differentiation and identification, and they played an essential part in the synthetic investigation of the sugars.

Since it was found that glucose, galactose and mannose are structurally identical aldo-hexoses, it is clear that the isomerism must be due to differences in the spatial arrangement of the atoms; and such isomerism is to be expected since these compounds contain four asymmetric carbon atoms. In fact, if the theory of van't Hoff be accepted, it is to be expected that the number of optically active isomers should be $2^4 = 16$; and the number of optically active keto-hexoses should be $2^3 = 8$, isomeric with fructose and sorbose. Guided by the theory of van't Hoff, therefore, Fischer set himself the task of synthesizing as many of these isomers as possible, and by studying the chemical reactions and the inter-conversions of one sugar into another, of determining the relative molecular configurations of the different isomers. No fewer than fourteen of the predicted sixteen aldo-hexoses and five of the eight keto-hexoses were thus synthesized and their relative configurations determined. The remaining two aldo-hexoses were later synthesized by W. C. Austin and F. L. Humoller in 1934. These configurations Fischer represented as projections of the space-formulae, of which the following may be given as illustrations:

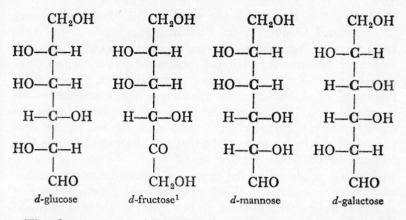

The first sugar to be synthesized by Fischer, in 1887, was optically inactive and proved to be racemic fructose (*dl*-fructose), When a solution of this sugar was acted on by yeast, the naturally-occurring, laevo-rotatory component of the synthetic sugar was destroyed by the action of the ferment, and there was left the hitherto unknown dextro-rotatory fructose. From this first synthetic sugar, *dl*-fructose, Fischer was able, by making use of reactions introduced by himself, as well as of reactions discovered by Kiliani, Alfred Wohl,[2] Otto Ruff[3] and others, to transform aldoses to ketoses, ketoses to aldoses, hexoses to pentoses ($C_5H_{10}O_5$), and pentoses to hexoses. In this way, therefore, Fischer was able to synthesize a large number of different sugars (only a few of which occur in nature), and to determine their relative configurations.

By the end of the nineteenth century the genius of Fischer seemed to have solved the riddle of the simpler sugars (the monosaccharides), and for a considerable time the stereochemical formulae deduced by him met with general acceptance. Interest in the carbohydrates now waned. In the early years of the present century, however, weapons were being forged, more especially by Purdie, Irvine and Haworth at the University of St. Andrews, which were later to be of great importance not only in a further attack on the problem of the structure of the monosaccharides, but also in an attack on the structure of the more complex carbohydrates, the disaccharides (cane sugar, etc.), the starches and cellulose. This attack, associated more especially with the names

---

[1] Although the naturally-occurring fructose is laevo-rotatory, the prefix *d* is used in order to indicate a genetic relationship with *d*-glucose.
[2] *Ber.*, 1893, **26**, 730.
[3] *Ber.*, 1898, **31**, 1573.

of Irvine and of Haworth but in which others (among whom may be mentioned Claude Silbert Hudson (1881–1952), of the U.S. Department of Agriculture), also participated, was carried out with an energy, experimental skill and intellectual acumen which place these later investigators on an equality with the great pioneer, Emil Fischer. Just as Fischer made special use of the osazones in his investigations of the sugars, so these later workers made use especially of the methyl ethers—compounds formed by the conversion of the hydroxyl or OH-groups of the carbohydrates into methoxy- or OCH_3-groups.[1]

The first challenge to the open-chain formulae of the monosaccharides, suggested by Fischer, came from a recognition that although the general properties of the aldoses are those of hydroxyaldehydes, these compounds do not show the chemical activity which might be expected or the properties which could be ascribed to the presence of an unmodified aldehydic group. Even as early as 1883, the German chemist, Bernhard Christian Gottfried Tollens (1841–1918), Professor of Agricultural Chemistry in the University of Göttingen, suggested that the molecule of glucose, for example, contains a butylene oxide, or *furan*, ring structure, represented by the formula

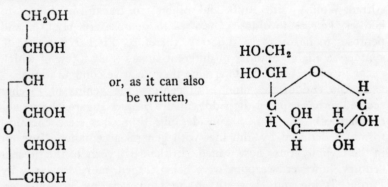

Since, in this formula, the aldehydic carbon atom becomes asymmetric, the existence of two isomeric glucoses, α- and β-glucose, which had long been known, can be accounted for by the interchange of the H and OH attached to that carbon atom; and the phenomenon of mutarotation can be explained as due to the conversion of the one form of glucose into the other, leading to a

[1] See Irvine, *Chemical Reviews*, 1927, **4**, 203; Haworth, *British Association Reports*, 1935, p. 31.

state of equilibrium.[1] When α-D-glucopyranose is made into an aqueous solution this first shows $[\alpha]_D = + 113°$, but the value gradually falls to $+ 52°$.

While later investigations confirmed the presence of a ring structure in the sugars, the view is now held that the normal pentoses and hexoses contain not a five-membered butylene oxide ring but a six-membered amylene oxide or *pyran* ring, so that glucose is represented by the formula

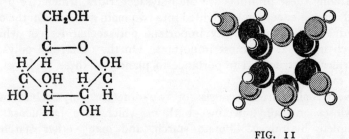

FIG. 11

a model of which is shown in Fig. 11.[2] Besides these normal sugars, however, there also exist labile γ-sugars, which contain a five-membered furan ring.

The elucidation of the structure and configuration of the simpler sugars, important as it is in itself, acquires a wider significance from the fact that these simpler sugars may be regarded as the structural units of which are built up the more complex natural products, the disaccharides, the starches and cellulose; and it is in the elucidation of the structure of these more complex carbo-hydrates that British chemists made their most valuable contributions. By a study of the methyl ethers of the simpler sugars and of the products of hydrolysis of the methyl ethers of the higher carbohydrates, a fairly complete picture has now been obtained of the structure of these more complex compounds. From these investigations it would appear that cane sugar (sucrose) is built up of the two units, glucose and fructose, as has, in fact, long been known; but while the α-glucose unit is present in the normal form with a pyran ring, the β-fructose unit exists in sucrose not with its normal pyran ring but with a furan ring.

Maltose, moreover, an isomer of sucrose, is found to be built

[1] Lowry, *J. Chem. Soc.*, 1903, **85**, 1314; E. F. Armstrong, *ibid.*, p. 1306.
[2] Haworth, *B.A. Reports*, 1935. In the model the black spheres represent carbon atoms, the larger white spheres oxygen atoms, and the smaller white spheres hydrogen atoms.

up of two normal α-glucose molecules; and cellobiose, similarly, of two β-glucose molecules. Starch and cellulose, in turn, are built up of hundreds of maltose units and cellobiose units respectively.[1] Another naturally-occurring disaccharide is trehalose, found in fungi and yeasts. Step by step, knowledge has been advanced and insight obtained, by chemical and physico-chemical methods, into the intimate molecular architecture of some of nature's most important products.

Among these products are the polysaccharides, which, for practical convenience, can be divided into two main groups. On the one hand are the nutritionally important polysaccharides, of which starch is by far the most important. On the other, are polysaccharides of structural importance in plants; of these, cellulose is the most important.

Nearly all starches consist of two different polysaccharides—amyloses and amylopectins—both of which give D-glucose on complete hydrolysis. Potato starch, and many other starches, consists of about one-fifth amylose and four-fifths amylopectin. The glucose units of amylose are evidently joined together in the same way as they are in maltose, for if an amylose is treated with the enzyme amylase the product consists almost wholly of maltose. The number of sugar units is large; potato amylose, for example, consists of polymers with molecular weights in the range 4000–150,000.

In the amyloses the sugar units are joined end to end; the amylopectins also consist of chains of sugars but these are relatively short and there is much cross-linking, mainly through 1,6-glycosidic bonds. The molecule of amylopectin derived from rice starch, for example, appears to consist of about ninety cross-linked glucose chains, each containing about thirty glucose units.

Glycogen, of great importance in animal nutrition because it forms the normal reserve of carbohydrates, resembles amylopectin. The molecule is very large, having a molecular weight measured in millions, and apparently consists mainly of complexly cross-linked D-glucopyranose units.

In 1936, G. Blix[2] of Uppsala obtained from the mucin of the sub-maxillary gland of cattle a crystalline substance which he called sialic acid. This proved to be a member of an important group of

[1] K. H. Meyer, *Naturwiss.*, 1940, **28**, 397; *Cellulose and Cellulose Derivatives*, edited by E. Ott.
[2] G. Blix, *Z. Physiol. Chem.*, 1936, **240**, 43.

natural substances—the nonulosaminic acids[1]—which clinically are hybrids of sugars and amino-acids. They are, therefore, in a sense intermediate between carbohydrates and proteins, which we shall now go on to consider. The biochemical significance of the nonulosaminic acids is still obscure, but they appear to play an important part in the defence mechanisms of the body.

If it is to Emil Fischer that we owe our first clear understanding of the constitution of the sugars, it is also to him that we owe chiefly the elucidation of the nature and structure of other important groups of compounds occurring in both the plant and animal organism. These compounds, all of them containing nitrogen as well as carbon, hydrogen and oxygen, are the *proteins* and the *purines*. Belonging to the latter group are such plant products as caffeine, theophylline and xanthine, which are found in tea, and theobromine, which is present in cocoa, as well as uric acid, which forms the main end-product of the decomposition of proteins in certain animal organisms; the excrement of snakes, for example, consists of almost pure uric acid.

These compounds had for long been the subject of investigation by chemists. The presence of uric acid in urinary calculi was discovered by Scheele in 1766, and its medical importance led to its further study by chemists; in gout, sodium urate accumulates in the tissues. The first real contribution to our knowledge of the structure of uric acid was made in 1838 by Liebig and Wöhler[2] who studied the decomposition products of uric acid and established the presence in it of residues of urea and mesoxalic acid, $CO(COOH)_2$; and other important relationships were later worked out by Baeyer.[3] Important advances had thus been made in our knowledge of uric acid and attempts also had been made at fitting together the various data into a chemical formula, but it was only when this compound and various derivatives had been synthesized, more especially by Emil Fischer,[4] that certainty was gained with regard to the structure of uric acid and confirmation obtained of the correctness of the formula suggested[5] in 1875 by Ludwig Medicus (1847–1915), who in 1900 became Professor of Chemistry and Pharmacy in the University of Würzburg. By his syntheses, moreover, of xanthine, theobromine, adenine, caffeine and purine, in

[1] M. Stacey, *Endeavour*, 1960, **19**, 43.
[3] *Annalen*, 1861, **117**, 178; 1864, **130**, 158.
[4] *Ber.*, 1897, **30**, 559; 1899, **32**, 436.
[2] *Annalen*, 1838, **26**, 245.
[5] *Annalen*, 1875, **175**, 236.

the years 1897–9, Fischer was able to show that all these compounds are derivatives of purine (from *purum uricium*) to which he assigned the formula

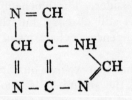

In more recent years other compounds belonging to the same group have been discovered and synthesized.

Among these compounds, the nucleic acids are of outstanding importance. The pioneer in this field was F. Miescher (1844–95), who in 1869 obtained from pus cells an acidic phosphorus-containing substance that was not derived from ordinary protein. Later workers who made outstanding contributions were A. Kossel (1853–1927), W. Jones (1865–1935), and P. A. Levene (1869–1940). As their name implies, nucleic acids are present in the nuclei of cells and for this reason tissues—such as glandular tissue—that contain relatively large nuclei are rich in it. The normal source of nucleic acid for experimental purposes is, however, yeast. There is at the present time intense interest in the nucleic acids because they form part of the nucleoproteins which, in the chromosomes of cell nuclei, appear to transmit hereditary characteristics.

When yeast nucleic acid is hydrolysed, each molecule yields four substances known as nucleotides. These nucleotides are known respectively as adenylic, guanylic, cytidylic, and uridylic acids. Each of these four nucleotides can be further hydrolysed, yielding in each case phosphoric acid, furfural (a degradation product of the sugar ribose) and either a purine or one of the closely related substances known as pyrimidine. Nucleic acids can therefore be described chemically as polynucleotides. They all appear to fall into two main classes, distinguished by the nature of the sugar unit; they are known respectively as ribonucleic acid (RNA) and deoxyribonucleic acid (DNA).[1] A related substance, adenosine triphosphate (ATP)—an adenine nucleotide containing three phosphoric acid groups—has been shown to play a vital role in many important biochemical reactions.

[1] The molecule of DNA has a helical structure. See F. H. C. Crick and J. D. Watson, *Proc. Roy. Soc.*, 1954, A, **223**, 80.

The elucidation of the structure of the nitrogenous purine derivatives led naturally to the study of the somewhat ill-defined but biologically very important group of substances, the *proteins*, which are essential constituents of plant and animal organisms and of which, for the most part, living tissues are built up. The investigation of these substances, however, encounters greater difficulties perhaps than that of any other group of naturally-occurring compounds, because being, for the most part, non-crystalline, they are difficult to obtain in a state of purity. Even those obtained in crystalline form are not necessarily homogeneous. In 1948, for example, β-lacto-globulin, long supposed to be a pure substance, was shown to be a mixture. They are, moreover, very sensitive to the action of acids, of alcohol and of heat, and are therefore liable to undergo change in the course of their investigation. The proteins occur mainly as lyophilic colloids (Chapter XII), and exist in the form of molecules of great complexity or of multi-molecular aggregates to which Karl Wilhelm von Naegeli (1817–91), of the University of Zurich, gave the name of *micelles*. The molecular or micellar weight of proteins has been determined by osmotic pressure measurements[1] and by means of the ultra-centrifuge, the introduction of which we owe to The Svedberg and R. Fåhraeus,[2] of the University of Uppsala. The molecular or micellar weight of haemoglobin was found by both methods to be in the neighbourhood of 68,000; and determination of the molecular weights of other proteins by means of the ultra-centrifuge[3] led to the suggestion that the proteins fall into groups in each of which the molecular weight is approximately equal to 17,600, or to an even multiple of this value.[4]

The development of other methods of determining the molecular weights of protein has shown that while many do in fact seem to be approximate multiples of 17,600—especially if one selects the results from different methods of measurement—this is not a universal rule and it certainly seems to have no particular significance. Among the newer methods introduced is one dependent on the long-known Tyndall effect, that is, the scattering of light by

[1] G. S. Adair, *Proc. Roy. Soc.*, 1915, A, **108**, 627; **109**, 292; 1928, A, **120**, 573; Peters and Saslow, *J. Gen. Physiol.*, 1939, **23**, 177.
[2] *J. Amer. Chem. Soc.*, 1926, **48**, 430; Svedberg and Pedersen, *The Ultracentrifuge*.
Theodor (The) Svedberg was awarded the Nobel Prize for Chemistry in 1926.
[3] Svedberg, *Kolloid-Z.*, 1930, **51**, 10; *J. Amer. Chem. Soc.*, 1930, **52**, 241, 279, 701, 5187; *Nature*, 1937, **139**, 1051.
[4] See E. E. Broda and C. F. Goodeve, *Nature*, 1941, **148**, 200.

suspended particles.[1] The molecular weight of proteins may also be deduced from osmosis experiments, or computed directly in the case of crystalline proteins by means of the electron microscope. In the main, these different methods give fairly consistent results and show that proteins have an extremely wide range of molecular weights. They may be as low as 13,000 (ribonuclease) or as high as 10,000,000 (tomato bushy stunt virus).

As early as 1820 Henri Braconnot (1781–1855),[2] by boiling glue with dilute sulphuric acid, had obtained a substance which he called glycocoll;[3] and this substance was later obtained by the decomposition of other plant and animal proteins. In the same year, also, Braconnot obtained leucine from gelatine, and in the following decades a number of other substances of a similar character were discovered—tyrosine and sarcosine by Liebig (1846–7), alanine and guanidine by Strecker (1850 and 1861), serine by Cramer (1865), etc. The chemical similarity of these compounds was shown to be due to the fact that they are all α-amino-acids, or derivatives of α-amino-acids. These acids, therefore, the simpler as well as the more complex, acquired much interest and importance, more especially as it was found that they are formed as the products of the hydrolytic decomposition of proteins, and may be regarded as the units of which the proteins are built up. Some of these amino-acids are open-chain, aliphatic compounds, while in the molecular structure of others one or more rings may be present. One of the amino-acids, cystine, discovered in 1810 by Wollaston, contains sulphur.

The first notable approach to an elucidation of the manner in which the amino-acids are built up into the complex protein molecules was made by Fischer in the course of a series of investigations commenced in 1899. Insight into the nature of the proteins and an understanding of their mutual relationships were gained, in the first instance, by analysis. By esterifying the mixture of amino-acids produced in the hydrolytic decomposition of the proteins and separating the esters by fractional distillation, Fischer greatly increased the effectiveness with which the different amino-acids could be isolated and characterized; and he was thereby able very greatly to extend and to render more precise our knowledge of the units of which the different proteins are built up.

[1] P. Doty and J. T. Edsall, *Advances in Protein Chem.*, 1951, 6, 35.
[2] Professor of Natural History, Lyceum, Nancy.
[3] From γλυκύς (glukus), sweet, and κόλλα (kolla), glue.

Having thus obtained a knowledge of the mono-amino acids which are formed as decomposition products of the proteins, Fischer was led to what must be regarded as his greatest achievement in this domain, the building up of these mono-amino acids (many of which had already been synthesized by others) into more and more complex structures which he called *polypeptides*. This name Fischer applied to the compounds formed by combining two or more amino-acid molecules with elimination of one or more molecules of water. Thus, two molecules of glycine (glycocoll), $NH_2 \cdot CH_2 \cdot COOH$, give one molecule of the dipeptide glycylglycine, $NH_2 \cdot CH_2 \cdot CO \cdot NH \cdot CH_2 \cdot COOH$, and one molecule of water. Similarly, this compound will react with other amino-acids, reaction taking place between the carboxyl group of one acid and the amino-group of another acid with elimination of water, and so more and more complex polypeptides can be formed.[1] Many polypeptides, have now been prepared artificially by various methods; among early examples were an octadecapeptide (with eighteen amino-acid residues), synthesized by Fischer in 1907, and a nonadecapeptide, $C_{54}H_{91}O_{20}N_{19}$, synthesized by Emil Abderhalden and Andor Fodor[2] in 1916. Although the first products of synthesis are inactive (racemic) compounds, these racemic forms have been resolved into their optically active antipodes, and in this way substances have been obtained which exhibit all the characteristics of the polypeptides derived from proteins. As the number of amino-acids in the molecule of a polypeptide increases, the number of possible isomers, due to a variation in the arrangement of the amino-acids, becomes greater and greater; and it may be that this fact affords an explanation of the great diversity of proteins in nature and of the differences which are found even among proteins of the same class, *e.g.*, the proteins of different bloods.

Although much is still to be learned about the structure of proteins, considerable progress has been made in recent years. In 1955, F. Sanger and his co-workers at Cambridge elucidated, by essentially classical chemical methods, the structure of the important natural hormone insulin.[3] This is a relatively simple protein, containing some fifty amino-acid residues; its structure is shown below. For this elegant piece of research Sanger received

[1] See E. Fischer, *Faraday Memorial Lecture, J. Chem. Soc.*, 1907, **91**, 1749.
[2] *Ber.*, 1916, **49**, 561.
[3] F. Sanger and L. F. Smith, *Endeavour*, 1957, **16**, 48.

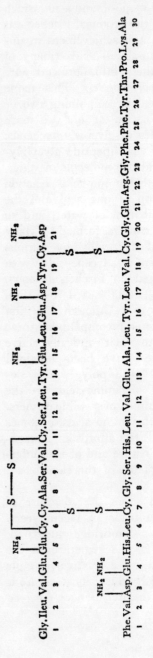

a Nobel Prize in 1958. Four years later, two of his colleagues at Cambridge—J. C. Kendrew and M. Perutz—followed in his footsteps to Stockholm to receive the same award. Their contribution had been respectively to unravel the structure of the proteins myoglobin and haemoglobin (p. 184), using an X-ray crystallographic method.[1] The molecular weight of myoglobin is approximately 17,000 and of haemoglobin 66,000. Their method was in principle the same as that developed by von Laue, the Braggs, and others some fifty years earlier but with two major innovations. The first was the introduction of a heavy-metal atom as a marker in the molecule. The second was the extensive use of computers to handle the data accumulated—an amount so enormous that it could not have been processed by ordinary conventional methods. Even so, the investigation was a very long and laborious one, and it seems clear that still other techniques must be learnt if we are to elucidate the structure of the hundreds of biologically important proteins.

Among the naturally-occurring plant products, an important place is taken by the fragrant "essential oils" which are present in the sap and tissues of certain plants and which give to them their characteristic odour. These essential oils, valued as perfumes and extracted from plants even from very early times, differ from the ordinary fixed or "fatty oils" (olive oil, linseed oil, etc.) by their

[1] *Nature*, 1961, **190**, 666.

fragrance or aroma and by being volatile in steam. Of the substances present as constituents of the essential oils, some of the most important are the various hydrocarbons and oxygenated compounds designated generally as *terpenes* and *camphors*. The work of the earlier chemists of the nineteenth century, which was restricted to the isolation and purification of the constituents of the crude oils, to the determination of their composition and the study of their characteristic reactions, showed that there was apparently an indefinitely large number of hydrocarbons of the same composition, a composition represented by the formula $C_{10}H_{16}$. In 1878 (Sir) William Tilden[1] for the first time attempted a classification of these diverse compounds, but this classification, which was based on a study of the compounds formed by the action of nitrosyl chloride (a valuable terpene reagent introduced by Tilden in 1874), was later superseded by a classification based on molecular structure. Since about 1885, down to the present day, the investigation of the terpenes has been vigorously prosecuted by an increasing number of chemists in all countries, but it is more especially to the German chemist, Otto Wallach (1847–1931), Professor of Organic Chemistry, University of Göttingen and Nobel Prizeman (1910), that we owe, in the first instance, the introduction of order into the chaotic mass of hydrocarbons of diverse origin;[2] and it is to him, as well as to Adolf von Baeyer, Friedrich Wilhelm Semmler, Georg Wagner, W. H. Perkin, junr., and others, that we owe the elucidation of the constitution of the terpenes. The molecular structures which chemical analysis and a study of the chemical transformations suggested have, in most cases, been confirmed by the partial or complete synthesis of the compounds.

Although the terpenes proper are hydrocarbons of the formula $C_{10}H_{16}$, it came to be recognized that the essential oils also contain hydrocarbons of a more complex structure, the composition of which could be represented by the general formula $(C_5H_8)_n$. In addition to the terpenes proper, therefore, one now recognizes the existence of sesquiterpenes, $C_{15}H_{24}$, of diterpenes, $C_{20}H_{32}$, and of triterpenes, $C_{30}H_{48}$. Although some of the terpenes belong to the open-chain, aliphatic compounds, most of them have a monocyclic or (less often) a polycyclic structure. Most of them, moreover, are optically active.

[1] *J. Chem. Soc.*, 1878, **33**, 80.
[2] *Annalen*, 1884, **225**, 314, etc. See also Wallach, *Terpene und Campher*; J. Simonsen and W. C. J. Ross, *The Terpenes*.

That some, at least, of the terpenes are related in structure to the aromatic hydrocarbon, *p*-cymene, had already been recognized[1] as early as 1860; and the relationship became clear when Wallach showed that the reactions of the oxygenated terpene, carvone, the dextro-rotatory form of which constitutes about 50 per cent of caraway oil, indicate that this compound has the structure

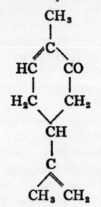

and that, similarly, the compound *a*-terpineol has the structure

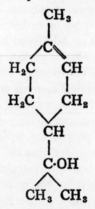

This compound was, at a later time, synthesized by W. H. Perkin, junr., and its structure thereby completely confirmed.[2] By the dehydration of terpineol, however, the hydrocarbon dipentene is formed[3] and, as Wallach showed in 1888, dipentene is the inactive

[1] C. Greville Williams, *Proc. Roy. Soc.*, 1860, **10**, 516; *Phil. Trans.*, 1860, **150**, 245.

[2] *J. Chem. Soc.*, 1904, **85**, 654. An optically active terpineol was also obtained by Fisher and Perkin (*J. Chem. Soc.*, 1908, **93**, 1871).

[3] The production of a dipentene from synthetic terpineol by Perkin was the first complete synthesis of a naturally-occurring terpene to be effected.

(racemic) form of limonene, a terpene which, in its dextro-rotatory form, is the main constituent of bitter orange oil.[1] Limonene was thereby shown to have the constitution

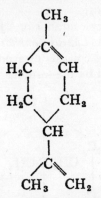

This constitution was confirmed by the synthesis of limonene from isoprene, C_5H_8, by Wladimir N. Ipatieff (1897), and by Wilhelm Euler (1898).

Other important terpenes, e.g., a-phellandrene, which occurs, in its l-form, in the essential oils from many species of *Eucalyptus*, and in its d-form in bitter fennel oil, have structures which differ from that of limonene only in the position of the two double linkings.

In the case of the terpene, a-pinene, the structure has been shown to be dicyclic. This terpene is one of the commonest constituents of the essential oils obtained from leaves, barks and woods, and occurs naturally in the inactive (racemic) as well as in the active forms. Its chief source is oil of turpentine.

From a detailed study of its oxidation products, Wagner suggested the formula

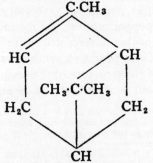

[1] Laevo-rotatory limonene also occurs naturally, but is not so common as the dextro form.

from which it will be seen that there is present not only the six-carbon ring of the hydrogenated benzene nucleus, but also a four-carbon ring.[1] A number of other terpenes have also been found to have this dicyclic structure.

In the case of ordinary camphor or "Japan camphor," obtained from the camphor laurel (*Laurus camphora*) and known from a very early time, the molecule consists of a polycyclic, "para-bridged," ring system, represented by the formula

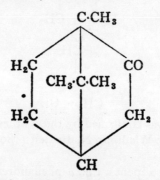

The composition of camphor was determined by Dumas in 1840, and the above formula was first suggested by the German chemist, Conrad Julius Bredt, in 1893. The correctness of this formula was fully confirmed by the synthesis of the oxidation products of camphor, camphoronic acid[2] and camphoric acid,[3] and by the synthesis of camphor from camphoric acid.[4]

Camphor, identical chemically with the natural Japan camphor but optically inactive, is now produced commercially from pinene which, in turn, is obtained from turpentine.

Unlike ordinary camphor, the oxygenated terpene, menthol or "mint camphor," $C_{10}H_{20}O$, the laevo-rotatory form of which is the chief constituent of peppermint oil, is a monocyclic alcohol, with a structure analogous to that of terpineol (p. 176). Optically inactive (racemic) menthol is also produced synthetically.

The elucidation of the structure of the sesquiterpenes and diterpenes was also effected in later years more especially by Leopold

---

[1] The presence of a stable 4-carbon ring in a naturally-occurring compound is an illustration of the stabilizing and ring-forming effect of the two methyl groups attached to a carbon atom in the ring (p. 130). Many other examples of this are to be found among the terpenes.

[2] Perkin and Thorpe, *J. Chem. Soc.*, 1897, **71**, 1169.

[3] Komppa, *Ber.*, 1901, **34**, 2472; 1903, **36**, 4332; Perkin and Thorpe, *Proc. Chem. Soc.*, 1903, **19**, 61; *J. Chem. Soc.*, 1906, **89**, 795.

[4] Haller, *Compt. rend.*, 1896, **122**, 448.

Ruzicka in Switzerland; and the triterpene, squalene, a hydrocarbon occurring in the liver of elasmobranch fishes, was investigated, from 1926, more especially by (Sir) Ian (formerly Isidor) Morris Heilbron, Professor of Organic Chemistry, Imperial College, London. Among these more complex terpenes are found derivatives of naphthalene and anthracene.

The terpenes, as has already been indicated, may be regarded formally as built up of units of C_5H_8; and an unsaturated hydrocarbon of this composition, known as *isoprene* $\left(\begin{array}{c} CH_3 \\ CH_2 \end{array}\!\!\!\diagdown C\cdot CH\!:\!CH_2\right)$ was obtained in 1860 by Charles Greville Williams (1829–1910),[1] by the destructive distillation of rubber. When this isoprene was kept in contact with air oxidation took place, and from the viscous liquid so formed Williams obtained "a pure white spongy elastic mass" which, on being burned, gave an odour of burning rubber.

A similar elastic mass, possessing the characteristics of rubber, was obtained in 1879 by the French chemist, Gustave Bouchardat (1842–1918), by allowing hydrochloric acid to act on isoprene,[2] and was found to have a composition agreeing closely with the formula $(C_5H_8)_n$. All the properties of the material, moreover, appeared "to identify the polymer of isoprene with the substance from which isoprene is formed, namely rubber."

In the experiments of Bouchardat the rubber was obtained by bringing about the polymerization of the isoprene which was formed as a decomposition product of rubber; but in 1882 (Sir) William Tilden[3] obtained isoprene by passing the vapour of turpentine through a red-hot tube. The isoprene obtained in this way was found to be converted into a rubber-like mass by the action of concentrated hydrochloric acid and of nitrosyl chloride. Moreover, it was found that when the limpid, colourless isoprene was kept for some years it passed into "a dense syrup in which were floating several large masses of solid, of a yellowish colour. Upon examination this turned out to be indiarubber."[4] This was the first real synthesis of rubber to be effected.

Although it was shown, at the end of last century,[5] that

---

[1] Consulting and industrial chemist. [2] *Compt. rend.*, 1879, **89**, 1117.

[3] *Chem. News*, 1882, **46**, 220.

[4] Paper read before the Philosophical Society of Birmingham in 1892; Tilden, *Chemical Discovery and Invention in the Twentieth Century*, 1916. Although, in 1892, the chemical identity of the synthetic and natural rubber could not be determined, the proof of this identity was obtained at a later time.

[5] I. L. Kondakoff, *J. prakt. Chem.*, 1900, **62**, 175; 1901, **63**, 113; **64**, 109. See *J. Ind. Eng. Chem.*, 1939, **31**, 941.

rubber-like materials can also be obtained from various hydrocarbons having a constitution similar to that of isoprene, it was not till 1910 that the commercial synthesis of rubber began to appear to be practicable. In that year Dr. F. E. Matthews, working for an Anglo-French syndicate, found that metallic sodium so greatly increases the rate of polymerization of isoprene that in the course of a few days, or even a few hours, depending on the temperature, conversion of the isoprene into pure rubber takes place.[1] The material obtained, however, is not identical in physical properties with natural rubber. In place of isoprene the hydrocarbon butadiene (CH_2:$CH\cdot CH$:CH_2), for the production of which acetylene, butane and butylene are the chief raw materials, is now largely employed.[2] At first the polymerization of butadiene was brought about by means of the catalytic activity of sodium (Na). Hence the name *Buna*, given to the material. Nowadays, benzoyl peroxide, or other substance, is used as catalyst. To improve the qualities of the buna rubber it is usual to mix the butadiene with another polymerizable substance, styrene ($C_6H_2\cdot CH$:CH_2) or acrylic nitrile (CH_2:$CH\cdot CN$). The mixed products so obtained are known as Buna-S (re-named in the U.S.A., GR.S.) and as Buna-N, and they are in some respects superior in properties to natural rubber.

In America, the important discovery was made by the chemists of the Du Pont de Nemours firm that the chlorinated compound, CH_2:$CCl\cdot CH$:CH_2, known as "chloroprene," readily undergoes polymerization with production of a rubber-like material called *neoprene*.[3] This material possesses a number of special and valuable qualities, and more than any other synthetic product, resembles rubber in its properties. Neoprene is at present used mainly as an addition to ordinary rubber, the properties of which are thereby improved.

Although the industrial production of synthetic rubbers must depend on the cost of plantation rubber and other economic considerations, the practicability of their production on a very large scale has been fully demonstrated. Their importance—apart from strategic considerations—lies mainly in the fact that they can be

---

[1] See Perkin, *J. Soc. Chem. Ind.*, 1912, **31**, 616. The discovery of the catalytic action of sodium was also made later in the same year by the German chemist, Carl Dietrich Harries.

[2] E. Konrad, *Angew. Chem.*, 1950, **62**, 491 (an account of the development of the synthetic rubber industry in Germany).

[3] W. H. Carothers, *J. Amer. Chem. Soc.*, 1931, **53**, 4303.

made to exhibit predetermined properties, *e.g.*, high resistance to oil and petrol, and low hysteresis loss.

In the preceding pages three large groups of naturally-occurring substances have been considered in some detail so as to emphasize the success which organic chemists have achieved in elucidating the molecular constitution of and in synthesizing some of nature's most important products, and for the purpose, also, of indicating in a general way the methods by which that success was attained. One must now refer, although in less detail, to other groups of compounds, the investigation of which has been vigorously pursued by a considerable number of chemists since 1870–80.

Although physiologically active principles had been extracted from various plants at a very early time, it was not till 1817 that Friedrich Wilhelm Sertürner (1783–1841), a German apothecary of Eimbeck, isolated from opium a crystalline "vegetable alkali," morphine.[1]

In the same and following years other substances of a similar character—narcotine (1817), strychnine (1818), brucine (1819), quinine (1820), etc.—were isolated and, owing to their basic character, which was ascribed by Liebig to the presence of nitrogen, were classed together as *alkaloids*.[2] Very many such compounds are now known, and their number continues to increase.

Insight into the chemical nature of the alkaloids was first obtained when Gerhardt, in 1842, showed that when strychnine, cinchonine and quinine are distilled with solid caustic potash, an oil is obtained which he called quinoleine (quinoline), and which was shown by Hofmann to be identical with a compound separated by Runge from coal tar (p. 148). At a later date it was found that other alkaloids, *e.g.*, nicotine, coniine, etc., are converted by chemical treatment into pyridine,

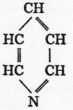

<hr />

[1] This had previously been obtained in an impure form in 1803 by Charles Derosne, a Paris apothecary, and in 1806 by Sertürner, who called it morphium. See H. M. Wuest, *A Hundred Years of Alkaloid Chemistry* (*Chem. and Ind.*, **1937**, 1084).

[2] This name originally included the compounds now called purines, but is now generally used to denote "a vegetable base which contains a cyclic nitrogenous nucleus."

or one of its derivatives, while still other alkaloids, *e.g.*, narcotine, are similarly related to iso-quinoline,

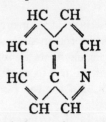

On the basis of this knowledge and as a result of the investigations carried out during the past half-century, the constitution of many of the more familiar alkaloids has been unravelled and a number of the simpler alkaloids have also been synthesized (*e.g.*, coniine and piperine by Albert Ladenburg in 1885–6, and nicotine and papaverine by Amé Pictet (1857–1937), University of Geneva, in 1904 and 1909). In 1885 W. Merck completed the synthesis of cocaine through tropine and ecgonine.

Although the investigation of this large and important class of naturally-occurring compounds has been actively prosecuted for many years and although a large number of alkaloids have been isolated and their chemical behaviour studied, doubt still exists concerning the molecular constitution of some of the more complex members of this group.[1] Notable advances, however, were made in more recent years towards the synthesis of the important alkaloid, quinine. In 1918 Paul C. L. Rabe,[2] later Professor of Chemistry, University of Hamburg, effected the conversion of the compound *d*-quinotoxine into quinine, and the first total synthesis of the alkaloid was realized when R. B. Woodward and W. E. Doering,[3] of Harvard University, successfully synthesized quinotoxine from 7-hydroxy*iso*quinoline. In 1954, Woodward announced the total synthesis of strychnine, in accordance with the structural formula advanced by Sir Robert Robinson.

The success which chemists achieved in unravelling the constitution of the essential oils has been paralleled, during the present century, by their success in elucidating the nature of the compounds which give to plants and flowers and fruits their peculiar and beautiful colours. In the prosecution of this work, the chemical

[1] See T. A. Henry, *The Plant Alkaloids*; *The Alkaloids*, edited by R. H. F. Manske and H. L. Holmes.
[2] *Ber.*, 1918, **51**, 466.   [3] *J. Amer. Chem. Soc.*, 1944, **66**, 849.

instability of many of these substances and the difficulty of obtaining them in a pure state have made the greatest demands on the skill and ingenuity of chemists.

The green substance, *chlorophyll*, which gives its colour to the leaves of plants and is now also used for colouring oils, soaps, salves, etc., is all-important for the building up of the carbohydrates in plants from the atmospheric carbon dioxide and moisture. Although chlorophyll had already been the subject of investigation in 1884 by Henry Edward Schunck (1820–1903) and others during the last quarter of the nineteenth century, it is to the work of Professor R. Willstätter and his collaborators, carried out during the years 1906–13, that our knowledge of the composition and constitution of that compound is mainly due.[1]

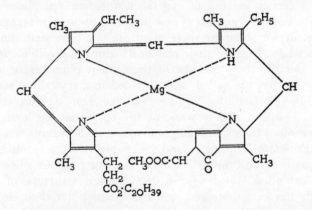

Having devised methods for the extraction of the green plant pigments without altering them chemically, Willstätter was able to show that there are two compounds, chlorophyll-A and chlorophyll-B, which occur together but not always in the same proportions in different plants, and which have the composition $C_{55}H_{72}O_5N_4Mg$ and $C_{55}H_{70}O_6N_4Mg$ respectively. The elucidation of the way in which these complex molecules are built up was a matter of extreme difficulty and necessitated a prolonged investigation of the chemical behaviour of chlorophyll and of the products which are formed by the breaking down of its molecule. Among the products of decomposition was a nitrogen-containing nucleus, known to chemists as pyrrole, and so, bit by bit, it became possible, from the experimental data, to fit the fragments together. In 1934

[1] See *Ber.*, 1914, **47**, 2831; K. F. Armstrong, *Chem. and Ind.*, **1933**, 809.

the labours of many years were gathered together in the formula[1] shown on p. 183. In 1960, Woodward synthesized chlorophyll.

It is of interest to note that the substance *haematin*, which when combined with the protein globin gives haemoglobin, the red colouring-matter of blood, has been found to be closely related chemically to chlorophyll, decomposition products of the same fundamental structure being formed by both substances. In haematin, however, an atom of iron takes the place of the magnesium in chlorophyll. It was synthesized by Fischer in 1929.

In the cell sap of plants there exist, also, a number of soluble plant pigments which give to flowers and fruits their great array of colours, and the study of these has engaged the attention of chemists throughout the whole of the past hundred years.[2] Although a certain amount of general information was gleaned and the existence of a large class of new pigments came to be recognized, the difficulty of obtaining these pigments in a pure state rendered any exact knowledge of their chemical nature impossible. It was, in fact, not till 1903 that one of these plant pigments or *anthocyanins*,[3] as they are called, was obtained in a crystalline state by Arthur Bower Griffiths, of the National Dental Hospital and College, London; and it was not till 1913 that the systematic investigation of the anthocyan pigments or anthocyanins was taken up by Professor Willstätter and his pupils.[4] As a result of the numerous investigations which have been carried out since then, a very extended knowledge of the chemical structure of these pigments has been obtained, and the methods for their isolation have been greatly improved.

In the plant the anthocyan pigment exists as a glucoside, formed by the combination of a coloured substance or *anthocyanidin* with glucose, or less often with the sugars rhamnose or galactose; and since the anthocyanidin may be combined with different sugars and with different numbers of sugar molecules, and since, also, the sugar molecule may be attached at different points of the anthocyanidin molecule, a large variety of colours may be obtained with one and the same anthocyanidin. Thus, it is found that the blue colour of the cornflower (*Centaurea cyanus*) and the red of the rose

[1] H. Fischer and J. Hasenkamp, *Annalen*, 1934, **513**, 107.

[2] See A. E. Everest, *Science Progress*, 1915, **9**, 597.

[3] The term anthocyan was introduced in 1835 by Ludwig Clamor Marquart (1804–81) to designate the blue pigment present in flowers. Later it was thought that the red and purple colours were due merely to the nature of the cell sap, and so the term anthocyan came to be applied to all.

[4] *Annalen*, 1913, **401**, 189, etc.

(*Rosa gallica*) are each due to the substance cyanidin combined with two molecules of glucose; while the colour of the cranberry is due to cyanidin combined with only one molecule of glucose. Moreover, the colour depends also on the nature of the cell sap and on the presence of iron salts, tannin, enzymes, etc. So far seven anthocyanidins have been obtained and all have been found, by the analytical and synthetic work of Willstätter and others and the remarkable syntheses of (Sir) Robert Robinson,[1] carried out since 1924, to be derivatives of the polycyclic substance flavonol:

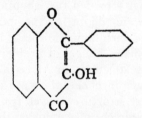

Besides the anthocyanins, other pigments of a yellow or red colour are found. Because of their relation to carotene, $C_{40}H_{56}$, a pigment first isolated from the carrot by Heinrich Wilhelm Ferdinand Wackenroder (1798–1854),[2] in 1831, they are called carotenoid pigments. Carotene and xanthophyll, a hydroxy-derivative of carotene, are yellow in colour and occur in many yellow flowers and fruits; lycopene, an isomer of carotene, is the red pigment of tomatoes, rose hips, etc. These and other similar colouring matters were obtained in crystalline form and their chemical structure elucidated more especially by R. M. Willstätter in Germany and Paul Karrer, Professor of Chemistry, University of Zurich, and Nobel Prizeman for Chemistry (1937).

The investigation of these and other natural pigments, some of which have been found to contain nitrogen, is still being actively prosecuted, and much has still to be learned regarding not only the nature of these substances but also the manner of their production and elaboration in the plant.

.   .   .   .   .   .

[1] Robinson occupied the Chair of Chemistry successively in the Universities of Sydney, Liverpool, St. Andrews and Manchester, and in University College, London. In 1930 he succeeded W. H. Perkin, junr., as Waynflete Professor of Chemistry in the University of Oxford. Created a Knight in 1939 and elected President of the Royal Society in 1945, from which he retired in 1955. Nobel Prizeman, 1947; O.M., 1949. He gives an account of the chemistry of the anthocyanin pigments in *Endeavour*, 1942, **1**, 92.

[2] Geiger's *Mag. für Pharm.*, 1831, **33**, 144.

One of the most notable features in the history of chemistry, more especially during the present century, is the manner in which the organic chemist has co-operated with the physician and physiologist in the study and elucidation of the processes of animal and human nutrition, and of the factors involved in the maintenance of health. Working in co-operation, the biologist and chemist have investigated the chemical substances involved in the processes taking place in the living organism, and have built up the important and rapidly growing branch of science known as *biochemistry*. The great success which has attended the labours of the modern biochemist is partly due, no doubt, to the introduction of special experimental methods, including the methods of micro-analysis[1] which have brought many biological problems within the range of exact chemical investigation, but it is also due, in greatest measure, to the advance in chemical theory and to the extensive knowledge of molecular structure derived from the great development of synthetic organic chemistry during the second half of the nineteenth century. The advances which have already been made in the domain of biochemistry are quite remarkable, but reference can be made here to only a few of these.

Since early in the eighteenth century, it was known that the disease scurvy, to which sailors absent on long voyages became subject, could be cured by the administration of lemon or orange juice; and for the general recognition of the antiscorbutic properties of fresh fruit and vegetables much is owed to Captain James Cook. Very gradually it came to be recognized that other diseases, *e.g.*, beriberi and rickets, could similarly be cured by dietary measures. In other words, it came gradually to be recognized that these diseases are due not to the presence of toxins or other substances in the diet, but to the absence of certain essential substances from the diet. By the work more especially of (Sir) F. Gowland Hopkins (1862–1947), of the University of Cambridge (1906), and of Elmer Verner McCollum and Marguerite Davis in America (1915), definite proof was obtained that health could not be maintained on a diet of fat, carbohydrate and protein only, but that small quantities of other substances—to which Casimir Funk, working at the Lister Institute in London, had, in 1912, given the name *vitamins*—are also necessary.[2] Thereupon there began a search for and an

[1] Introduced by Fritz Pregl (1870–1931), Head of the Institute of Medical Chemistry at Graz.

[2] In 1905 Pekelharing, of the University of Utrecht, claimed that there is present in milk an unknown substance which is essential to nutrition.

investigation of the sources and nature of the vitamins, and this investigation has been carried out in the different countries of the world with great vigour and success. The existence of a considerable number of different vitamins, designated at first by the letters of the alphabet, has been recognized, and most of them have even been prepared synthetically by the chemist. Their number is continually increasing.[1]

Vitamin-A, which is necessary for the growth of young animals and absence of which leads to the disease xerophthalmia in adults, is present in fish-liver oils, green plants, tomatoes, etc. It was isolated in 1931 by Paul Karrer and obtained in a crystalline form by H. N. Holmes and Ruth E. C. Corbet in 1937. Its composition $C_{20}H_{30}O$, was determined by Karrer, who also suggested the constitution

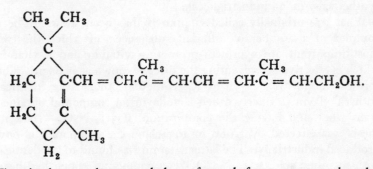

Vitamin-A may be regarded as formed from carotene by the cleavage at the double bond in the middle of the long hydrocarbon chain and the taking up of two molecules of water. Carotene, in fact, has been found to be the precursor of vitamin-A and is converted into it in the animal liver. If added, in any of its three isomeric forms, to the diet of an animal, its physiological effect is the same as that of the vitamin.

A synthesis of vitamin-A was announced in 1937 by Professor Richard Kuhn,[2] of the Kaiser Wilhelm Institute of Medical Research, Heidelberg, and C. J. O. R. Morris.[3] Two new syntheses of vitamin-A were later announced, one by J. F. Arens and D. A. van Dorp, of the laboratory of N. V. Organon, Holland, and one by Isler and his co-workers of the Hoffmann-La Roche

[1] H. R. Rosenberg, *Chemistry and Physiology of the Vitamins*; Leslie J. Harris, *Vitamins in Theory and Practice*.

[2] By instruction of the German Government Kuhn declined the Nobel Prize for Chemistry in 1938.

[3] *Ber.*, 1937, **70** [B], 853.

Laboratories, Switzerland. The latter closely follows a synthesis outlined by Heilbron, Johnson, Jones and Spinks[1] in 1942.

Investigation has shown that what was originally called vitamin-B is, in reality, a mixture of a number of different substances. Vitamin-B_1, deficiency of which results in the neuritic disease beriberi, was isolated from yeast by Professor Adolf Windaus (1876–1959), of the University of Göttingen, and Nobel Prizeman (1928), by whom, also, it was shown to be the hydrochloride of a base having the composition $C_{12}H_{17}ClN_4OS$, a substance to which the name *aneurine* (in America, *thiamin*) has been given. Its chemical structure was elucidated in 1936 by Robert R. Williams, of Bell Telephone Laboratories, New York, and his collaborators.[2] In the same year its synthesis was effected by Williams and J. K. Cline and by H. Andersag and K. Westphal.[3] It is now produced synthetically on an industrial scale.

What was originally called vitamin-B_2 has been shown to be a complex of a number of different substances, of which the two most important are a growth-promoting vitamin and a vitamin which is effective in preventing the disease known as pellagra. The former vitamin was shown by R. Kuhn and others to be identical with riboflavin (formerly called lactoflavin), a compound obtained from whey and having the composition, $C_{17}H_{20}N_4O_6$. Its synthesis was effected by Kuhn in 1934, and the compound is now produced industrially. The latter vitamin was found to be identical with nicotinic acid, $C_5H_4N \cdot COOH$, a compound which had been known to chemists long before its vitamin activity was recognized.

Other substances present in the B_2-complex are: pantothenic acid, $C_9H_{17}NO_5$; adermin or pyridoxin, $C_6H_{11}NO_3$; and *p*-amino-benzoic acid, $NH_2 \cdot C_6H_4 \cdot COOH$. The synthesis of all these compounds has been effected.

Vitamin-C, which was known to be present in fresh green vegetables and in citrus fruits, was, in 1932, isolated in a pure crystalline state and in relatively large amount from Hungarian red pepper (paprika) by Professor A. Szent-Györgyi,[4] of the University of

[1] *J. Chem. Soc.*, **1942**, 727.
[2] *J. Amer. Chem. Soc.*, **1936**, **58**, 1063, 1504; R. R. Williams, *Ind. Eng. Chem.*, **1937**, **29**, 980.
[3] *Ber.*, **1937**, **70** [B], 2035.
[4] Albert Szent-Györgyi, born in 1893, was Professor of Medical Chemistry, Szeged University, 1931–45; Visiting Professor, Harvard University, 1936; Professor of Biochemistry, University of Budapest, 1945–47; Director of Research, Institute of Muscle Research, Massachusetts, 1947– . Awarded Nobel Prize for Medicine, 1937.

Szeged. In 1933 Professor Haworth and his collaborators in the University of Birmingham succeeded not only in elucidating the constitution of this compound but also in effecting its synthesis. This substance, which is now known as *ascorbic acid*, has the comparatively simple structure

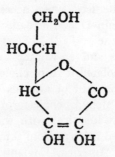

This structure is obviously related to that of the simple sugars (p. 167). The naturally-occurring ascorbic acid (vitamin-C) is laevo-rotatory, and this as well as the dextro-rotatory form have been prepared synthetically. The latter form, however, has no antiscorbutic effect.

Rickets, due to the imperfect calcification of the bones, is brought about by the absence of vitamin-D, which is present in various fish-liver oils, more especially halibut-liver oil.

The discovery made in 1919 that rickets are associated with absence of sunlight and can be cured by exposure to ultra-violet light was followed in 1924 by the discovery, by H. Steenbock, in Wisconsin, and by A. F. Hess, in New York, that foods were rendered antirachitically active by irradiation with ultra-violet light. In 1927 A. Windaus showed that when ergosterol, first isolated in 1889 by the French chemist, C. Tanret, from the ergot fungus of rye, was irradiated with ultra-violet light, an antirachitically active material was obtained, and from this material, in 1932, Windaus and workers at the National Institute for Medical Research, London,[1] isolated a compound which was named *calciferol*. It has the composition $C_{28}H_{43} \cdot OH$. This substance, the constitution of which was also determined, was thought at first to be identical with the natural vitamin-D (now called D₃) of fish-liver oils; but it was later found that this is not the case. In chemical

[1] *Proc. Roy. Soc.*, 1932, B, **109**, 488; *Annalen*, 1932, **492**, 226.

constitution the two substances are very similar and differ only in the hydrocarbon chain attached to a complex ring-system. In the natural vitamin this chain has the constitution $\cdot CH(CH_3) \cdot CH_2 \cdot CH_2 \cdot CH_2 \cdot CH(CH_3)_2$, whereas in calciferol the constitution is: $CH(CH_3) \cdot CH = CH \cdot CH(CH_3) \cdot CH(CH_3)_2$. Calciferol is now an article of commerce, and although not quite as potent as the natural vitamin, it can be used, when necessary, to supplement the natural supply of vitamin-D in the daily diet. It is now added, for example, to margarine.

In 1935 A. Windaus and co-workers[1] synthesized 7-dehydrocholesterol, and later isolated it from cholesterol, which is present in the fat glands of the skin of animals. When irradiated with ultra-violet light, the dehydrocholesterol is converted into an isomeric form which is identical with the natural vitamin—D_3. The vitamin was isolated in 1936 from tunny and halibut-liver oils by various workers in America, England and Germany, and in 1937 crystalline D_3 was obtained by F. Schenck[2] in Germany. It has the composition $C_{27}H_{43} \cdot OH$.

Other vitamins have also been the subject of intensive study. In 1936 H. M. Evans, O. H. Emerson and G. A. Emerson[3] isolated two different antisterility vitamins E, which have been named α- and β-*tocopherol*, and in 1938 the synthesis was effected[4] of *dl*-α-tocopherol, $C_{29}H_{50}O_2$, which can be used in place of the natural vitamin.

The two compounds present in vitamin-K, an antihaemorrhagic vitamin, have been isolated and their constitution as derivatives of 2-methyl-1 : 4-naphthoquinone, $C_{11}H_8O_2$, has been determined. The parent substance has itself been found to have antihaemorrhagic properties[5] and is used, under the name *menaphthone*, in place of the natural vitamin.

Biotin, originally designated vitamin-H, was isolated from egg yolk by Kögl in 1935. Its structure (2′-keto-3,4-imidazolidothiophane-2-n-valeric acid) was elucidated by du Vigneaud in 1942 and shortly afterwards it was synthesized. It seems to be of relatively minor importance in human nutrition but is an essential growth factor for many micro-organisms; it is identical with coenzyme R.

---

[1] *J. Physiol. Chem.*, 1936, **241**, 100, 104.   [2] *Naturwiss.*, 1937, **25**, 159.
[3] *J. Biol. Chem.*, 1936, **113**, 319.
[4] F. Bergel, A. Jacob, A. R. Todd and T. S. Work, *Nature*, 1938, **142**, 36; P. Karrer, H. Fritzsche, B. H. Ringier and H. Salomon, *Helv. Chim. Acta.*, 1938, **21**, 820.
[5] S. Ansbacher and E. Fernholz, *J. Amer. Chem. Soc.*, 1939, **61**, 1924.

In 1926 it was discovered that whole liver was effective in the dietary treatment of pernicious anaemia, but more than twenty years elapsed before the active principle (vitamin-B_{12}) was isolated independently by K. Folkers in the United States and E. Lester Smith in Britain. Its structure, which was elucidated in 1955 by Sir Alexander Todd (Lord Todd)[1] and his fellow-workers at the University of Cambridge, indicates that vitamin B_{12} is closely related to haem, chlorophyll and other natural porphyrin derivatives.[2] The compound is unique among vitamins in containing the element cobalt.

The collaboration between the organic chemist, the physiologist and the physician, which has proved so valuable in the study of the accessory food factors, the vitamins, has been of no less value in the study of the internal secretions of the ductless or endocrine glands, the so-called *hormones*,[3] the presence of which is essential for the health of the animal organism.

As early as 1849 it had been shown that the pathological condition known as Addison's disease is associated with degeneration of the suprarenal glands, and in 1874 it was also established that myxoedema is associated with degeneration of the thyroid gland. In 1891, further, it was found that injection of a glycerine extract of the thyroid gland led to the disappearance of the symptoms and signs of myxoedema; and cretinism, or arrested physical and mental development, could similarly be cured by means of the thyroid extract. It was thus demonstrated that the thyroid gland produces a substance which is passed into the blood-stream and which is essential for normal growth and for the maintenance of health. At this point the organic chemist enters and attempts to isolate and to determine the chemical nature of the active principle present in the secretions of the different endocrine glands.

The first hormone to be isolated and synthesized was *adrenalin*, the active principle of the suprarenal gland. Reference to the structure and synthesis of this compound has already been made

---

[1] Alexander Robertus Todd, Professor of Organic Chemistry, University of Cambridge, since 1944, was Reader in Biochemistry, University of London, 1937–38; Visiting Lecturer, California Institute of Technology, 1938; Professor of Chemistry, University of Manchester, 1938–44. He was created a Knight in 1954, was awarded the Nobel Prize for Chemistry in 1957, and created a Life Peer, with the title Baron Todd of Trumpington, in 1962.

[2] A. W. Johnson and Sir Alexander Todd, *Endeavour*, 1956, 15, 29.

[3] The name is due to the late Professor Ernest H. Starling, of University College, London, and is derived from the Greek ὁρμάω (hormaō), to excite or to arouse (*Lancet*, 1905, II, 339).

(p. 149). Adrenalin counters the hypo-glycæmic action of insulin, and is thus of great physiological importance.

In 1915 Professor Edward Calvin Kendall,[1] Professor of Physiological Chemistry, Mayo Foundation, Rochester, Minnesota, isolated from the thyroid gland a crystalline substance, *thyroxine*, which is the active principle of the thyroid gland. The hormone itself is a compound of thyroxine with a protein. In 1926 the constitution of thyroxine was determined by (Sir) Charles Robert Harington, clinical chemist, University College Hospital, London,[2] and George Barger (1878–1939), Professor of Medical Chemistry, University of Edinburgh, and was shown to be represented by the formula

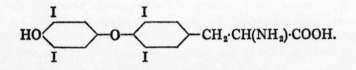

This compound was synthesized by Harington and obtained not only in the racemic but also in the optically active forms. It is the laevo-form, which is much more active than the dextro-form, that occurs in the gland.

The hormone insulin, which controls the level of sugar in the blood, has already been referred to (p. 173), in connection with its structure. It is a protein, as are at least two of the half-dozen hormones secreted by the pituitary gland (hypophysis), which plays a dominant role in controlling the vital functions. The follicle-stimulating hormone (FSH), which influences ovarian function, is a glycoprotein (molecular weight about 70,000) containing hexosamine and mannose units. Prolactin, which controls milk secretion, is also a protein, though its molecular weight apparently varies from one animal species to another.

In 1775, Conradi isolated from gallstones a substance to which Chevreul in 1815 gave the name cholesterol (Greek: chole, bile, stereos, solid). It is widely distributed in animal tissues, and its structure was shown by Rosenheim and Wieland (1932) to be

---

[1] Since 1951, Emeritus Professor. Awarded the Nobel Prize for Physiology and Medicine, 1950.
[2] Director of the National Institute for Medical Research, London, 1942–62.

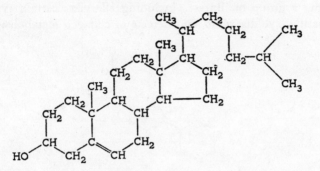

and in 1945 Dorothy Crowfoot confirmed this by X-ray analysis of the iodide.

Cholesterol is important less for its own properties—though it is currently believed to be implicated in the disease known as atherosclerosis (coronary thrombosis)—than because it is the parent substance of a wide range of substances that are physiologically very active; to this important chemical class Callow and Young,[1] in 1936, gave the name steroids. Of these the most important are the sex hormones, secreted by the ovaries and testes, which are associated with sex activity and with the appearance of the secondary sexual characteristics. The unravelling of the chemistry and intermediary metabolism of the steroid sex hormones —with the early phases of which the names of Butenandt in Germany and Ruzicka in Switzerland are particularly identified—have led to important advances in medical science. In recent years, following pioneer work by Sir Charles Dodds[2] and Professor J. M. Robson,[3] a number of synthetic steroids—such as stilboestrol, a derivative of stilbene—have been successfully introduced into medical practice.

The wide range of naturally occurring steroids is apparent from the fact that the cortex of the adrenal gland alone has so far yielded some thirty different steroids. Of these, cortisone, the formula of which is given on p. 194, has attracted particular attention. It is used medically for two main purposes. Firstly, for replacement therapy in disorders of the adrenal or pituitary glands. Secondly, for

[1] *Proc. Roy. Soc.*, 1936, **157A**, 194; E. C. Dodds and F. Dickens, *Chemical and Physiological Properties of Internal Secretions.*
[2] *Proc. Roy. Soc.*, 1939, **127B**, 140; 1940, **128B**, 253. See also E. C. Dodds and F. Dickens, *Chemical and Physiological Properties of Internal Secretions.*
[3] *Nature*, 1938, **142**, 292; 1942, **150**, 22; *J. Chem. Soc.*, 1943, 394.

treating a group of illnesses including allergies, certain types of dermatitis, eye disorders, and defects of collagen metabolism.

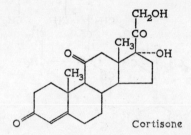

Cortisone

In a brief general work of this kind it is not possible even to mention all the many advances that have taken place in the chemistry of natural products, and this chapter must conclude with a brief reference to mevalonic acid, with which the name of K. Folkers[1] is particularly associated. This relatively simple substance ($C_6H_{12}O_4$, β-δ-dihydroxy-β-methylvaleric acid) was identified by degradation and synthesis in 1956–7. It is a key substance in the biosynthesis of a wide range of natural substances, including steroids, terpenes, and carotenoids. A complete biosynthetic pathway from acetic acid to cholesterol, involving mevalonic acid at an early stage, has been described.

In the preceding pages it has been possible only to indicate some of the main directions in which synthetic organic chemistry has moved during the past fifty or sixty years. From these main lines of advance, however, organic chemistry has branched and re-branched in innumerable directions. Not only have large numbers of natural compounds been synthesized, but countless other new compounds have been prepared, the number of which increases year by year by many hundreds. Through the unprecedented activity of the organic chemists of the past sixty years, great skill in the building up of even very complex organic compounds has been acquired and a very extended and deep insight into their molecular structure obtained.

[1] A. F. Wagner and K. Folkers. The organic and biological chemistry of mevalonic acid, *Endeavour*, 1961, **20**, 177.

CHAPTER TEN

THE DISCOVERY OF NEW ELEMENTS.
THE RARE GASES

IT is to the Hon. Robert Boyle, in the seventeenth century, that we owe the recognition of the importance of distinguishing between elements, compounds and mixtures, and it was he who introduced the term and encouraged the practice of *analysis* for the purpose of ascertaining the composition of materials. The teaching of Boyle, however, met with little response and his precepts were not generally accepted by chemists; and it was not until Lavoisier had initiated the "chemical revolution" and had reiterated[1] the teaching of Boyle that "the object of chemistry, in investigating the different natural bodies, is to decompose them and to learn how to examine separately the different substances which enter into their combination," that the work of analysing and of classifying the different kinds of matter was seriously undertaken by chemists. Those substances which had resisted all attempts to decompose them or to break them up into two or more substances simpler than themselves were, according to the practical definition due to Lavoisier, regarded as elements.[2]

In his *Traité élémentaire de Chimie* (1789) Lavoisier gave a list of substances, thirty-one in number (excluding *light* and *heat*), which were at that time regarded as elements, and of these, eight were later shown to be compounds. As chemists became possessed of more effective methods of decomposition and as the methods of analysis were improved and extended by men like Martin Heinrich Klaproth (1743–1817), in Germany, Louis Nicolas Vauquelin (1763–1829) in France, William Hyde Wollaston (1766–1828) in England, and Jöns Jacob Berzelius (1779–1848) in Sweden, the number of substances recognized as elements—

---

[1] *Œuvres*, I, 136.
[2] "We cannot be certain," wrote Lavoisier, "that what we think to-day to be simple is in fact simple; all that we can say is that such or such a substance is the limit at which chemical analysis has at the present moment arrived, and that in the existing state of knowledge it cannot be further subdivided." (*Œuvres*, I, 137.)

and still so recognized—rapidly increased. By 1800 the twenty-three proved elements in the list drawn up by Lavoisier in 1789 had increased in number to twenty-seven, through the discovery of uranium and tellurium (by Klaproth), titanium (by William Gregor) and chromium (by Vauquelin); and in the following three years no fewer than seven new elements were added to the list. In 1830 the number had grown to fifty-four, but in several cases the isolation of an element, in a reasonably pure state, followed the discovery of its existence only after a period of years.

The introduction in 1800 of the voltaic cell as a new and effective instrument for the decomposition of compounds led, in the hands of (Sir) Humphry Davy, to the isolation of sodium, potassium, barium, strontium and calcium in 1807; and in the following year, through the high chemical reactivity of the newly isolated potassium,[1] Davy[2] and also Gay-Lussac and Thenard[3] succeeded in isolating boron from boric acid. Later, the same metal was found effective in isolating silicon (Gay-Lussac and Thenard) and aluminium (Wöhler).

Of the important group of the halogen elements, chlorine had been prepared as early as 1774 by Scheele, and its elementary nature demonstrated by Humphry Davy in 1810. Iodine was obtained in 1811 by the French saltpetre manufacturer, Bernard Courtois (1777–1838), and bromine was discovered in 1826 by Antoine Jérome Balard (1802–76), who later succeeded Thenard at the Sorbonne and became Professor of Chemistry at the Collège de France. The existence of an element fluorine in the hydrofluoric acid described by Marggraf in 1768 was surmised by Ampère about 1816, but the element, chemically the most active of the halogens, was not known till 1886 when Henri Moissan succeeded in isolating it by the electrolysis of a solution of potassium hydrogen fluoride in liquid hydrogen fluoride.

As early as 1758 the German chemist, Andreas Sigismund Marggraf (1709–82), a pioneer of analytical chemistry, had shown that the salts of sodium and of potassium can be distinguished by the yellow and the lavender colour which they respectively impart to a flame; and the English astronomer, (Sir) John Frederick William Herschel (1792–1871), showed, in 1822, that when the coloured light is passed through a prism a spectrum of bright lines separated by dark spaces is obtained. Although, however,

---

[1] Also prepared by Gay-Lussac and Thenard by strongly heating caustic potash with iron.   [2] *Phil. Trans.*, 1808, 343.   [3] *Ann. Chim.*, 1808, **68**, 169.

it was known that different spectra are produced when the salts of different metals are vaporized and the vapour raised to incandescence in a flame, it was not till 1859 that the colour and position of the bright spectral lines were systematically investigated by the chemist, Robert Wilhelm Bunsen, and the physicist, Gustav Robert Kirchhoff (1824–87), at Heidelberg.[1] In the memoir in which they communicated the results of their investigation, Kirchhoff and Bunsen described the construction of a *spectroscope* for the examination of the coloured flames, and there was thereby introduced into practical chemistry a new instrument of chemical analysis and a powerful aid in the discovery of new elements.

The positions which the spectrum lines occupy were found by Kirchhoff and Bunsen to be definite (corresponding to definite wave-lengths of light), and to depend only on the metal present in the vaporized salt;[2] and the position of the lines was not altered by the presence of other salts. As Kirchhoff and Bunsen wrote: "The positions which they [the coloured lines] occupy in the spectrum are due to a chemical property as invariable and of as fundamental a nature as the atomic weight, and can therefore be determined with almost astronomical accuracy." It thus became possible, by means of the spectroscope, to analyse a mixture of salts by examining the colour given to a flame by the mixture. Moreover, the method of spectroscopic analysis showed itself to be superior to the ordinary method of chemical analysis, which depended largely on the production of differently coloured precipitates, since, in the latter case, the colours of the different precipitates may interfere with one another.[3]

[1] *Ann. Physik*, 1860, **110**, 161; *J. Chem. Soc.*, 1861, **13**, 270; Ostwald's *Klassiker*, No. 72. See also Roscoe, *Bunsen Memorial Lecture, J. Chem. Soc.*, 1900, **77**, 513. See also Miller, *Chem. News*, 1862, **5**, 201, 214; Kirchhoff, *Phil. Mag*, 1863 (IV), **25**, 250; Herschel, *Encycl. Metropol.*, IV, 438. (This was published in 1827. The encyclopaedia was issued in parts between 1817 and 1845, and the title page is dated 1845.)

[2] Although this is true in the case of the salts of the alkali metals, it was later found not to be true in the case of the salts of the alkaline earths. In the case of these, the spectrum obtained when the salt is heated in the Bunsen flame is different from that obtained when the salt is heated to the high temperature of the electric spark. In the latter case the line spectrum of the metal is obtained.

[3] According to Bunsen, the history of the discovery of spectroscopic analysis is as follows: For some time Bunsen had made use, for analytical purposes, of the different colours imparted by salts of different metals to the colourless Bunsen flame; and in order to distinguish between flames of similar appearance, he viewed the flames through coloured glasses or solutions. One day, when discussing this method with his colleague Kirchhoff, the latter pointed out to Bunsen that he could attain his ends more simply and accurately by passing the light of the coloured flames through a glass prism and so obtaining their spectra. This suggestion was acted upon, and Bunsen and Kirchhoff developed the method of spectrum analysis (Ostwald's *Klassiker*, No. 72, p. 71).

Although a number of salts can be vaporized at the temperature of the Bunsen flame, others require a higher temperature, *e.g.*, that of the oxy-hydrogen flame. Substances may also be heated in the electric arc or, in the case of metals, for example, by means of a spark discharge from an induction coil between wires or thin rods of the metal.[1] Since molecules as well as atoms can emit light and give rise to definite spectra, the spectrum obtained may differ considerably according to the method used in raising the substance to incandescence. Investigations carried out since Bunsen's time have revealed how complex spectra may become, and spectroscopy has now developed into a large and special field of study.

Spectroscopic analysis is of enormously greater delicacy than chemical analysis, and Kirchhoff and Bunsen showed that it is easily possible to detect spectroscopically the presence of less than one three-millionth of a milligram of sodium salt, and the lithium in twenty grams of sea water. Owing to this extreme sensitiveness spectroscopic analysis lends itself admirably to the discovery of elements which may be present in nature in very small amounts or in great dilution; in amounts, that is, which are too small to be detected or separated by the ordinary methods of chemical analysis. This fact was very soon demonstrated by Bunsen and Kirchhoff who, almost immediately after their invention of the spectroscope, discovered with its aid a new alkali metal, present in the mineral water of Dürkheim. In a letter to Sir Henry Roscoe, dated April 10th, 1860, Bunsen[2] wrote: "I have obtained full certainty, by means of spectrum analysis, that besides Ka, Na and Li, a fourth alkali metal must exist. . . . Where the presence of this body is indicated, it occurs in such minute quantity that I almost give up hope of isolating it, unless I am fortunate enough to find a material which contains it in larger amount." Later in the year, on November 6th, Bunsen again wrote to Roscoe: "I have been very fortunate with my new metal. I have got 50 grams of the nearly chemically pure chloro-platinic compound. It is true that this 50 grams has been obtained from no less than 40 tons of the mineral water. . . . I am calling the new metal *caesium* (from *caesius* = blue), on account of the splendid blue line in its spectrum."

The properties of caesium, the first element to be discovered by means of the spectroscope, were found to be very similar to those

---

[1] Bunsen, *Ann. Physik*, 1875, **155**, 230, 366.
[2] *J. Chem. Soc.*, 1900, 77, 513. The discovery was communicated to the Berlin Academy of Sciences on May 10th. See also *Chem. News*, 1860, **2**, 281.

of potassium; and the German mineralogist, Carl Friedrich Plattner (1800–58), of the Freiberg School of Mines in Saxony, had, in fact, in analysing the mineral pollux from Elba, mistaken the sulphate of caesium for the sulphate of potassium.[1]

The discovery of caesium was followed a few months later by the announcement, on February 23rd, 1861, of the discovery of still another alkali metal,[2] the presence of which was detected in the mineral lepidolite. The spectrum of this element was characterized by the presence of two red lines towards the extreme end of the visible spectrum. "The magnificent dark-red colour of these rays," wrote Bunsen and Kirchhoff, "led us to give to this element the name *rubidium*, from *rubidus*, which, with the ancients, served to designate the deepest red."

The application of the newly invented method of spectrum analysis soon led to the discovery of a number of other elements. Thus, in 1861 (Sir) William Crookes, when examining spectroscopically some residues from a sulphuric-acid plant, after removal of the selenium which they contained, observed in the spectrum a hitherto unknown beautiful green line. This line, he concluded, must be due to a new element to which he gave the name *thallium*.[3] This new metallic element was isolated in 1862 by Crookes and also by the French chemist, Claude Auguste Lamy (1820–78), at that time Professor of Physics at Lille.[4] It was, in fact, to Lamy, who also prepared and studied the properties of its salts, that a committee of the French Academy gave the credit of first isolating the metal.[5]

A further triumph attended the application of the spectroscope, for, in 1863, Ferdinand Reich (1799–1882), Professor of Physics at the Freiberg School of Mines, and his assistant, Hieronymus Theodor Richter (1824–98), who later became Director of the School, discovered, in some zinc ores, a new element which, on account of the brilliant indigo line occurring in its spectrum, was named *indium*.

[1] Plattner, *Ann. Physik*, 1846, **69**, 443; F. Pisani, *Compt. rend.*, 1864, **58**, 714.
[2] Bunsen and Kirchhoff, *Berlin Monatsh.*, 1861, **6**, 273; Bunsen, *Annalen*, 1861, **119**, 107.
[3] From the Greek θαλλός (thallos), a green twig. See *Chem. News*, 1861, **3**, 193.
[4] In 1865 Lamy became Professor of Chemistry at the *École Centrale des Arts et Manufactures*, Paris.
[5] *Compt. rend.*, 1862, **55**, 866; *Ann. Chim.*, 1863, **67**, 418.
At the International Exhibition in London in 1862 Crookes exhibited a specimen of thallium in the form of a black powder, and Lamy, who had independently discovered the element a year later than Crookes, exhibited a solid ingot of the metal. Crookes was awarded a medal for the discovery of the element and Lamy was awarded a medal for his production of a solid ingot.

The spectroscope has been found of inestimable value for the purpose of chemical analysis, not only in the case of terrestrial materials, but also in the case of the sun and the stars. As early as 1814 Josef Fraunhofer, a German glassmaker and physicist, observed that the spectrum of sunlight is traversed by a large number of dark lines; and, although he could not explain their presence, Fraunhofer mapped many of them and designated the more prominent ones by letters. The explanation of these lines, now known as Fraunhofer's lines, was, however, given in 1860 by Kirchhoff, who showed that they are due to the absorption of certain wave-lengths of light by vapours which, in a state of incandescence, emit those wave-lengths. Thus, the Fraunhofer D-line coincides exactly with the bright yellow line[1] of the spectrum of incandescent sodium vapour; and the existence of the Fraunhofer D-line points to the presence of sodium in the sun's atmosphere. Similarly with the other Fraunhofer lines. Thus, as Bunsen wrote to Roscoe, "a means has been found to determine the composition of the sun and fixed stars with the same accuracy as we determine sulphuric acid, chlorine, etc., with our chemical reagents."

In the periodic classification of the elements which he drew up in 1872, Mendeléeff left (amongst others) three vacant spaces for undiscovered elements, which he designated as eka-boron (atomic weight about 44), eka-aluminium (atomic weight about 68), and eka-silicon (atomic weight about 72). These three elements, the properties of which Mendeléeff had predicted in considerable detail,[2] were discovered within a period of fifteen years. The first of the three missing elements to be discovered was gallium, which Mendeléeff had designated as eka-aluminium; and its discovery, as Mendeléeff had predicted as probable, was made with the aid of the spectroscope.

In 1874 Paul Émile Lecoq de Boisbaudran, who had devoted himself for many years to spectroscopic investigations, began the examination of a zinc blende from the Argelès Valley in the Pyrenees. Having dissolved the ore in acid, he placed metallic zinc in the solution, when a deposit was formed on the zinc. When this was heated in the oxy-hydrogen flame a spectrum was obtained which showed the presence of two hitherto unobserved lines. To the element, the existence of which was thus indicated, Lecoq de

---

[1] With sufficient dispersion this line is resolved into a doublet.
[2] *Annalen*, 1872, Suppl. VIII, 196.

Boisbaudran, in honour of his native country, gave the name *gallium*.[1] Later, starting with several hundred kilograms of the zinc ore, he was able to prepare gallium hydroxide, and by electrolysing a solution of this in caustic potash the metal itself was obtained. The properties of the metal and of its compounds were found to be in remarkable agreement with the predictions of Mendeléeff.

Gallium melts at the low temperature of 30° C., but boils only at 1,700°. Thermometers constructed of fused quartz and filled with molten gallium, which readily remains superfused, may therefore be used for the measurement of high temperatures.

The second of the three missing elements to be discovered was *scandium*, the properties of which showed that it was the eka-boron of Mendeléeff. This element was discovered in 1879 by Lars Fredrik Nilson in the rare earth ytterbia which the Swiss chemist, Jean Charles Galissard de Marignac (1817–94), Professor of Chemistry in the University of Geneva, had isolated in the preceding year from the mineral gadolinite.[2] The identity of scandium with the hypothetical eka-boron was pointed out by the Swedish chemist Per Theodor Cleve (1840–1905), the discoverer of the rare-earth elements thulium and holmium.

In 1885 a new ore, argyrodite, was found in the Himmelsfürst mine near Freiberg, and when this was subjected to analysis by Clemens Alexander Winkler, the results always came out too low. From this, Winkler concluded that there was some unknown element present, and after many attempts and failures he succeeded in obtaining the sulphide of a new metallic element from which the metal itself was isolated by heating the sulphide in a current of hydrogen. To this element Winkler gave the name *germanium*, and he was later able to show that it was the element eka-silicon predicted by Mendeléeff.

As far back as 1794 the Finnish chemist, Johann Gadolin 1760–1852), Professor of Chemistry in the University of Åbo, on examining a black mineral found at the small town of Ytterby, near Stockholm—a mineral to which later the name gadolinite was given—isolated from it the oxide of a hitherto unknown metal,[3] to which later investigators gave the name *yttrium*, the oxide being

[1] *Amer. Chemist*, 1875, **6**, 146; *Compt. rend.*, 1875, **81**, 493, 1100; 1876, **82**, 168.
[2] *Compt. rend.*, 1879, **88**, 642, 645; *Ber.*, 1879, **12**, 550, 554.
[3] *Svenska Akad. Handlingar*, 1794, **15**, 137.

known (according to the usage of the time) as *yttria*. Nine years later, in 1803, Klaproth[1] and also Hisinger and Berzelius[2] discovered the oxide of another new metal which they called *cerium*, the mineral from which the oxide was obtained receiving the name cerite. These two oxides, yttria and ceria, were the first members of a large group of closely related oxides or earths, as they were then termed, and as these were found in certain complex and, it was thought, not very abundant minerals, they were called *rare earths*.[3] So similar are these oxides in their properties that their separation proved to be extremely difficult. Their discovery was spread over a large number of years and came as the reward of careful and skilful chemical analysis and as the result of elaborate and laborious fractionation. Only recently, with the advent of new techniques (p. 205) has the isolation of the metals become relatively easy.

After a period of many years the investigation of the earths yttria and ceria was taken up, appropriately enough by one of Berzelius's assistants, Carl Gustav Mosander (1797–1858), who also, for a time, lived with Berzelius and was affectionately called by him Pater Moses. From the nitrate of ceria Mosander,[4] in 1839, succeeded in obtaining the oxide of a new metal which he called *lanthanum* (= hidden); but this new oxide was shown, by Mosander[5] himself three years later, not to be a pure substance, but to contain the oxide of another metal which he called *didymium*, "an inseparable twin brother of lanthanum."

Although the simplicity of the earth didymia was doubted by a number of chemists, it was not until 1885 that the Austrian chemist, Carl Auer (1858–1929), who was ennobled by the Emperor Franz Josef in 1901 and became Freiherr Auer von Welsbach, resolved didymia into two different oxides, *praseodymia* and *neodymia*.[6] These oxides could not be further resolved and were therefore taken as the oxides of two elements, praseodymium and neodymium.

Stimulated by the success of his investigation of the ceria of Klaproth, Berzelius and Hisinger, and the discovery of lanthana and didymia, Mosander turned his attention to the oxide yttria, discovered by Gadolin. This yttria, as he showed in 1843, could be resolved by fractional precipitation with ammonium hydroxide

[1] *Gehlen's Annalen*, 1804, **2**, 203.
[2] *Gehlen's Annalen*, 1804, **2**, 303, 397; *Nicholson's J.*, 1804, **9**, 290; 1805, **10**, 10.
[3] The name is no longer quite appropriate. For a more detailed account of the rare earths see J. F. Spencer, *The Metals of the Rare Earths*; Levy, *The Rare Earths*.
[4] *Ann. Physik*, 1839, **46**, 648.   [5] *Ann. Physik*, 1842, **56**, 503.
[6] *Ber.*, 1885, **18**, 605.

into three different oxides; one for which the name yttria was
retained and two others which he called *erbia* and *terbia*. This
separation of the original yttria into three oxides was confirmed by
a number of later workers, but the names erbia and terbia were
interchanged, the name erbia being applied to the oxide called
terbia by Mosander.

There was now a lull in the rush of discovery of new elements
belonging to the group of the rare earths, but thirty-five years later,
in 1878, the rush started once more. In that year Marignac was
able to show that the earth known as erbia, obtained from gadolinite,
could be decomposed into two oxides, one for which the name erbia
was retained and a second which was called *ytterbia*.[1] From this
ytterbia, as we have seen, Nilson, in 1879, isolated the oxide of the
element scandium; and, in the same year, erbia, from which
ytterbia and scandia had been removed, was again resolved by
Cleve into three oxides—*erbia, holmia* and *thulia*.[2]

While the earth erbia was thus being resolved into different
oxides, attention was turned to the didymia which Mosander had
obtained sixty years before. Suspicions regarding the purity of this
earth had arisen and these suspicions were confirmed by a spectroscopic examination; and in 1879, by fractional precipitation with
ammonium hydroxide, Lecoq de Boisbaudran separated from
didymia a new earth which he called *samaria*. In 1886 still another
earth was found which, however, had already been obtained, in
1880, by Marignac. This earth was now named *gadolinia*.

And still the laborious process of resolution of the earths went
on, for, as we have seen, Carl Auer, in 1885, resolved didymia
(from which samaria and gadolinia had been removed), into two
earths, praeseodymia and neodymia; and in 1886 Lecoq de Boisbaudran separated Cleve's holmia into two earths which he called
holmia and *dysprosia*. Carrying the tale on into the early years of
the present century, we find that, in 1901, the French chemist,
Eugène Anatole Demarçay (1852–1904), working in his own well-
equipped laboratory, succeeded in isolating from samarium magnesium nitrate a new earth which, with a somewhat wider patriot-
ism than was shown by most of the discoverers of rare earths, he
called *europia*. Lastly, in 1907, Georges Urbain (1872–1938), Professor of Chemistry at the Sorbonne, obtained from ytterbia a new

---

[1] The names yttria, erbia, terbia and ytterbia are all derived from Ytterby, the
town where the mineral gadolinite, the source of these earths, was found.

[2] Holmia and Thulia were named after Stockholm and Thule, the ancient name
of Scandinavia.

earth which he called *lutecia* (the Latin name for Paris). The same resolution was carried out independently, but somewhat later, by Auer von Welsbach, who called the new elements aldebaranium and cassiopeium. The names ytterbia and lutecia, given to the earths by Urbain, have, however, been universally accepted.[1]

Between 1830 and 1907 no fewer than fifteen rare-earth elements had been discovered, namely, cerium, dysprosium, erbium, europium, gadolinium, holmium, lanthanum, lutecium, neodymium, praseodymium, samarium, terbium, thulium, ytterbium, and yttrium. The discovery of so many elements with closely related properties, "meta-elements" as Crookes called them,[2] created great difficulties in classification; and for long these elements were huddled together in the periodic table, occupying an ill-defined area with a somewhat uncertain tenure. These difficulties, however, have in more recent years been overcome and the numerous elements mentioned in the preceding pages, along with others discovered at a later time, now fall into an orderly array in the periodic classification of the elements based on atomic numbers. (Chapter XI.)

It was in 1884, while investigating the rare earths in the laboratory of Professor Bunsen at Heidelberg, that Carl Auer was impressed by the brilliance of the light emitted by certain mixtures of the rare earths when heated to incandescence in a Bunsen flame. He therefore conceived the idea of turning this fact to practical account for the purpose of securing increased illumination with coal-gas; and after some time he was able to make the incandescent gas-mantle a commercial success. An industry of very considerable magnitude and importance was thus brought into being. Auer found that the most brilliant light was emitted by a mixture consisting of 99 per cent of thorium oxide and 1 per cent of cerium oxide.[3]

The metal cerium, which is obtained in large amount as a byproduct of the incandescent gas-mantle industry, has also found widespread use in the production of pyrophoric alloys. An alloy of iron and cerium, for example, when rubbed against a rough surface of steel, emits a shower of sparks, and if these are allowed

[1] The earth lutecia had actually been prepared by Charles James (1880–1928), Professor of Chemistry in the University of New Hampshire, before he received news of Urbain's discovery. The results of his work, however, had not then been published.

[2] *Chem. News*, 1887, **55**, 83, 95.

[3] The main source of these oxides is now the mineral known as monazite sand found, more especially, in Brazil and at Travancore, India.

to come in contact with tinder, with the vapour of petrol, or with low-boiling hydrocarbons such as propane or butane, the tinder or the vapour is ignited. This property of iron-cerium alloys is therefore made use of in petrol and similar lighters. The pyrophoric iron-cerium alloy has also found use in tracer shells and tracer bullets.

Compounds of cerium and of some of the other rare-earth metals have found application for the colouring of glass and enamels and for the production of heat-resisting pigments for porcelain, etc. Cerium also has been found to be a valuable catalyst in a number of reactions of industrial importance.

The fundamental difficulty in the study of the rare earths is that their close chemical similarity makes exceedingly difficult their separation by chemical methods. Research workers who entered this field had to be prepared to spend months, or even years, in repetitive crystallizations and other refining processes; a single slip, and all might be lost. Within the last twenty years, however, new methods of analysis have transformed the situation and most of the rare-earth metals are now comparatively easily obtainable. This transformation was a by-product of the wartime Manhattan project, when it was necessary to isolate some of the fission products produced in very small quantities in the uranium piles. For this purpose a form of chromatography based on ion-exchange proved exceedingly valuable; the details need not be discussed here as chromatography is to be discussed later in the book as a separate subject. In the present context it suffices to say that chromatographic methods, pioneered particularly by F. H. Spedding and E. R. Tompkins and their colleagues,[1] made it possible to obtain substantial quantities of rare-earth metals quite quickly and easily. In 1949, for example, Spedding reported the isolation of 300 g of pure ytterbium and about 15 g each of lutecium and thallium. At a Faraday Society meeting held in Reading in that year he exhibited bars of pure lanthanum, cerium, praseodymium, neodymium, and yttrium derived from chromatographically purified compounds of the metals.

The discovery of a number of the elements described in the preceding pages came, as it were, in satisfactory fulfilment of expectations, and filled the gaps in the table of the elements drawn

[1] E. R. Tompkins, J. X. Khym and W. E. Cohn, *J. Amer. Chem. Soc.*, 1947, **69** 2769; F. H. Spedding, *Gen. Disc. Faraday Soc.*, 1949, **7**, 214.

up by Mendeléeff in 1872 (p. 53), although, it is true, the multitude of the rare-earth metals was something of an embarrassment. In the middle of the nineties of last century, however, a group of elements was discovered, the existence of which could not be predicted on the basis of the periodic law and for which, in fact, there appeared to be no room at all in Mendeléeff's table. The discovery of these elements forms one of the most romantic episodes in the history of chemistry.

As far back as 1785 the Hon. Henry Cavendish (1731–1810), grandson of the second Duke of Devonshire, investigated the properties of what was then called "phlogisticated air" (nitrogen) with the object of deciding "whether there are not in reality many different substances confounded together under the name of phlogisticated air." Repeating experiments first carried out by Priestley, Cavendish showed that when electric sparks are passed through air, combination of nitrogen and oxygen takes place, and the oxides of nitrogen formed can be removed by a solution of caustic potash. By adding further quantities of oxygen and continuing to pass the sparks, the nitrogen could be entirely removed, and any excess of oxygen remaining could be absorbed by a solution of *liver of sulphur* (potassium sulphide). As a result of his experiments, Cavendish found that no matter how long the passage of electric sparks was continued, there always remained a small bubble of gas, the volume of which was "certainly not more than $\frac{1}{120}$ of the bulk of the phlogisticated air let up into the tube; so that if there is any part of the phlogisticated air of our atmosphere which differs from the rest . . . we may safely conclude that it is not more than $\frac{1}{120}$ part of the whole."[1] Unfortunately, it was scarcely possible for Cavendish, at that time, to push the investigation further, and for over a hundred years the importance of the little bubble of gas was overlooked. Until 1894 chemists everywhere believed that dry air, freed from carbon dioxide and incidental impurities, consisted solely of oxygen and nitrogen; and when, at the meeting of the British Association at Oxford in 1894, Lord Rayleigh and Professor Ramsay announced the discovery of a new element in the air, the announcement was received with incredulity. The story of the discovery is full of interest and illustrates the importance of accurate experiment and logical reasoning.

The Hon. John William Strutt, third Baron Rayleigh (1842–1919)

[1] See Ramsay, *The Gases of the Atmosphere* (1896).

Professor of Experimental Physics in the University of Cambridge, while engaged in the accurate determination of the density of different gases—carried out, it would seem,[1] as part of a wider investigation, for the purpose of testing Prout's hypothesis regarding the constitution of the elements (p. 213), and for the purpose of determining atomic weights by a volumetric method—found that the density of "atmospheric nitrogen," the gas left after removal of the oxygen from purified air, is greater than the density of nitrogen obtained from one of its compounds by about 1 part in 230; a difference which, although small, was much greater than the possible errors of the determination. Various hypotheses were put forward by Lord Rayleigh to account for this difference, but none withstood the test of experiment, In a letter to *Nature*, dated September 29th, 1892, Lord Rayleigh drew attention to the results he had obtained and invited the help of chemists in explaining the divergent results. Although letters were received containing "useful suggestions, but none going to the root of the matter," Ramsay, early in 1894, "asked and received permission to make some experiments on the nitrogen of the atmosphere, with the view of explaining its anomalous behaviour."[2]

Alive even then to the possibility that there might be present in "atmospheric nitrogen," some inert gas which had hitherto escaped notice, and to the possibility that he might, in fact, discover a new element, Ramsay passed "atmospheric nitrogen" repeatedly over red-hot magnesium (which combines with nitrogen to form a nitride), and obtained a quantity of residual gas which did not combine with the metal. At the same time Lord Rayleigh repeated the experiment of Cavendish and obtained a small quantity of the new gas. Later, he carried out the experiment on a much larger scale and obtained no less than two litres of gas which did not combine with oxygen.[3]

The density of the residual gas obtained by Ramsay and by Lord Rayleigh was found to be about 20, compared with the value 14 for nitrogen; and the spectrum, as Crookes showed, was a very definite and characteristic one, and different from that of nitrogen.[4] Many attempts were made by Ramsay to bring the new gas into combination with other elements, but entirely without success; and

---

[1] See M. W. Travers, *The Discovery of the Rare Gases*, 1928; *A Life of Sir William Ramsay*, 1956.

[2] Ramsay, *The Gases of the Atmosphere*, 1896.

[3] Lord Rayleigh and W. Ramsay, *Phil. Trans.*, 1895, A, **186**, 187.

[4] *Phil. Trans.*, 1895, A, **186**, 243.

the new element, on account of its chemical inertness, was christened
argon (Greek ἀργόν, inert).

With the discovery and isolation of this new element, the task
was imposed of determining its atomic weight, and since the
element is chemically inactive, chemical methods of determining
the atomic weight were ruled out. From determinations of the
density, however, the molecular weight can be calculated, and if
the number of atoms in the molecule is known, the atomic weight
can be obtained. For a considerable time a number of chemists
held the view that the new gas was probably an allotrope of nitrogen, N_3, and comparable therefore with ozone, O_3; but, on determining the ratio of the specific heats, Ramsay found the value 1·653,
a value which indicated that the gas is monatomic. Its atomic
weight and molecular weight, therefore, are the same and equal, as
later determinations have shown, to 39·9, on the basis of $O = 16·0$.

In 1868 the French astronomer, Pierre Jules César Janssen, and
the English astronomer, Sir Norman Lockyer, while examining the
spectrum of the sun's photosphere during an eclipse observed a
yellow line which they could not find in the spectrum of any
known terrestrial substance. Lockyer concluded, therefore, that
it must be due to some extra-terrestrial element to which he gave
the name helium (Greek ἥλιος—helios, the sun). On March 26th,
1895, the discovery of this element on the earth was announced
to the Royal Society and to the French Academy by Ramsay.[1]
The history of the discovery is given by Ramsay in these words:[2]

"In seeking for a clue which would guide to the formation of
some compounds of argon, Mr. Miers [1858–1943], of the British
Museum,[3] kindly informed me that a gas, supposed to be nitrogen,
was obtainable from certain minerals containing the metal uranium,
and notably from cleveite, a Norwegian mineral discovered by
Nordenskjöld. The gas evolved from a number of such minerals
had been examined by Dr. W. F. Hillebrand, of the United States
Geological Survey, and was pronounced by him to be nitrogen."
Ramsay, therefore, boiled some cleveite, "about a gram of which
[he] bought for 3s. 6d.," with dilute sulphuric acid, and obtained
a small quantity of a gas which, on spectroscopic examination,
he saw at once was neither nitrogen nor argon. A specimen of the
gas, provisionally named krypton (hidden), was sent to Sir William
Crookes for spectroscopic examination, and on Saturday, March

[1] *Proc. Roy. Soc.*, 1895, **58**, 65.   [2] *Proc. Roy. Soc.*, 1895, **58**, 81.
[3] Afterwards Sir Henry Miers, Vice-Chancellor of the University of Manchester

23rd, Ramsay received in reply a telegram, "Krypton is helium." In other words, the gas which Ramsay had obtained from the mineral cleveite was none other than the helium of Janssen and Lockyer.[1] Investigation of this new gas, undertaken with the assistance of John Norman Collie and Morris William Travers[2] (1872–1961), showed that it is an element with the atomic weight 4·00, and that, like argon, it is chemically inert. As in the case of argon, also, the molecule is monatomic.[3]

With the discovery of argon, the difficulty arose of fitting that element into the periodic classification. At first Ramsay suggested that it might fill one of the gaps in column VIII, forming a transition between chlorine and potassium.[4] But this suggestion encountered the difficulty that the atomic weight of argon is greater than the atomic weight of potassium; and although there already existed the parallel case of tellurium and iodine, many chemists at that time thought that this was due to experimental error. The discovery of helium did not immediately help in the removal of the difficulty; and the suggestion could still be entertained that argon and helium were mixtures. These suggestions, however, were disproved by the diffusion experiments of Ramsay and Collie,[5] and also by the experiments of Lord Rayleigh.[6] The elementary

---

[1] Regarding his failure to discover helium, Dr. Hillebrand wrote to Ramsay: "The circumstances and conditions under which my work was done were unfavourable; the chemical investigation had consumed a vast amount of time, and I felt strong scruples about taking more from regular routine work. I was a novice at spectroscopic work of this kind and was thereby led to attach too little importance to certain observations which in the light of your discoveries deserved the utmost consideration. . . . It doubtless has appeared incomprehensible to you, in view of the bright argon and other lines noticed by you in the gas from cleveite, that they should have escaped my observation. *They did not.* Both Dr. Hallock and I observed numerous bright lines on one or two occasions, some of which apparently could be accounted for by known elements—as mercury, or sulphur from sulphuric acid; but there were others which I could not identify with any mapped lines. The well-known variability in the spectra of some substances under varying conditions of current and degree of evacuation of the tube, led me to ascribe similar causes for these anomalous appearances, and to reject the suggestion made by one of us in a doubtfully serious spirit, that a new element might be in question." (*Proc. Roy. Soc.*, 1895, **58**, 81.) And so Hillebrand failed to discover helium, just as Cavendish failed to discover argon.

[2] Afterwards Professor of Chemistry, University College, Bristol, 1904–6; Director of the Indian Institute of Science, Bangalore, 1906–14; Hon. Professor of Applied Chemistry, University of Bristol, 1927–37.

[3] *J. Chem. Soc.*, 1895, **67**, 684.

[4] According to the views of Mendeléeff there was really no gap in column VIII in the position suggested by Ramsay. Ramsay and Rayleigh at first entertained the idea that argon was probably a mixture of three gases, all of which possessed nearly the same atomic weight, like iron, cobalt and nickel (see *British Association Reports*, 1897, p. 593).

[5] *Compt. rend.*, 1896, **123**, 214; *Nature*, 1896, **54**, 546; *Proc. Roy. Soc.*, 1896, **60**, 206.   [6] *Proc. Roy. Soc.*, 1896, **59**, 198.

nature of argon and helium was therefore accepted, and Ramsay, in discussing the problem in his *Gases of the Atmosphere*, inserted a new group (later known as the zero group), in which helium falls between hydrogen and lithium, and argon between chlorine and potassium. No explanation of what was then regarded as an anomaly in the position of argon could be given, although Ramsay made the highly speculative suggestion that the atomic weight value might be affected by the power of combination possessed by an element. This, it was suggested, might cause a transposition of the elements and interfere with the strict arrangement according to ascending atomic weights.

With the discovery of argon and helium, two elements which are chemically inert and which may therefore be regarded as having zero valency, a new group—a zero group—had to be inserted in the Mendeléeff classification of the elements. Moreover, since helium must precede the element lithium, and argon the element potassium, the probability of the existence of other similar elements of zero valency was at once indicated. These elements would then fall quite naturally into the zero group in front of the other univalent alkali metals, sodium, rubidium and caesium. The possibility of the existence of a new element between helium and argon was discussed by Ramsay in his presidential address to the Chemistry Section of the British Association at Toronto in 1897.

During 1895, and also from time to time during 1896 and 1897, the gases from a large number of minerals and from other sources were examined in the hope of obtaining the unknown gas, but no new gas was found. It seemed, therefore, as if the atmosphere must be regarded as the one source from which the new gas or gases might be obtained. Fortunately, the development by Olszevski of a method of liquefying air made available a new method of investigation, the fractional distillation of liquid air. On May 24th, 1898, Dr. W. Hampson, the inventor of a simple apparatus for the liquefaction of air, brought to Ramsay a small quantity of liquid air, the greater part of which was allowed to evaporate. The more volatile constituents of the air therefore escaped and a concentration of the less volatile constituents took place. The small residue which was left was freed from oxygen and nitrogen and examined spectroscopically. At once it was seen that a new gas was present. This gas, called *krypton*, was found to have a density of about 41 and to be monatomic. Its atomic weight, therefore, was about 82, and this new element fell into its place in the zero group in front of rubidium.[1]

A relatively large quantity of air was now freed from oxygen and nitrogen and the residual gas (argon, etc.) was liquefied in fair quantity and fractionally distilled. On June 12th, the most volatile fraction was examined spectroscopically, and the "blaze of crimson light" which was emitted by the incandescent gas showed that another new element had been discovered.[2] It was christened *neon*, and investigation of the density showed that it was the "undiscovered gas," the existence of which Ramsay had foretold the previous year at Toronto. It now took its place in the periodic classification immediately before sodium.

Soon thereafter, in July 1898, the last of the rare gases of the atmosphere, *xenon* (ξένος (xenos), strange), was discovered in the least volatile fractions of liquid air. The discovery of the last as of the first of these new elements was announced at a meeting of the British Association;[3] and in this case appropriately enough at Bristol, where for a number of years Ramsay had occupied the Chair of Chemistry. With an atomic weight of $131\cdot3$ xenon takes its place in the periodic table immediately before caesium.

The discovery and isolation, within the space of four years, of five hitherto unknown elements in the air is one of the great epics in the history of scientific investigation; and the series of gases belonging to the zero group was completed later by the discovery of *radon*, as will be recounted later (p. 219).

Recently, the notable discovery has been made that xenon, which had hitherto been regarded as an inert gas, can form compounds with fluorine (XeF_2, XeF_4, XeF_6) and with oxygen (XeO_3).[4]

The rare gases, the discovery of which has just been described, have, in several cases, found important applications. Thus, argon, which forms rather less than 1 per cent of the air by volume, finds an important industrial application in filling incandescent electric lamps ("gas-filled") of the "half-watt" type. Owing to the presence of the gas the tungsten filament of the lamp can be heated to a much higher temperature, and a brighter light can therefore be obtained, without dispersion of the metal and darkening of the bulb taking place. It is also made use of for filling tubes for

[1] Ramsay and Travers, *Proc. Roy. Soc.*, 1898, **63**, 405 (Paper received June 3rd).
[2] Ramsay and Travers, *Proc. Roy. Soc.*, 1898, **63**, 437.
[3] *British Association Reports*, 1898, p. 830.
[4] Bartlett, *Proc. chem. Soc. Lond.* 218, 1962.

illuminated signs. Another important use of argon is in arc-welding of metals the reactivity of which demands an inert atmo-sphere. Neon, similarly, which occurs in the air to the extent of only about 15 parts per million, and which emits a beautiful rose-coloured light when raised to incandescence by the passage of an electric discharge, is also used for illuminated signs. Neon lamps, moreover, are used as lighthouse lamps at airports owing to the great power which the neon light has—because of its long wave-length—of penetrating mist and fog. The widespread use of these two gases has been made possible through the extensive production at the present day of liquid air.

Helium, which was originally obtained only in small quantities from certain minerals, being present there, as we shall see, as a disintegration product of radioactive substances, is now obtained, to the extent of many millions of cubic feet, from the natural gas which escapes from the earth in certain districts of Canada and of the United States. It is used mainly for filling airships and for the production of artificial breathing-atmospheres for caisson workers and for divers working at great depths. With such atmo-spheres the period of decompression can be greatly reduced.

Krypton and xenon may be used in place of argon for filling electric lamp bulbs.

The discovery of other new elements will be discussed in the following chapter.

CHAPTER ELEVEN

RADIOACTIVITY AND ATOMIC CONSTITUTION

THE introduction of the atomic theory by Dalton early in the nineteenth century marked an important stage in the development of chemistry, but during the closing years of that century, and, more especially, during the present century that theory has undergone a very remarkable evolution. To Dalton, as to Newton, the atoms were "solid, massy, hard, impenetrable particles," forming distinct and separate individuals, primordial and unalterable. Even in the early decades of the century, however, the old Aristotelian idea of a unique matter, a *prima materia*, of which all forms of matter are built up, began to be revived; and throughout the century investigations and discussions took place regarding the nature of the atom and the problem of the constitution of matter. In the last quarter of the century interest in such investigations and discussions was stimulated by the recognition of the essential validity of the periodic law and by the indications which that law gave of a common origin and parentage of the elements. To the theoretical speculations of the nineteenth century there then succeeded the experimental investigations of the twentieth, which have shown that the atom of the chemist can no longer be regarded as the unit of subdivision of matter, but that it is a complex structure built up of smaller particles, the nature of which is the same for all elements. The chemical revolution at the end of the eighteenth century was thereby paralleled by the physico-chemical revolution at the end of the nineteenth.

One of the first to revive the ancient idea of a *prima materia* was William Prout (1785–1850), who graduated Doctor of Medicine at Edinburgh in 1811 and afterwards lectured on chemistry in London. On the basis of an examination of the early determinations of atomic weights, from which it appeared that the atomic weights of the elements are whole multiples of the atomic weight of hydrogen, taken as unity. Prout conjectured that this element might be identified with the primordial matter of the ancient philosophers.[1] This view, which was widely accepted in England, was seized upon by Thomas Thomson (1773–1852), Professor of Chemistry in the University of Glasgow, and accounted one of the leading chemists of the time, who, dominated by a belief in the "sagacious conjecture" of Dr. Prout, put speculation before experiment and changed the atomic weight numbers determined by himself and by his predecessors into the nearest whole multiples of the atomic weight of hydrogen.[2] Against such a method of procedure, Berzelius, whose analytical skill, according to Stas, "has never been surpassed, if indeed it has ever been equalled," vigorously protested; and the justice of Berzelius's protest and the inaccuracy of Thomson's atomic weight values were later recognized by Edward Turner who found that "as far as experimental evidence at present goes, the hypothesis above alluded to is unsupported."[3] Nevertheless, although interest in atomic weight determinations greatly abated during the period of rapid development of organic chemistry, the hypothesis of Prout was not suffered to die; and its captivating simplicity stimulated a number of chemists to undertake determinations of atomic weights in order to test its validity. Among these, Jean Charles Galissard de Marignac, an accurate and skilful analyst, accepted Prout's hypothesis of a primordial matter but found it necessary to halve the value of Prout's unit;[4] and Dumas, also, convinced of the exact validity of Prout's principle, believed that all atomic weights are multiples of the atomic weight of hydrogen, or of half or quarter of this value.[5] This subdivision of the unit, which Marignac was prepared to reduce to the hundredth or even the thousandth part of the atom of hydrogen, robbed Prout's hypothesis of much of its interest and value; and Stas,[6] who began

[1] *Annals of Philosophy*, 1815, 6, 321; 1816, 7, 111. These papers were published anonymously, but the identity of the author was revealed by Thomson (*ibid.*, 1816, 7, 343). The same hypothesis was put forward, somewhat later, by Johann Ludwig Georg Meinecke (1781–1823), Professor of Technology, University of Halle (*J. Pharm.*, 1816, 25, 72, 200; *Ann. Physik*, 1816, 24, 162).

[2] *An Attempt to Establish the First Principles of Chemistry by Experiment*, 1825. With reference to this, Berzelius wrote in his *Jahresbericht* for 1827: "The greatest consideration which contemporaries can show to the author is to treat his book as if it had never appeared."

[3] *British Association Reports*, 1833, p. 399; *Phil. Trans.*, 1833, p. 523.

[4] *Bibliothèque Univ. Archives*, 1860, 9, 97; *Œuvres*, I, 693; *Alembic Club Reprints*, No. 20, p. 48.

[5] Even as early as 1831, Prout, in a letter to Professor Daubeny of Oxford, had suggested the possible existence of substances with atomic weights smaller than that of hydrogen (*Alembic Club Reprints*, No. 20, p. 14).

[6] *Bull. Acad. Roy. Belg.*, 1860, 10, 208; *Œuvres*, I, 308; *Alembic Club Reprints*, No. 20, p. 41.

his researches "with an almost absolute confidence in the exactness of Prout's principle," but placing his reliance only on experiment, "reached the complete conviction, the entire certainty, if it is humanly possible that certainty can be attained on such a subject, that Prout's law . . . is nothing but an illusion, a mere hypothesis which is definitely contradicted by experiment." Not only did Stas consider Prout's law to be illusory, but he discarded the conception of the unity of matter and felt compelled to "regard the undecomposable bodies of our globe as distinct entities having no simple relation by weight to one another." This view was adopted also by Mendeléeff.

While the deviations from Prout's law were undoubtedly greater than the experimental errors, Marignac suggested that Prout's law embodied a real truth and that "the unknown cause which has determined certain groupings of the atoms of the unique primordial matter so as to give rise to our simple chemical atoms . . . might have . . . exercised an influence . . . in such wise that the weight of each group might not be exactly the sum of the weights of the primordial atoms composing it."

Finality, however, had not yet been reached and although, during the last two or three decades of the nineteenth century, little interest was shown in Prout's hypothesis or law, as Marignac called it, many chemists, while not fully accepting the validity of the law, were not quite prepared to subscribe to Stas's declaration that it was *une pure illusion*. In 1886 Crookes,[1] in his presidential address to the Chemical Section of the British Association, definitely suggested an evolution of the elements from some few or even from one antecedent form of matter, which he called *protyle*;[2] and the feeling persisted that Prout's hypothesis enshrines a truth "masked by some residual or collateral phenomena which we have not yet succeeded in eliminating." This feeling was strengthened when, in 1901, the Hon. Robert John Strutt,[3] fourth Lord Rayleigh (1875–1947), pointed out that "a calculation of the probabilities involved confirms the verdict of common sense that the atomic weights" (referred to that of oxygen equal to 16·00), "tend to approximate to whole numbers far more closely than can reasonably be accounted for by any accidental coincidence. The chance of any such coincidence being the explanation is not more than 1 in 1,000,

---

[1] *British Association Reports*, 1886, p. 558; *Genesis of the Elements*, 1887.
[2] From the Greek πρώτη ὕλη (prōtē hylē), first matter.
[3] *Phil. Mag.*, 1901 (6), **1**, 311. Compare J. W. Mallet, *Phil. Trans.*, 1881, **171**, 1003.

so that . . . we have stronger reasons for believing in the truth of some modification of Prout's law than in that of many historical events which are universally accepted as unquestionable."

The methods of chemical analysis were incapable of solving the problem of the constitution of the atom, and it was from an unexpected direction, from a study of the discharge of electricity through highly rarefied gases, that a new and powerful light was thrown on the problem. In 1859 the German physicist, Julius Plücker (1801–68), showed that when an electric discharge is allowed to take place in a highly evacuated glass tube, a peculiar radiation is projected from the negative electrode or *cathode*; and this radiation excites a phosphorescent light in the glass of the tube opposite to the cathode. Moreover, this radiation, known as *cathode rays*, travels in straight lines and casts a shadow of an opaque object placed in its path.

Although, in 1879, (Sir) William Crookes advanced the view that the cathode rays are composed of small particles or corpuscles, it was the discovery of X-rays, in 1895, by the German physicist, Wilhelm Konrad von Röntgen (1845–1923), that awakened interest in the cathode rays and stimulated an inquiry into their nature and into the origin of the X-rays. In 1897 (Sir) Joseph John Thomson,[1] then Cavendish Professor of Experimental Physics in the University of Cambridge, succeeded in proving that the cathode rays consist of particles carrying a charge of negative electricity and moving with a velocity of 10,000 to 100,000 miles per second; and the mass of these negatively charged particles or *electrons*,[2] as they are called, was found to be about one-thousandth of the mass of the hydrogen atom.[3]

In this way, the existence of particles very much smaller than the chemical atom was discovered. Moreover, since investigation showed that electrons can be produced in various ways—by the electric discharge as cathode rays, by the action of ultra-violet light on metals and by raising metals to incandescence—and that the electrons are the same in all cases, the conclusion was reached that the electron must be a constituent of different substances.

[1] Joseph John Thomson, O.M. (1856–1940), Cavendish Professor of Experimental Physics, Cambridge (1884–1918); Master of Trinity College (1918–40); Nobel Prizeman, 1906.
[2] The term *electron* was introduced by G. Johnston Stoney to designate the unit or atomic charge of electricity.
[3] The mass was later found to be about $\frac{1}{1840}$ of the mass of the hydrogen atom. That the cathode rays are negatively electrified had already been proved by the French physicist, Jean Perrin, in 1895.

There thus arose the idea of the electronic constitution of the atom.[1]

Startling as this conception at first seemed to be, it was not long before further evidence of an impressive and almost unbelievable character was obtained, which showed that the atom could no longer be thought of as an indivisible, unchangeable particle, but must rather be regarded as a complex system, capable of undergoing spontaneously, in certain cases at least, an extraordinary transformation into atoms of a different kind.

It has already been mentioned that when cathode rays strike against the walls of the discharge tube, the glass is rendered fluorescent. In 1896, therefore, the French physicist, Antoine Henri Becquerel (1852–1908), thinking that the X-rays, discovered in the previous year by Röntgen, might have their origin in the fluorescence produced by the impact of the cathode rays, placed a salt of uranium, which fluoresces when exposed to light, on a photographic plate protected from the action of ordinary light. After some time, on developing the photographic plate, Becquerel found that a dark patch had been produced; and this result appeared to confirm his view regarding the origin of X-rays. While preparing to repeat the experiment, Becquerel was called away, and he placed the photographic plate with uranium salt in a dark cupboard, where it remained some time. On placing the plate in the developing solution, Becquerel was surprised to find that the plate had been acted on in the same way as before; but since the uranium salt does not fluoresce in the dark, the action on the photographic plate could not be due to fluorescence. By this happy chance Becquerel was led to the discovery that uranium and its compounds continuously and spontaneously emit "rays" or radiations which have the power of passing through wood, paper and other materials opaque to ordinary light, and of affecting a photographic plate in a manner similar to X-rays. In this way was discovered the property of *radioactivity*,[2] a property which was later shown by Mme Curie to be a property not of the salt as a whole but of the uranium *atom*.[3]

The investigation of radioactivity was greatly facilitated by the discovery by Becquerel that radioactive substances have the property of rendering the air a conductor of electricity, the conductivity being due, as Lord Rutherford explained,[4] to an ionization of

---

[1] On theoretical grounds the possibility of an electronic constitution of matter had already been recognized by the physicists, Hendrik Anton Lorentz, of the University of Leyden, and (Sir) Joseph Larmor, of the University of Cambridge.
[2] *Compt. rend.*, 1896, **122**, 420, 501, 689, 1086.   [3] *Compt. rend.*, 1898, **126**, 1101.
[4] *Phil. Mag.*, 1899, **134**, 109; *Conduction of Electricity through Gases*, 1903.

the air or the production of electrically charged ions in the air. The intensity of radioactivity could therefore be measured by the rate of discharge of an electroscope or by the current produced between the two charged plates of a condenser. By this electrical method Professor and Mme Curie undertook a systematic investigation of different elements and minerals, and they found, as also Gerhardt Carl Schmidt, Professor of Physics in the University of Münster, had previously found,[1] that not only uranium but also thorium is radioactive. They also found that certain minerals containing uranium, such as the pitchblende from Joachimstal in Czechoslovakia, had a radioactivity very much greater than could be explained by the amount of uranium present;[2] and they were thus led to the discovery of a hitherto unknown element which, on account of its great radioactive power—more than a million times greater than that of uranium—received the name *radium*. By a long and laborious process Mme Curie succeeded in separating this element from pitchblende, in which it is present to the extent of only 1 part in 10,000,000 and obtained it in the form of its chloride. In 1910 by the electrolysis of a solution of radium chloride, using a mercury cathode, radium amalgam was formed from which, by volatizing away the mercury, radium was obtained as a brilliant white metal. In chemical properties it resembles the alkaline earth metals, calcium, strontium and barium.[3] In addition to radium Mme Curie also obtained from uranium minerals another radioactive element which, in honour of her native country, she named *polonium*; and in 1900 André Debierne, Professor of Physical Chemistry and Radioactivity in the University of Paris, detected the presence of still another radioactive element, which he called actinium.

The close connection which exists between the radiations emitted by radioactive substances and the cathode and canal rays produced by the discharge of electricity through gases, at that time also the subject of investigation, led to a rapid advance in our knowledge of radioactive change. Thus, the investigation of the radioactive elements, uranium, thorium and radium, showed that they are continuously and spontaneously emitting positively and negatively charged particles, the so-called alpha (α) and beta (β) rays;[4] and

[1] *Ann. Physik*, 1898, **65**, 141.
[2] P. Curie, M. Curie and G. Bémont, *Compt. rend.*, 1898, **127**, 1215; M. Curie, *Chem. News*, 1903, **88**, 85 and later.
[3] M. Curie and A. Debierne, *Compt. rend.*, 1910, **151**, 523; E. Demarçay, *ibid.*, 1898, **127**, 1218; 1899, **129**, 716; 1900, **131**, 258.
[4] E. Rutherford, *Phil. Mag.*, 1899, **134**, 109; 1903, **5**, 441; *Physikal. Z.*, 1902, **4**, 235; F. Giesel, *Ann. Physik*, 1899, **69**, 834; H. Becquerel, *Compt. rend.*, 1900, **130**, 809.

they are the source, also, of highly penetrating, ethereal vibrations, the gamma (γ) rays, which are similar in character to the X-rays.[1] The beta rays, which are of the same nature as cathode rays and consist of negatively charged electrons shot out from the atom with a velocity approaching that of light, also have a considerable penetrating power; but the alpha rays, which were later found to be positively charged helium atoms, can be stopped by a few centimetres of air or a thin sheet of paper. By reason of their mass, however, these alpha rays possess much more energy than the beta rays and are much more effective in ionizing the molecules of gases and so of rendering the air a conductor of electricity. By the impact of an alpha particle on zinc blende, moreover, a flash of light is produced,[2] and so, by examining these flashes of light, as in Crookes's *spinthariscope*,[3] one can witness effects due to single individual atoms of matter. The atomic hypothesis, or the hypothesis of the discontinuous structure of matter, thereby gains an objective reality which was undreamt of by the chemists of the nineteenth century.

The investigation of the radioactive properties of uranium, thorium and radium was undertaken with great energy by a number of workers in the early years of the present century, the intensity of radioactivity being determined by means of electrical measurements. As a result of these investigations it was found that uranium, thorium and radium give rise to a large number of other radioactive elements which have a more or less transitory existence.[4] Thus, radium gives rise to a gas or "emanation" to which the name *radon* has been given.[5] This gas, when removed from the original radium salt, gradually loses its radioactivity and disappears. As fast, however, as the separated radon disappears, fresh radon is produced, at exactly the same rate, by the original radium. These

[1] P. Villard, *Compt. rend.*, 1899, **130**, 1010, 1178. The gamma rays, it is supposed, are produced by the β-rays as they escape from the atom.

[2] F. Giesel, *Ann. Physik*, 1899, **69**, 91, 834; J. Elster and H. Geitel, *Physikal. Z.*, 1903, **4**, 439.

[3] *Proc. Roy. Soc.*, 1903, **71**, 405.

[4] For a detailed discussion of radioactivity and the radioactive elements the reader may consult, for example, Soddy, *The Interpretation of Radium and the Structure of the Atom*, or *The Chemistry of the Radio-elements*. See also Lord Rutherford, *Faraday Lecture, J. Chem. Soc.*, **1936**, 508; S. J. Bresler, *Die Radioaktiven Elemente*.

[5] The production of a radioactive gas, or "emanation" as it was called, was first observed in the case of thorium (Rutherford, *Phil. Mag.*, 1900, **49**, 1). The properties of this gas, *thoron*, which is formed from thorium-X and not directly from thorium, were investigated by Rutherford and Soddy (*J. Chem. Soc.*, 1902, **81**, 321, 837), who found that it behaved like a member of the argon group. The formation of an emanation from radium was discovered by Friedrich Ernst Dorn, in Germany, in 1900, and of an emanation from actinium by A. Debierne in 1903.

changes are spontaneous and are unaffected by any alteration of the external conditions. It was found, moreover, that not only radon but quite a number of other radioactive substances are produced successively: radium, radon, radium-A, radium-B, etc., down to radium-G, which is lead. Radium-F, the second-last member of the series, was found identical with the element to which Mme Curie, herself a Pole, had given the name polonium.

Post-war developments in atomic physics have made it possible to obtain certain radioactive isotopes of the elements in substantial quantities, but in the early days of radiochemistry the members of a radioactive series were, for the most part, obtained only in un-weighable amounts and were distinguished from each other by the rate at which the radioactivity diminishes or decays. Thus, the radioactivity of radium falls to half its initial value in 1730 years. By contrast, the corresponding "half-life" for radon and radium-A is 3·85 days and 3 minutes respectively.[1]

The fact that in old uranium minerals the ratio of uranium to radium is constant, indicated that uranium might be the parent of radium. Further investigation showed that this is the case, and that from uranium there are formed, successively, uranium-X_1, uranium-X_2, uranium-II and ionium. As Professor Bertram Borden Boltwood (1870–1927), of Yale University, showed, it is from ionium, by the loss of an alpha particle, that radium is formed.[2]

Besides giving rise to ionium and the radium series of elements, uranium-II may give rise also to another series of radioactive substances, uranium-Y, protoactinium, actinium, etc.

From thorium, as from radium, a series of radioactive substances is formed: thorium, mesothorium-I, mesothorium-II, radio-thorium, thorium-X, thoron, thorium-A, etc., the last member of this series, thorium-D, also being lead. Of these products meso-thorium-I is now obtained commercially in considerable quantities as a by-product of the incandescent gas-mantle industry. It has a half-life period of only five and a half years, and its activity is there-fore fairly great. For this reason mesothorium is employed in the manufacture of luminous paints (though the possible health hazards of all such substances are now being closely scrutinized) and

---

[1] The rate at which radioactivity decays is in accordance with the law of mass action applied to a unimolecular reaction (p. 88), and is given by the expression $\log_{10} I_t - \log_{10} I_0 = 0·4343\lambda t$. Here I_t and I_0 are the intensities of radioactivity after a given interval of time, t, and initially. λ is the reaction constant or "dis-integration constant," and is characteristic of the given change. The "period of half-life" is given by the expression $\log_{10} 0·5 = 0·4343\lambda t$, or $t = 0·6932 \, (1/\lambda)$.

[2] *Amer. J. Sci.*, 1905, **20**, 239, 253; 1907, **24**, 370.

as a disperser of static electricity, for example, in textile manufacture.

In 1902 the complex and bewildering phenomena of radioactivity found their explanation on the basis of the theory of atomic disintegration put forward by Lord Rutherford and Frederick Soddy (1877–1956), of McGill University, Montreal.[1] According to this theory the atom of a radioactive element is a complex system of particles which undergoes spontaneous change into a different atom or system of particles, with the emission at a very high velocity, sometimes of negatively charged electrons (the beta rays) and sometimes of positively charged particles (alpha rays). Not all but only a certain definite fraction of the atoms reach the condition of instability at the same time, and the fraction is definite for each element and characteristic of the element. In the case of radium one atom out of every hundred thousand million disintegrates each second; and the greater the fraction the more powerfully radioactive is the substance. The spontaneous disintegration of the atom is accompanied by emission of heat, as was observed by P. Curie and A. Laborde.[2]

The disintegration theory of radioactivity received powerful support from the work of Sir William Ramsay and Lord Rutherford who showed, by different experimental methods, that the positively charged alpha particle which is expelled from radioactive substances is a positively charged atom of the element helium, which had been discovered only a few years before. In the course of their investigation of the properties of radon, Ramsay and Soddy found that this radioactive gas, which undergoes transformation with expulsion of alpha particles, disappears on keeping and in its place helium is found.[3] It was therefore clear that the radioactive process was accompanied by an actual transformation of matter and that one of the products of this transformation is helium. Rutherford, also, was able to isolate the alpha particles and to show that these particles, on losing their positive charge by taking up electrons, form ordinary helium.[4] Rutherford, moreover, succeeded in counting the number of alpha particles emitted by a

[1] *J. Chem. Soc.*, 1902, **81**, 837; *Phil. Mag.*, 1903, **5**, 576.
Ernest Rutherford (1871–1937), born near Nelson, New Zealand, and Professor of Physics, successively, at McGill University, Montreal and the University of Manchester, was Cavendish Professor of Experimental Physics, University of Cambridge, 1919–37. He was awarded the Nobel Prize for Physics in 1908. Created a Knight in 1914, he had the Order of Merit conferred on him in 1925. In 1931 he was raised to the Peerage as Baron Rutherford of Nelson.
[2] *Compt. rend.*, 1903, **136**, 673.
[3] *Proc. Roy. Soc.*, 1903, **72**, 204; 1904, **73**, 341.
[4] Rutherford and T. Royds, *Phil. Mag.*, 1909, **17**, 281.

given weight of radium by allowing the particles to pass through a small hole and counting the number of flashes produced by their impact on a fluorescent screen. Since each alpha particle becomes a helium atom, it is possible, by measuring the volume of helium produced by a given weight of radium in a given time, to calculate the number of helium atoms (or molecules) in a given volume of the gas.[1] The number obtained in this way, namely, $2 \cdot 75 \times 10^{19}$ in one cubic centimetre at $0°$ C., is in good agreement with the value deduced from measurements of an entirely different kind (p. 75).

The phenomena of radioactivity and those accompanying the discharge of electricity through gases afforded a striking proof of the complexity of the chemical atom; and the enunciation of the disintegration theory of radioactivity gave a great impetus to the formation of definite views regarding the constitution or inner structure of what had hitherto been thought of as an indivisible particle. That negatively charged electrons constitute a part of the atom of different substances had already been inferred (p. 217), and since the atom as a whole is electrically neutral there must also be within the atom an amount of positive electricity equal to the total negative charge carried by the electrons. It was therefore suggested by Lord Kelvin,[2] and the suggestion was further elaborated by Sir J. J. Thomson,[3] that the atom might be conceived of as a uniform mass of positive electricity throughout which the electrons are distributed. Such an atomic structure, however, was found not to be in harmony with the properties of matter, and the investigation, more especially, of the passage of alpha rays through matter led to the conclusion that the positive electricity in the atom must be highly condensed.

This conclusion was based partly on the beautiful experiment devised, in 1897, by C. T. R. Wilson (1869-1959), by which the path of an alpha particle may be traced, and partly on the deflection of alpha particles on passing through thin metal foil. Alpha particles expelled from radium were passed through air saturated with moisture, and when the air was cooled by rapid expansion the ions into which the molecules of the atmospheric gases are broken by the alpha particles acted as condensation nuclei for the moisture, and the paths of the rays were traced in fog.[4] On examination, it

[1] Rutherford and H. Geiger, *Proc. Roy. Soc.*, 1908, A, **81**, 141.
[2] *Phil. Mag.*, 1904, **8**, 528; 1905, **10**, 695.   [3] *Ibid.*, 1903, **6**, 673; 1904, **7**, 237.
[4] *Phil. Trans.*, 1897, A, **189**, 265.
Charles Thomson Rees Wilson was Jacksonian Professor of Natural Philosophy, University of Cambridge, 1925-34. He was awarded the Nobel Prize for Physics in 1927 and created Companion of Honour (C.H.) in 1937.

was found that, in most cases, the track of the particle is a straight, or nearly straight, line; but occasionally a very sharp deflection is shown. A similar behaviour is found, as Lord Rutherford and others showed,[1] when alpha particles are passed through gold leaf or thin metal foil; that is, most of the alpha rays pass undeflected through the metal foil, but others are deflected to a greater or less extent. Occasionally an alpha particle was deflected to such an extent that it did not pass through the metal foil at all but emerged on the side on which it entered.

From these experimental data Lord Rutherford made the brilliant inference that the atom is an open structure composed of a relatively massive, positively charged nucleus surrounded by comparatively very light, negatively charged electrons.[2] For the most part the alpha particles can pass unimpeded through the spaces between the electrons, and any deflection caused by these will be comparatively slight. When, however, the positively charged alpha particle comes near to the relatively massive, positively charged nucleus of the metal atom, electrical repulsion may cause a large deflection, or the alpha particle may even be stopped and driven back along its path. The general correctness of the atomic model which was thus suggested by Lord Rutherford has been confirmed by all later work.

From the angles of deflection observed in the above experiment, Rutherford was able also to calculate the amount of positive charge on the nucleus and therefore the number of "planetary" electrons, or electrons surrounding the nucleus. In all cases the number was found to be equal to about half the atomic weight of the element, and this conclusion was in harmony with that reached by the English physicist and Nobel Prizeman, Charles Glover Barkla (1877–1944), from a study of the scattering of X-rays.[3]

One of the most important advances in connection with the problem of atomic structure was the identification, in 1913, of the number of excess positive charges on the atomic nucleus with the serial number, or *atomic number* as it is now called, of the element in the periodic classification.[4] The number of excess positive

[1] E. Rutherford, *Phil. Mag.*, 1906, **11**, 166; H. Geiger, *Proc. Roy. Soc.*, 1908, A, **81**, 174; 1910, A, **83**, 492; H. Geiger and E. Marsden, *ibid.*, 1909, A, **82**, 495.

[2] *Phil. Mag.*, 1911, **21**, 669. In 1904 the Japanese physicist, H. Nagaoka, had shown mathematically that a system constructed on the principle of the Rutherford model would be stable, provided the attraction of the nucleus was great and that the electrons were in rapid motion (*Phil. Mag.*, 1904, **7**, 445).

[3] *Phil. Mag.*, 1911, **21**, 648.

[4] A. van den Broek, *Physikal. Z.*, 1913, **14**, 32; Niels Bohr, *Phil. Mag.*, 1913, **26**, 476; J. Chadwick, *ibid.*, 1920, **40**, 734.

charges on the nucleus, or the number of extranuclear electrons in the atom could, therefore, be obtained from determinations of the serial number of the element, and such determinations were first made by a young English physicist, Henry Gwyn Jeffreys Moseley (1887–1915), working in the Physics Laboratory of the University of Manchester.[1]

When cathode rays are allowed to impinge on a solid target X-rays are produced; and it was found by Moseley in 1913 that when these X-rays are analysed by means of a crystal, in which the layers of atoms cause it to act as a diffraction grating, each substance used as a target emits X-rays of definite wave-length characteristic of the material of the target. The X-ray spectra which are thus obtained are very simple and consist of groups of lines, known as the K, L, M, etc., series (p. 228). Moreover, when one compares the corresponding lines in the different spectra it is found that there is a regular shift as one passes from element to element, in accordance with the equation, $\sqrt{\frac{1}{\lambda}} = a(Z - b)$. Here λ denotes the wave-length of the particular line in the spectrum, a and b are constants and Z is an integer which is characteristic of the element and was called by Moseley the *atomic number*. It follows from this that the square root of $1/\lambda$, for corresponding lines in the X-ray spectra of the elements, increases by *equal amounts* as one passes from element to element; and from this it may be concluded that, as Moseley wrote, "there is in the atom a fundamental quantity which increases by regular steps as we pass from one atom to the next. This quantity can only be the charge on the central positive nucleus."

From the investigation of the X-ray spectra of the elements, the important conclusion can be drawn that as one passes from element to element, ascending the series, there is, at each step, an addition of one unit of positive electricity to the nucleus, and a corresponding increase in the number of extranuclear electrons; and one can give to each element a whole number which represents its serial number in the list of the elements, starting with hydrogen, which has the atomic number 1, and passing on to uranium, which has the atomic number 92.

When the elements are arranged in the order of their atomic numbers, as in the following table, it is found that the anomalies of the Mendeléeff classification according to atomic weights disappear, and argon and potassium, telluruim and iodine, cobalt and

[1] *Phil. Mag.*, 1913, **26**, 1024; 1914, **27**, 703.

nickel follow regularly in the order of their atomic numbers, as given by their X-ray spectra, and as required by their chemical properties. Moreover, since it is now recognized that the chemical, and most of the physical, properties of an element depend on the number and arrangement of the extranuclear electrons rather than on the atomic weight, the periodic law may be restated in the form: The properties of the elements are a periodic function of the *atomic number*. The recognition that the atomic number is more

PERIODIC CLASSIFICATION OF THE ELEMENTS
ACCORDING TO ATOMIC NUMBERS

					55.Cs	87.—
					56.Ba	88.Ra
					57.La	89.Ac
					58.Ce	90.Th
					59.Pr	91.Pa
					60.Nd	92.U
					61.Il	
			19.K	37.Rb	62.Sa	
			20.Ca	38.Sr	63.Eu	
			21.Sc	39.Yt	64.Gd	
			22.Ti	40.Zr	65.Tb	
			23.V	41.Nb	66.Dy	
1.H	3.Li	11.Na	24.Cr	42.Mo	67.Ho	
	4.Be	12.Mg	25.Mn	43.Ma	68.Er	
	5.B	13.Al	26.Fe	44.Ru	69.Tm	
	6.C	14.Si	27.Co	45.Rh	70.Yb	
	7.N	15.P	28.Ni	46.Pd	71.Lu	
	8.O	16.S	29.Cu	47.Ag	72.Hf	
	9.F	17.Cl	30.Zn	48.Cd	73.Ta	
2.He	10.Ne	18.A	31.Ga	49.In	74.W	
			32.Ge	50.Sn	75.Re	
			33.As	51.Sb	76.Os	
			34.Se	52.Te	77.Ir	
			35.Br	53.I	78.Pt	
			36.Kr	54.Xe	79.Au	
					80.Hg	
					81.Tl	
					82.Pb	
					83.Bi	
					84.Po	
					85.—	
					86.Rn	

definite and fundamental than the atomic weight as an index of the properties of an element constitutes one of the most important advances in chemical theory.

The existence of a series of integral atomic numbers not only shows that the elements are limited in number but also throws light on the number of elements still to be discovered. This fact is of especial importance in the case of the elements of the rare-earth group, which, as was pointed out, occupied an ill-defined position in the periodic table of Mendeléeff and offered no internal evidence regarding the possible number of elements belonging to the group.

The work of Moseley was confirmed by the numerous determinations of X-ray spectra carried out by workers in different countries. Many of these workers, moreover, used the method of Moseley as an aid in the search for missing elements, for, since the atomic numbers of these missing elements were known, the wave-length of the X-ray spectral lines could be calculated. A knowledge of these gave a means of identifying and also guided the quest of hitherto unknown elements. By means of their X-ray spectra the identity of the rare-earth metals, which were so difficult to separate and distinguish by chemical methods (p. 205), could readily be confirmed.

At the time when Moseley published his determinations of X-ray spectra and atomic numbers, there were seven missing elements with atomic numbers below that of uranium. Of these *proto-actinium*, a radioactive element, was discovered[1] in 1917; and in 1923, by the method of X-ray analysis, the presence of the element of atomic number 72 was discovered in zirconium ores, and to it was given the name *hafnium* (from Hafnia, an old name for Copenhagen).[2] Two years later X-ray analysis was thought to reveal the existence of two other elements with atomic numbers 43 and 75, which were christened *masurium* and *rhenium*;[3] and in 1926 Professor B. Smith Hopkins of the University of Illinois, announced the discovery of element No. 61, which he called *illinium*.[4] The existence of masurium and of illinium in nature has, however, not been confirmed.

In 1943 (Mrs.) Alice Leigh-Smith and Walter Minder, working

[1] Otto Hahn and Lise Meitner, *Physikal. Z.*, 1918, **19**, 208; F. Soddy and J. A. Cranston, *Proc. Roy. Soc.*, 1918, A, **94**, 384.

[2] Dirk Coster and Georg von Hevesy, *Nature*, 1923, **111**, 79, 182.

[3] W. Noddack, J. Tacke and O. Berg, *Naturwiss.*, 1925, **13**, 567; *Z. angew. Chem.*, 1927, **40**, 250; *Nature*, 1928, **116**, 54.

[4] J. A. Harris, L. F. Yntema and B. S. Hopkins, *Nature*, 1926, **117**, 792; B. Smith Hopkins, *J. Franklin Inst.*, 1927, **204**, 1; *J. Amer. Chem. Soc.*, 1926, **48**, 1585, 1594.

in the Radium Institute in Berne, discovered element 85, an analogue of iodine, as a branch product in the radium and thorium series;[1] and element 87, to which the name *francium* (Fr) was given, was discovered in 1939, as a branch product of the actinium series, by (Miss) M. Perey,[2] in Paris. While the existence in nature of elements 43 and 61 is doubtful, isotopes of these two elements[3] and also of element 85 have been produced artificially, in recent years, in the United States, with the aid of the cyclotron (p. 238). They can be produced much more readily in the uranium pile, developed in connection with the utilization of nuclear energy and the production of the atomic bomb, owing to its far larger output of neutrons. The chemical consequences of atomic energy, and in particular the production of elements beyond uranium which terminates the natural series, will be discussed later.

Although the investigations of Moseley enabled one to determine the number of electrons surrounding the central nucleus of the atom, the question of the arrangement of these electrons, on which the chemical behaviour and main physical properties of the atom depend, gave rise to much discussion. So far as the interpretation of chemical properties, more especially the valence, is concerned, a static model, in which the electrons are immobilized, may suffice; and such models had been suggested as early as 1902 and had been elaborated in 1916 by Gilbert Newton Lewis, Professor of Chemistry in the University of California,[4] and also by Walther Kossel, Professor of Theoretical Physics, University of Kiel.[5] For an explanation of the dynamic stability of the atom, however, and of the optical properties, it is necessary to assume that the electrons are in motion and that they are revolving round the nucleus in circular or elliptical orbits. According to the theory put forward by the Danish physicist, Niels Bohr,[6] and now generally accepted, the electrons in the outer orbits are regarded as responsible for the visible spectra, those in the inner orbits for the X-ray spectra.

---

[1] *Nature*, 1942, **150**, 767.
[2] *Compt. rend.*, 1939, **208**, 97.
[3] The isotope of element 61 is radioactive with a half-life of 3·7 years.
[4] *J. Amer. Chem. Soc.*, 1916, **38**, 762; *Valence and the Structure of Atoms and Molecules.*
[5] *Ann. Physik*, 1916, **49**, 229; *Valenzkräfte und Röntgenspektren.*
[6] Niels Henrik David Bohr (1885–1962) was Reader in Mathematical Physics, University of Manchester, 1914–16; Professor of Theoretical Physics, University of Copenhagen, since 1916, and Director of the Institute for Theoretical Physics, Copenhagen, since 1920. He was awarded the Nobel Prize for Physics in 1922, the Atoms for Peace Award in 1957, and the Sonning Prize in 1961.

In Bohr's theory of atomic structure,[1] published at a time when he was working under Lord Rutherford in the University of Manchester, each extranuclear electron is considered as being able to revolve in one of a definite number of possible orbits; but it is postulated that the electron, while revolving in one orbit, does not emit radiation.[2] Energy of radiation may, however, be absorbed by the electron, which is thereby raised to a higher energy level, or is caused to pass into a larger orbit in which the electron has a greater potential energy. Bohr then invokes the *quantum theory* which was enunciated by the German mathematical physicist, Max Planck, and assumes that the radiation taken up by the electron is not absorbed continuously, but as a quantum of energy or in integral multiples of energy quanta.[3] When an electron falls back again from an orbit of higher energy level, E_2, to an orbit of lower energy level, E_1, a quantum of energy is emitted, and the frequency of the radiation so emitted (spectral line) is given by the expression, $E_2 - E_1 = h\nu$, where ν is equal to $1/\lambda$.

Without attempting to enter into details or to discuss the modifications which physicists have found it necessary to introduce into the theory of Bohr, it may be said that, for a given nucleus, there are only certain orbits or energy levels which an electron can occupy; and these are defined by what are known as the *principal quantum numbers*, $n = 1, 2, 3$, etc., which also denote the number of the electron sheath or shell. The maximum number of electrons which can exist at these different energy levels is given by the series, $2 \times 1^2, 2 \times 2^2, 2 \times 3^2$, etc.[4]—that is, the number is equal to twice the square of the principal quantum number. The radiations emitted by electrons for which $n = 1, 2, 3$, etc., correspond to the groups of X-ray spectral lines previously referred to as the K, L. M, etc., series respectively.

So far as the number of electrons in different orbits is concerned, the Bohr arrangement of electrons is similar to that put forward on

---

[1]*Phil. Mag.*, 1913, **26**, 1, 476, 857; **27**, 506.

[2] This very important postulate, which represents a breakaway from classical electrodynamics, was made in order to take into account the permanence of the atom. If energy were emitted as radiation, the electron orbit would become smaller and smaller and the electron would finally fall into the nucleus.

[3] The quantum of energy of radiation is not a constant but depends on the frequency of radiation (ν) and is equal to $h\nu$, where ν is equal to $1/\lambda$ and h is a constant, known as *Planck's constant*. This constant has the dimensions of energy multiplied by time, and has the value $6 \cdot 547 \times 10^{-27}$ erg-second.

[4] This is known as Rydberg's series (*Phil. Mag.*, 1914, **28**, 144).

chemical grounds by various chemists.[1] In all cases, the maximum number of electrons existing in an outer layer was taken to be eight, forming what is called an *octet*. This octet of electrons is particularly stable and is found in the case of all the inert gases except helium. The arrangement of the extranuclear electrons, therefore, which has been adopted as expressing most completely the behaviour of the inert gases, is given in the following table:

ARRANGEMENT OF EXTRANUCLEAR ELECTRONS IN THE INERT GASES

Element	Atomic Number	Number of Electrons in Quantum Group					
		$n = 1$	$n = 2$	$n = 3$	$n = 4$	$n = 5$	$n = 6$
Helium	2	2					
Neon	10	2	8				
Argon	18	2	8	8			
Krypton	36	2	8	18	8		
Xenon	54	2	8	18	18	8	
Radon	86	2	8	18	32	18	8

Hydrogen, with the atomic number 1, has only one planetary or extranuclear electron. Helium has two extranuclear electrons at $n = 1$, and when a third electron is introduced, in the case of lithium, it does not enter into the already existing shell but begins to form a new shell with quantum number $n = 2$. For each successive element one electron is added to this second shell until the octet is completed with neon, and a system of maximum stability is obtained. (See table on p. 225.) On passing to sodium, a new shell of electrons, with quantum number $n = 3$, begins to be formed, and this shell becomes complete in the case of argon, which has the electronic arrangement 2, 8, 8.

After argon comes the first long period and the elements acquire four quantum orbits, potassium having the electronic configuration 2, 8, 8, 1. This fourth shell, however, is not built up continuously,

[1] More especially by C. R. Bury, of University College of Wales, Aberystwyth (*J. Amer. Chem. Soc.*, 1921, **43**, 1602). See also J. D. Main Smith (*Chem. and Ind.*, 1923, **42**, 1073; 1924, **43**, 323; *Phil. Mag.*, 1925, **50**, 878); E. C. Stoner (*Phil. Mag.*, 1924, **48**, 719); R. Abegg (*Z. anorgan. Chem.*, 1904, **39**, 330); G. N. Lewis (*J. Amer. Chem. Soc.*, 1916, **38**, 762); I. Langmuir (*J. Amer. Chem. Soc.*, 1919, **41**, 868, 1543; 1920, **42**, 274).

but after calcium, with the configuration 2, 8, 8, 2, the added electrons go to build up the incomplete shell with quantum number $n = 3$, so that for scandium and the following elements we have the electronic configurations, 2, 8, 9, 2; 2, 8, 10, 2, etc., until one comes to copper[1] with the configuration, 2, 8, 18, 1. The third shell of electrons is now complete, and in the succeeding elements, the outer layer is built up, unit by unit, until the stable configuration of the inert gas, krypton, is reached with the electronic arrangement 2, 8, 18, 8.

Without entering into details, it may be said that the same general principles, regarding the arrangement of the electrons, are followed in the case of the other long series; and, although various slight modifications have been introduced, the Bohr theory has, in general, received confirmation from a study of the visible and X-ray spectra, and from the chemical behaviour of the elements.[2]

The Rutherford-Bohr conception of the atom as a complex structure in which the planetary or extranuclear electrons are built up into shells with a maximum of eight electrons in the outer shell furnishes a physical basis for the doctrine of valency which was first put forward as an empirical doctrine descriptive of the combining capacity of the atoms; and a discussion of valency from this point of view was first most fully undertaken by Gilbert N. Lewis[3] and by Walther Kossel[4] in 1916. Many of the consequences of the theory were worked out by Irving Langmuir (1881–1957), of the General Electric Co., in America.[5]

Chemical combination may be regarded as brought about by the transfer of electrons from one atom to another with the production of electrostatic charges which bind the atoms together. In the case of such reactions, involving only the outer layer of electrons, the valency may be regarded as indicating the number of electrons which an atom must gain or lose in order to form a system with the structure of the nearest inert gas, or a system with an outer shell of eight electrons. Thus, sodium has only one electron in the outer shell, and it can give up this electron so as to form a positively charged stable system having the electronic constitution of neon. On the other hand, chlorine has seven electrons in its outermost shell and can take up one electron so as to form a negatively charged

[1] There are one or two irregularities. Chromium, for example, has the arrangement 2, 8, 13, 1.
[2] D. W. Pearce, *Anomalous Valence and Periodicity within the Rare Earth Group* (*Chem. Rev.*, 1935, **16**, 121).
[3] *J. Amer. Chem. Soc.*, 1916, **38**, 762.   [4] *Ann. Physik*, 1916, **49**, 229.
[5] *J. Amer. Chem. Soc.*, 1919, **41**, 868, 1543; 1920, **42**, 274.

system with the stable configuration of argon. By the transfer, therefore, of an electron from a sodium atom to a chlorine atom, two oppositely charged systems are formed, and union can take place by electrostatic attraction. In this case both the sodium and the chlorine act as univalent elements, each giving or receiving only one electron, the sodium being electropositive and the chlorine electronegative.

Calcium, barium, etc., which have two electrons in the outer shell, can give up two electrons and act as bivalent electropositive elements; and oxygen, which has six electrons in the outer shell, can take up two electrons and so act as a bivalent electronegative element, and so on. Valency which is thus manifested as a transfer of one or more electrons from one atom to another is known as *electrovalency*.

According to the electronic conception of valency, however, chemical combination can take place not only by electrostatic attraction produced by the transfer of electrons, but also by the sharing of electrons by atoms, as when two atoms of chlorine each share an electron with the other and so give to each atom a completed

octet. Thus, $: \overset{..}{\underset{..}{Cl}} \cdot + \cdot \overset{..}{\underset{..}{Cl}} : \rightarrow : \overset{..}{\underset{..}{Cl}} : \overset{..}{\underset{..}{Cl}} :$

Valency due to the sharing of electrons is spoken of as *co-valency*. The sharing of one pair of electrons, as in chlorine, corresponds to what is usually represented by a single bond, whereas the sharing of two pairs of electrons corresponds to a double bond. The sharing of three pairs of electrons corresponds to a triple bond. No cases are known where four pairs of electrons are shared, nor are they to be expected on stereochemical grounds.

Ordinarily, one would expect that the maximum co-valency of a non-metal would be equal to the number of electrons in its outer shell in defect of eight, and in the case of the elements of the first short period this is true, since there cannot be more than eight electrons in the second shell or group with the principal quantum number 2. In the third shell, however, a larger number of electrons may be present, and, therefore, as pointed out in 1923 by Professor Nevil Vincent Sidgwick (1873–1952), of the University of Oxford, to whom much of the development of the electronic theory of valency was due, the maximum co-valency in the second short period, and in the first long period, is six, *e.g.*, sulphur hexafluoride. For the remaining elements, the maximum co-valency is eight. In the case of hydrogen, where one might expect a co-valency of only one, a maximum co-valency of two is found, indicating the existence of bivalent hydrogen.[1]

In cases of what is known as *co-ordinate co-valency*, the shared electrons are supplied by only one of the two atoms concerned. In such cases the molecule formed becomes polar, or forms a *dipole*, the atom which supplies the electrons (the "donor") becoming positive while the other atom (the "acceptor") becomes negative. These compounds, however, unlike the polar compounds formed by electrostatic attraction (*e.g.*, sodium chloride), are not ionized and do not conduct the electric current in the liquid state.[2]

The electronic theory of valency has received important application to the "co-ordination complexes," the existence of which was assumed by Alfred Werner (1866–1919), Professor of Chemistry at Zurich, in the case of the so-called "molecular compounds" like the complex ammines.[3] For the formulation of such compounds as $CoCl_3, 6NH_3$, the classical doctrine of valency is inadequate, and Werner put forward the view that the metal (cobalt) atom has the power, through "subsidiary valencies," of attaching to itself atoms or molecules up to the number of six, thereby forming a stable "co-ordination complex." The above compound was therefore formulated as $[Co,(NH_3)_6]Cl_3$, the square brackets enclosing the co-ordination complex. Other compounds belonging to this series —$CoCl_3,5NH_3$; $CoCl_3,4NH_3$; $CoCl_3,3NH_3$, etc.—were formulated $[Co,(NH_3)_5Cl]Cl_2$; $[Co,(NH_3)_4Cl_2]Cl$; $[Co,(NH_3)_3Cl_3]$, etc. The maximum number of atoms or groups that an atom can hold within its co-ordination complex is known as its "co-ordination number." In the case of certain metals (*e.g.*, cobalt, platinum, etc.) the co-ordination number is 6, but in other cases it may be 4, 2 or 8.

While the system of formulation suggested by Werner showed itself to be of very great value as a basis of classification and enabled one to explain not only the structure isomerism which is met with here, but also, as we shall see later, the stereo-isomerism (p. 266), the introduction of the assumption of different kinds of valency was felt to be a serious weakness and one which required explanation and support. This was given in 1923 by Sidgwick, who showed

---

[1] See Sidgwick, *Annual Reports of the Chemical Society*, 1933, **30**, 112.
[2] For a full discussion of modern views, see L. Pauling, *The Nature of the Chemical Bond and the Structure of Molecules and Crystals* (3rd edn.), 1959; C. A. Coulson, *Valence*.
[3] *Z. physikal. Chem.*, 1892, **3**, 267; *Neuere Anschauungen auf dem Gebiete der anorganischen Chemie*; *New Ideas on Inorganic Chemistry* (trans. by E. P. Hedley).

that the radicals within the co-ordination complex were attached to the metal atom by co-ordinate co-valencies. According to the electronic theory of valency, the nitrogen atom of ammonia has an unshared pair of electrons (a so-called "lone pair")[1] and the nitrogen atom of each of the six NH_3-molecules, according to Sidgwick, donates its lone pair of electrons to the cobalt atom, or cobalt ion, which thus acquires an outer shell of twelve electrons, all shared with nitrogen atoms. The co-ordination number, discovered by Werner, becomes in fact a special case of the co-valency rule put forward by Sidgwick.

If one accepts the general correctness of the "Rutherford atom," one must consider not only the arrangement of the extranuclear electrons but also the nature of the nucleus. This question has occupied the attention of chemists and physicists since 1911, and it is one of their main preoccupations at the present day.

According to the disintegration theory of radioactivity, the atoms of the radioactive elements emit, in some cases, alpha particles or positively charged helium nuclei, and, in other cases, beta rays or electrons. Since these alpha and beta rays are produced by a disruption of the atomic nucleus, one would appear to be justified in drawing the conclusion not only that the nucleus is complex but also that the positively charged nucleus of the helium atom is one of the units of nuclear structure. The helium nucleus, however, must itself be regarded as complex, its mass being about four times that of the hydrogen atom, and so one is led to the conclusion that the positively charged nucleus of the hydrogen atom is the structural unit of positive charge. To this unit the name *proton* has been given.

Although the nucleus of the hydrogen atom can be regarded as consisting of a single proton, the nuclei of the other elements must consist of both protons and electrons. In the case of helium, for example, which has an atomic mass about four times as great as hydrogen, the nucleus must contain four protons, but since it is found that the helium nucleus carries only two unit charges of positive electricity, the four protons must be associated with two negative charges, or two electrons. Until recently it was considered that the protons and electrons existed free in the nucleus, but the proof obtained by (Sir) James Chadwick,[2] in the Cavendish

[1] Maurice L. Huggins, *J. Phys. Chem.*, 1922, **26**, 601.
[2] Professor of Physics, University of Liverpool, 1935–48; Master of Gonville and Caius College, Cambridge, 1948–58; Nobel Prizeman, 1935.

Laboratory, Cambridge, that *neutrons*, or stable uncharged particles consisting of a firm combination of a proton and an electron, are also present in atomic nuclei,[1] has now led to the view that the nuclei consist of neutrons and protons. Although the nature of the nucleus remains obscure, recent research begins to make it possible to discern some pattern within it. Its extreme smallness does not simplify the problem; the radius of the nucleus is about 10^{-12} cm., that is, smaller than the radius of an atom by a factor of about 10^{-5}. In more concrete form, the size of the nucleus to the size of the atom is roughly as a sphere of 100 metres diameter is to the Earth. Within the nucleus, the protons and neutrons appear to move in definite orbits much as the electrons revolve about the nucleus itself.[2] Quantitatively, however, there are great differences; whereas the energy differences between orbits in nuclei are of the order of 1 MeV those between electronic orbitals are (in light atoms) about 10 cV.

The present-day conception of an atom, therefore, is that of a very open-spaced system of particles, the diameters of which are very small compared with that of the system as a whole, or the so-called atomic domain.

In 1906 Professor Boltwood, of Yale University, pointed out that the radioactive element, ionium, is so similar in chemical properties to thorium that if salts of these two elements are mixed it is impossible to separate them again by any chemical process.[3] Similar chemical identities were met with in the case of other radioactive substances.[4] Moreover, in the case of ionium and thorium not only the chemical properties but also the spectra were found to be identical,[5] and, similarly, the X-ray spectrum of radium-B was found to be identical with that of lead.[6] For these astounding facts a basis of interpretation was first found in the Rutherford theory of the nuclear atom.

According to this theory, as we have learned, the nucleus of an atom is made up of protons and neutrons and, consequently, if one

[1] *Nature*, 1932, **129**, 674; *Proc. Roy. Soc.*, 1932, A, **136**, 692.

[2] For a survey of modern theories of nuclear structure see J. Hamilton, *Endeavour* 1960, **19**, 163.

[3] *Amer. J. Sci.*, 1906, **22**, 537; 1907, **24**, 370. The chemical similarity of ionium and thorium was confirmed by W. Marckwald and B. Keetman (*Jahr. Radioaktivität*, 1909, **6**, 269), and by Auer von Welsbach (*Sitzungsber. K. Akad. Wiss. Wien*, 1910, **119**, 1011).

[4] For a detailed discussion see F. W. Aston, *Mass Spectra and Isotopes* (Arnold).

[5] A. S. Russell and R. Rossi, *Proc. Roy. Soc.*, 1912, A, **87**, 478.

[6] Rutherford and E. N. da C. Andrade, *Phil. Mag.*, 1914, **27**, 854.

neutron is removed from the nucleus the atomic weight of the element will be reduced by unity, but the positive charge on the nucleus, or the atomic number of the element, will be unchanged. Chemical and optical properties, however, depend on the atomic number (number of extranuclear electrons), and so the possibility is given for the existence of elements which have the same chemical properties but different atomic weights.

It is not surprising that the existence of such elements was first realized in the case of the radioactive elements, since radioactive change is associated with a disintegration of the nucleus; and for such substances the law was first stated by Soddy, in 1911, and later and more completely in 1913, by Soddy as well as by A. S. Russell and by Kasimir Fajans,[1] that a radioactive element moves back two places in the periodic table when it loses an alpha particle, and moves forward one place when it loses a beta particle. With the introduction of the atomic number in place of atomic weight as the basis of the periodic classification, the law can be stated that the emission of an alpha ray from the nucleus of a radioactive element corresponds with a reduction of the atomic number by two units and a diminution of the atomic weight by four units, while the emission of a beta ray from the nucleus increases the atomic number by one unit but leaves the atomic weight unchanged (the mass of the electron being negligible). In the disintegration of radium, for example, to form radium-B, one has the series:

	Radium	$\xrightarrow{\alpha\text{-ray}}$	radon	$\xrightarrow{\alpha\text{-ray}}$	radium-A	$\xrightarrow{\alpha\text{-ray}}$	radium-B
Atomic weight	226		222		218		214
Atomic number	88		86		84		82

The atomic number of radium-B, therefore, is the same as that of lead, and radium-B gives, as we have seen, the same X-ray spectrum as lead, but its atomic weight is greater than the atomic weight of lead (207·2).

On pursuing the investigation of the radium disintegration series further, one finds the following series of changes:

	Radium-B	$\xrightarrow{\beta\text{-ray}}$	radium-C	$\xrightarrow{\beta\text{-ray}}$	radium-C_1	$\xrightarrow{\alpha\text{-ray}}$	radium-D
Atomic weight	214		214		214		210
Atomic number	82		83		84		82

[1] Soddy, *Chemistry of the Radio-elements*, Part II (Longmans, 1911); Fajans, *Physikal. Z.*, 1913, **14**, 136, 257; A. Fleck, *J. Chem. Soc.*, 1913, **103**, 381, 1052; A. S. Russell, *Chem. News*, 1913, **107**, 49.

	Radium-D	→	radium-E	→	radium-F	→	radium-G
		β-ray		β-ray		α-ray	
Atomic weight	210		210		210		206
Atomic number	82		83		84		82

Radium-D and radium-G, therefore, although they have different atomic weights have the same atomic number as radium-B and lead, and consequently occupy the same place in the periodic classification. For this reason they were called by Soddy *isotopes*.[1]

The final disintegration product of radium and the final disintegration product of thorium were found to have the same atomic number as lead, but whereas the calculated value of the atomic weight of the lead formed by the disintegration of radium is 206, the atomic weight of the lead formed from thorium is 208·4. It was therefore predicted by Soddy that the atomic weight of lead will depend on its origin, and this prediction was confirmed experimentally. One therefore reaches the conclusion that ordinary lead, with the atomic weight 207·2, is a mixture of isotopes.

Although the occurrence of isotopes was first met with in the case of the radioactive elements, it was soon demonstrated very clearly by Sir J. J. Thomson and by F. W. Aston, of the University of Cambridge, that many, and in fact most, of the ordinary elements are mixtures of isotopes.[2]

The cathode rays produced by an electric discharge through a rarefied gas ionize the atoms and molecules of gas present in the tube by splitting off electrons from them and so producing positively charged particles, the mass of which depends on the nature of the atoms and molecules present. Under the fall of potential existing in the discharge tube, these "positive rays" pass to the cathode and, if the latter is perforated, pass through the perforations, forming what were originally called "canal rays." These positively charged rays or particles can be deflected by a magnet, and from the amount of the deflection it is possible to draw conclusions regarding the mass of the positively charged particles.

---

[1] From the Greek ἴσος (isos), the same, and τόπος (topos), place.
The term isotope was suggested to Soddy by Dr. Margaret Todd, an Edinburgh graduate, who wrote under the name of Graham Travers (*Endeavour*, 1964, **23**, 54).
[2] Thomson, *Rays of Positive Electricity and their Application to Chemical Analyses* (1913); Aston, *Mass-spectra* and *Isotopes*.
Francis William Aston (1877–1945), worked in the Cavendish Laboratory and devoted himself mainly to the development of positive ray analysis. Awarded the Nobel Prize for Chemistry in 1922 and the Royal Medal of the Royal Society in 1938.

By this method of *positive ray analysis*, Aston, more especially, determined, with a high degree of accuracy, the mass of the positively charged nuclei of different elements.

As a result of the first determinations of mass spectra the very important generalization was made that if the atomic weight of oxygen be represented by 16, the atomic weights of all the other elements can be represented by *whole numbers*.[1] In most cases the elements have been found to consist of a mixture of isotopes, the atomic masses of which are whole numbers, but in some cases the elements were found to be simple and to consist entirely of atoms having the same mass. With the increasing accuracy of investigation, however, the number of elements regarded as simple is diminishing, and possibly all the elements may be found to exist as isotopes; at present, some twenty seem to be anisotopic.[2]

The discovery, in 1932, of an isotope of hydrogen having an atomic mass 2, to which the name *deuterium* was given, aroused very great interest,[3] for in the case of an element of such simplicity the introduction of a neutron into the nucleus is accompanied by appreciable changes in the chemical properties.

The fact that so many of the elements exist in isotopic forms gives at once an explanation of the fractional atomic weights obtained by chemical methods. Thus, in the case of chlorine with the chemical atomic weight of 35·46, investigation has shown that this element is a mixture of isotopes having atomic weights of 35 and 37, mixed together in the proportion of 3 to 1. The two isotopes have the same number of extranuclear electrons (17), and are therefore chemically identical, but in one case the nucleus consists of 18 neutrons and 17 protons, while in the other case it consists of 20 neutrons and 17 protons. Similarly with other elements.

Modern views regarding the constitution of matter throw a new light on the question of the transformation or transmutation of elements which so greatly attracted the interest of the medieval alchemists. According to modern views, transformation of an

[1] This rule is now known to be only approximately correct, for when protons and electrons unite to form the nucleus of an element energy is liberated. According to the theory of Einstein, however, mass and energy are interrelated, and emission of energy corresponds to a reduction of mass. Nearly all isotopes, therefore, deviate to a slight extent from the whole-number rule.
[2] The existence of isotopes has been confirmed by a study of the band spectra of compounds of the isotopic element, which gives a more sensitive means of detecting isotopes than does the mass-spectrograph.
[3] H. C. Urey, F. G. Brickwedde and G. M. Murphy, *Phys. Rev.*, 1932, **40**, 1.

element can be brought about only by altering its nuclear mass or charge; that is, by altering its atomic mass or its atomic number. Theoretically, this is a simple matter, for the atomic mass could be altered by adding an alpha particle to, or subtracting it from, the nucleus of an element, and such transformations have actually been effected by a bombardment method.

The first case of atomic transformation was effected in 1919 by Lord Rutherford[1] who bombarded nitrogen by fast alpha particles. About one alpha particle in a hundred thousand enters a nitrogen nucleus which then breaks up into an *isotope* of oxygen of mass 17, and a proton. Transformation of a number of other light elements was similarly effected by bombardment with helions or alpha particles, protons and deuterons, and such transformations were greatly facilitated by means of the *cyclotron*, developed by the Nobel Prizeman E. O. Lawrence (1901–58), of the University of California at Berkeley, in which the charged particles, used as missiles, have their motions greatly accelerated by a powerful electric field.

Following on the work of W. Bothe and H. Becker, who showed, in 1930, that when beryllium, boron, or lithium is bombarded by very energetic alpha particles emitted by polonium, an unusually penetrating radiation is produced, Chadwick, in 1932, was able to prove that this radiation consists of neutrons.

In 1933 Frédéric Joliot and his wife, Irène,[2] succeeded in producing a new radioactive substance by bombarding boron with fast alpha particles. The radioactive substance thereby formed lost its activity in the same way as the well-known radioactive elements, but during the process high-speed positive electrons or *positrons* were emitted, and not alpha or beta particles. The radioactive substance would appear to be an isotope of nitrogen, of atomic mass 13, the activity of which decays to half-value in about 14 minutes, and passes into an isotope of carbon, of atomic mass 13. Other radioactive substances have also been produced by bombardment with neutrons[3] which, owing to their being electrically

[1] See Faraday Lecture, *J. Chem. Soc.*, **1936**, 508.
[2] *Compt. rend.*, 1933, **196**, 1105.
   Jean Frédéric Joliot (1900–58), after his marriage to Irène Curie (1897–1956), daughter of Pierre and Marie Curie, assumed the name Joliot-Curie. He was Professor of Physics, Collège de France, Paris, 1937–58, and Director of the Laboratory of Nuclear Physics and Chemistry; Director of the Curie Laboratory of the Radium Institute, University of Paris, 1956–58. He was awarded, jointly with his wife, the Nobel Prize for Physics in 1935.
[3] E. Fermi, *Nature*, 1934, **133**, 757, 898.

neutral, are able to enter the nucleus of even heavy elements, such as uranium;[1] and also by bombardment by fast protons.[2]

As a result of experiments carried out since 1934, and especially since the advent of atomic piles, radioactive isotopes of nearly every element can now be produced. These radioactive isotopes are of great value as indicators and as tracer elements in chemical and biological research and are also finding important applications in medicine.[3] And so, as Lord Rutherford said:[4] "We can build heavier elements from lighter and break up other atoms into fragments and produce novel radioactive elements by the score. This new field of what may be called nuclear chemistry is opening up with great rapidity. . . . Future work may disclose many surprises. . . . we are entering a no-man's-land with the ultimate hope of throwing light on the way atoms are built up from simpler particles."

Future work has indeed disclosed many surprises.

When protons and neutrons combine to form a stable nucleus, the mass is reduced and energy is evolved; for according to the theory of relativity, energy and mass are interrelated as given by the equation $E = mc^2$, where m is mass and c is the velocity of light.[5] When an unstable radioactive nucleus breaks up, a reduction of mass also takes place with consequent emission of energy (p. 221). It was for long realized, therefore, that by combining protons and neutrons, or by transmuting one kind of nucleus into another, energy might be obtained; and since in the Einstein equation c^2 is very large, a very small change of mass corresponds to a very large amount of energy. Until 1939, however, although many nuclear transmutations had been effected, the energy employed in effecting the transmutations was very much greater than the energy yielded by the small number of nuclei which, in any given case, were transmuted. There seemed, therefore, to be no prospect of successfully utilizing the enormous amounts of energy stored in atomic nuclei.

---

[1] Otto Hahn and Lise Meitner, of the Kaiser Wilhelm Institute for Chemistry, Berlin, obtained four radioactive elements, the properties of which correspond to elements of atomic numbers 93, 94, 95 and 96 (*Naturwiss.*, 1935, **23**, 37, 230; 1936, **24**, 158).

[2] J. D. Cockroft and E. T. S. Wilson, *Proc. Roy. Soc.*, 1932, A, **136**, 619; **137**, 229.

[3] F. Joliot and Irène Curie, *Radioactivité Artificielle*; F. Paneth, *Radio-elements as Indicators*; C. Rosenblum, *Chem. Rev.*, 1935, **16**, 99; G. T. Seaborg, *ibid.*, 1940, **27**, 199.

[4] *Nature*, 1935, **135**, 289.

[5] The mass of 1 gram is equivalent to about 2×10^{13} g-cal.; or 1 ounce of matter if transformed entirely into heat energy would be sufficient to convert about one million tons of water into steam.

Early in January 1939 Otto Hahn and F. Strassmann in Germany showed that when uranium is bombarded by neutrons an isotope of barium is produced, and this result gave rise to the hypothesis, put forward by O. R. Frisch and Lise Meitner, that the absorption of a neutron by a uranium nucleus causes the nucleus to split into two approximately equal parts with the release of very large amounts of energy. During 1939 this hypothesis of "nuclear fission" was investigated by a number of workers in different countries and confirmation of it was obtained. It was also found that in the process of nuclear fission neutrons were emitted which could bring about the fission of neighbouring uranium nuclei. It followed, therefore, that once the fission of uranium nuclei had been initiated in a few atoms, by means of neutrons liberated, for example, by the action of alpha particles on beryllium, a "chain process" would be established which would propagate itself throughout the mass of uranium with liberation of a large amount of energy.[1]

Whereas the nuclear fission of U^{238} requires bombardment by neutrons of high velocity, fission of the isotope U^{235} is brought about by slow neutrons. Moreover, when the isotope U^{238} is bombarded by neutrons which have a velocity intermediate between that required to bring about the fission of U^{238} and U^{235}, a new nucleus is produced with a mass number 239. This nucleus is radioactive and passes in two stages, in each case with emission of an electron, into *neptunium* (Np) with atomic number 93 and mass 239, and *plutonium* (Pu) with atomic number 94. This element, not known in nature, when bombarded by neutrons of low velocity also undergoes fission with emission of neutrons and release of a large amount of energy.

Recent developments in atomic physics have opened up a completely new chapter in chemistry; no less than ten new elements—represented by about 80 isotopes—have been prepared with atomic numbers greater than that of uranium, the last member of the natural series.[2] One of these, plutonium, has been isolated in substantial quantities. Some others have been obtained only in microgram quantities but with modern techniques this does not preclude detailed investigation of their chemical and physical properties on a quantitative as well as a qualitative basis. Indeed, even in the

[1] For an authoritative account of the early work done in liberating atomic energy, see *Statements Relating to the Atomic Bomb*, H.M. Stationery Office, 1945; *Atomic Energy*, published by the U.S. Government Printing Office and reprinted by H.M. Stationery Office, 1945.

[2] G. T. Seaborg, 'The eight new synthetic elements', *Kagaku*, 1949, **19**, 211.

case of plutonium experiments are commonly made with milligram quantities or less to diminish the hazards resulting from its
very high α-radioactivity.

The first transuranium element (neptunium (93)) was discovered
by E. M. McMillan and P. H. Abelson in 1940, though Fermi and
his co-workers believed they had prepared it some six years
earlier. In fact, however, the radioactive products Fermi obtained
by bombarding uranium with slow neutrons were fission products
with atomic numbers lower than that of uranium.

For practical purposes, all the transuranium elements are made
synthetically by bombardment processes but two of them—neptunium (93) and plutonium (94)—are in fact present in minute
traces in uranium ores as a result of the neutrons present. The other
new elements have been given the following names:

Americium	(95)	Einsteinium	(99)
Curium	(96)	Fermium	(100)
Berkelium	(97)	Mendelevium	(101)
Californium	(98)	Element 102	(102)

Element 102 was at one time given the name Nobelium but it subsequently appeared that this was a case of mistaken identity. The
supposed nobelium isotope had a half-life of about ten minutes;
a specimen prepared by bombarding californium with C^{12} ions
had a half-life of three seconds.

How far does the transuranium series extend? The answer to
this question depends to some extent on our definition of an
element, for with increasing atomic number the half-lives of the
isotopes seem to diminish steadily; by the time elements 104 or
105 are reached the half-life is probably too short to allow chemical
identification. It is conceivable that this trend is not a steady one
and that as the atomic number increases still further we might
enter a region of greater stability. Whether or not such a region of
stability exists, there are no present means of attaining it.

The transuranium elements are by no means the only products of
the uranium pile that have been of immense chemical interest. In
addition, it has been the source of a great many radioactive isotopes
of the commoner elements, and these have been immensely valuable
as so-called tracers. The theory underlying the use of these is very
simple. As we have already seen, the isotopes of any given element
are identical in their chemical behaviour, and this is true whether
they are radioactive or not. But because of their radioactivity the

isotopes can be followed throughout a series of chemical transformations, and thus show the behaviour of the particular element
as a whole. The supply of radioactive isotopes for research purposes is an important function of the United Kingdom Atomic
Energy Authority, which has established a special centre for the
purpose at Amersham. The United States, Russia, France, and
other countries with appropriate atomic energy programmes also
supply radioactive isotopes.

Carbon 14 is, perhaps, the radioactive isotope of greatest chemical and biochemical interest, for it can be used to throw light on a
vast range of organic reactions. In the study of photosynthesis, for
example, $^{14}CO_2$ has been used, particularly by M. Calvin in the
United States, to discover the nature of the organic molecules into
which atmospheric carbon dioxide is first elaborated. Carbon 14
is particularly convenient to use; its half-life is about 6,000 years
and its radiation consists only of low-energy β-particles.

Atmospheric carbon dioxide contains traces of $^{14}CO_2$, the carbon
14 resulting from the bombardment of nitrogen with neutrons
produced by cosmic radiation.

$$N + {}^{14}N^r = H + {}^{14}C.$$

All plant life—and therefore all animal life, since animals ulti-
mately depend on plants for their food—therefore contains a tiny
proportion of ^{14}C. When the plant dies, assimilation of carbon
dioxide—both $^{12}CO_2$ and $^{14}CO_2$—ceases, and the radioactivity due
to ^{14}C begins to decay according to the half-life of $5,568 \pm 30$ years.
Although the radioactivity is minute, it is measurable and thus we
have a means of dating dead organic matter by measuring its ^{14}C
activity. There are very considerable technical difficulties to be
overcome, but following the pioneer work of W. F. Libby considerable success has been achieved in dating a wide range of
archaeological material, especially when the method can be used in
conjunction with others. For example, charcoal samples from
Stonehenge have been estimated to be $3,798 \pm 275$ years by this
method; the age of one of the Dead Sea Scrolls, believed on internal
evidence to date from the first or second century B.C., was estimated
by radiocarbon dating to be $1,917 \pm 200$ years old. In biochemical
research ^{32}P and ^{131}I (half-life respectively 14 and 8 days) are in
great demand for metabolic studies and, to some extent, thera-
peutically. Radioactive iodine becomes concentrated in the
thyroid, manufacturer of thyroxin, and can be used for irradiation

of that gland. ^{32}P is made by irradiation of sulphur and ^{131}I by irradiation of tellurium 131.

As a result of the investigations and deductions which have been discussed in this chapter, an entirely new idea is given of the order and unity which run through the whole series of diverse elements known to the chemist. The ultimate components of matter now appear to be three in number, positively charged protons, negatively charged electrons, and neutrons; and the atomic model which has been constructed out of these primordial elements has made it possible to co-ordinate the varied physical and chemical properties of the elements. It must be borne in mind, however, that any model of atomic structure which may be formed reflects only the state of knowledge at a particular time, and must therefore be modified or replaced as knowledge widens and deepens. New types and methods of investigation are introduced, new phenomena are observed and new interpretations are advanced in order to explain and co-ordinate these phenomena. Thus the purely corpuscular theory of matter is passing into a wave theory, and the difficulties of reconciliation which are thereby produced give a stimulus to further advance and a fuller and richer significance to what is already known.

CHAPTER TWELVE

PHYSICAL CHEMISTRY AND CHEMICAL THEORY IN THE
TWENTIETH CENTURY

IN 1861 Thomas Graham, while investigating the diffusion of
dissolved substances through parchment paper or animal
membrane, found that while certain substances in solution pass
through such membranes, other substances do not do so. Since the
substances which passed through the membranes generally crystallized well (*e.g.*, sugar, salt, etc.), whereas those which did not
pass through the membranes were thought to be non-crystallizable
(*e.g.*, gelatin, silicic acid, etc.), Graham divided substances into two
classes, *crystalloids* and *colloids*.[1]

Although the experiments of Graham did not contribute much
to the elucidation of the fundamental nature of colloidal systems,
they were of great practical value; for the process of *dialysis* which
Graham introduced gave a means of separating and distinguishing
between what Graham regarded as two different kinds of matter,
but which are now recognized merely as different states of matter.

Although colloidal sols—the term *sol* was introduced by Graham
—have long been known,[2] and although, after Graham, the behaviour of a number of colloids was studied by others, it was not
till the end of the nineteenth and the beginning of the twentieth
century that the systematic investigation of colloids was undertaken.
During the present century the investigation of matter in the
colloidal state, to which some of the most important as also some of
the most complex naturally-occurring substances belong, has been
carried out by a large and increasing army of workers, whose
enthusiasm and energy have given rise to a voluminous literature
and have called into being such special journals as the *Zeitschrift*

---

[1] *Phil. Trans.*, 1861, **151**, 183; *Annalen*, 1862, **121**, 1. The term colloid was
derived from the Greek κόλλα (kolla), glue.

[2] A colloidal gold sol, obtained by the reduction of gold chloride by ethereal oils
and called *aurum potabile*, was known to the medieval alchemists, and was later
prepared by Faraday in 1857 (*Phil. Trans.*, 1857, **147**, 154); and colloidal sols of
arsenious sulphide, sulphur, etc., were well known in the nineteenth century.
They were termed *pseudo-solutions* by F. Selmi in 1844.

für Chemie und Industrie der Kolloide (1906), the *Revue Générale des Colloides* (1923) and the Journal of Colloid Science (1946).

In the theory of solutions that was put forward by van't Hoff (p. 106), the solute or dissolved substance is regarded as existing in the molecular state, or as being molecularly dispersed in the solution. A solution is homogeneous. A colloidal sol, however, although apparently homogeneous, is in reality heterogeneous, the heterogeneity being made evident, as Faraday showed,[1] by the scattering of the waves of light by the colloidal particles;[2] and the introduction, in 1903, of the *ultramicroscope*[3] by Richard Adolf Zsigmondy (1865–1929) made it possible not only to detect the particles but to count their number in a given volume of the sol and, indirectly, to estimate their magnitude. As a result of these and other experiments it came to be recognized, and the view was most definitely expressed in 1907 by Professor Peter Petrovitsch von Weimarn (1879–1935), of the Empress Catharina Mining Institute, St. Petersburg (Leningrad), that the colloidal state is a particular state of subdivision of matter, in which the particles of matter are intermediate in size between the microscopically visible and the molecular; that is, between one ten-thousandth of a millimetre (0·1 micron) and one millionth of a millimetre (1 milli-micron). A colloidal system, therefore, was defined by Wolfgang Ostwald (1883–1943), of the University of Leipzig, as a hetero-geneous system in which a substance is colloidally dispersed in a dispersion medium;[4] and the so-called disperse phase as well as the dispersion medium may be solid, liquid or gaseous (colloidal sols, smokes, fogs, foams, films, etc.). Moreover, since matter in the colloidal state is very finely subdivided, the extent of surface exposed is very large relatively to the total volume of the matter, and surface forces, therefore, play a predominant part.

One of the most important properties of a surface is that known as *adsorption*, or the production of an increase of concentration of matter at the surface. Silica gel or charcoal, for example, porous materials with a highly developed surface, are very effective agents for the removal of noxious gases from the air or for decolorizing

---

[1] *Phil. Trans.*, 1757, **147**, 154.

[2] The phenomenon was later studied more fully by the English physicist, John Tyndall (*Phil. Mag.*, 1869, **37**, 384) and is now known as the Tyndall pheno-menon.

[3] H. Siedentopf and R. Zsigmondy, *Ann. Physik*, 1903, **10**, 1. Zsigmondy, at that time with the firm of Schott and Genossen, Jena, was, in 1908 appointed Professor of Physics in the University of Göttingen. Nobel Prizeman, 1925.

[4] *Kolloid-Z.*, 1907, **1**, 291.

liquids. The extent of adsorption varies with the pressure of the gas or the concentration of the dissolved substance, and expressions for the adsorption isotherm have been deduced by H. F. Freundlich (1880–1941), Irving Langmuir and others,[1] which take the form $x/m = ap/1 + ap$ or $x/m = ac/1 + ac$, where x/m is the amount of substance adsorbed per gram of the adsorbent (the surface being taken as proportional to the mass), p is the pressure of the gas, and c the concentration of the solute. Adsorption from solution, it has been found, depends not only on the adsorbent and adsorbate but also on the solvent, adsorption being, in general, greater from liquids of high surface tension than from liquids of low surface tension.

Differences in adsorptive properties have long been the basis of analytical and preparative techniques; today, these are known as chromatography. That separations may occur when solutions are passed through columns of adsorptive powder was well known in the middle of the nineteenth century,[2] for example to the British soil chemists H. S. Thomson and J. T. Way. The first to apply the method systematically, however, seems to have been the American petroleum chemist D. T. Day, who gave a detailed account of his experiments at the First International Petroleum Congress held in Paris in August, 1900. The Russian petroleum chemist, S. K. Kvitka, was doing similar work at about the same time. This work was quickly followed up by C. Engler and E. Albrecht (1901) and later by the Polish botanist, M. S. Tswett (1903) working with plant pigments. Despite this considerable amount of early work, adsorption chromatography did not come into general use until very much later. In 1931, R. Kuhn and E. Lederer showed its immense value in the carotenoid field. Thereafter, it quickly came into general use, particularly for the isolation of natural products from the complex mixtures in which they commonly occur.

Ordinary column chromatography is based on distribution between liquid and solid phase, but other forms of chromatography have been developed in which other phase differences are used. Paper partition chromatography, for example, is based upon distribution between two liquid phases. The beginnings of this method can be seen in the experiments of F. F. Runge, whose *Der Bildungstrieb der Stoffe*, illustrated with twenty-two original paper chromatograms, appeared as early as 1855. Runge was closely

[1] Langmuir, *J. Amer. Chem. Soc.*, 1916, **38**, 2221; 1918, **40**, 1361; R. H. Fowler, *Proc. Camb. Phil. Soc.*, 1935, **31**, 260.

[2] For a critical study of the early history of chromatography see Trevor I. Williams and Herbert Weil, *Arkiv für Kemi*, 1953, **5**, 283.

followed by F. Goppelsroeder, author of several works on what he called capillary analysis, and G. F. Schoenbein. The full flowering of paper chromatography into one of the most powerful of all methods of chemical analysis did not occur until the last war, however, when two-dimensional paper chromatography was introduced; pioneers in this field were R. E. Liesegang and A. J. P. Martin. Later still (1952) gas-phase chromatography was introduced[1] and this has proved immensely valuable over a very wide field of chemistry, but especially in the fractionation of hydrocarbon.

Today, chromatography in all its aspects is undoubtedly one of the most important—if not the most important—of all methods of chemical analysis.[2]

The systematic investigation of the colloidal state may, perhaps, be dated from about 1898, when a number of colloidal sols of metals were prepared by Georg Bredig (1868–1944),[3] at that time Privat-Dozent in the University of Leipzig and later Professor of Physical Chemistry at the Technical High School, Karlsruhe, and by Zsigmondy,[4] at Jena. At first attention was directed mainly to the investigation of colloidal sols in which water or some other liquid was the dispersion medium; and of these sols two groups were recognized. These were called by Wolfgang Ostwald[5] *suspensoids* and *emulsoids*, but the terms *hydrophobe* and *hydrophile* colloids,[6] introduced by the French physicist, Jean Perrin, are perhaps more suitable and more generally employed. The hydrophile or lyophile colloids are characterized especially by their adsorption of the dispersion medium.

The stability of the lyophobe colloids (*e.g.*, metal sols, arsenious sulphide sol, etc.), is mainly due to the fact that the colloid particles carry an electric charge, due to the adsorption of ions from the dispersion medium, and, as S. Ernest Linder and Harold W. Picton showed in 1892, they move in an electric field towards the cathode or anode, according to the sign of the charge on the particles.[7] This process is known as *electrophoresis*, and is made use of industrially, for example for the electrodeposition of rubber from latex.

Owing to the fact that colloids have different mobilities, electrophoresis has also been applied to the investigation of the

---

[1] A. T. James, *Endeavour*, 1956, **15**, 73.
[2] For a comprehensive review see E. Lederer and M. Lederer, *Chromatography*.
[3] *Z. angew. Chem.*, 1898, p. 951; *Z. Elektrochem.*, 1898, **4**, 514.
[4] *Annalen*, 1898, **301**, 29.   [5] *Kolloid-Z.*, 1912, **11**, 230.
[6] *J. Chim. Phys.*, 1905, **3**, 84. The more general terms, lyophobe and lyophile, are also used.
[7] *J. Chem. Soc.*, 1892, **61**, 161.

homogeneity of colloidal material[1] and for the separation and purification of colloids, especially of biocolloids.[2]

By the neutralization of the electric charge on the particles, *e.g.*, by the addition of an electrolyte, flocculation and precipitation are brought about; and the effectiveness of an ion in causing precipitation is, as Hans Oscar Schulze, of the Academy of Mines, Freiberg, showed, dependent in the first instance on the valency of the ion of opposite electric charge to that on the colloid particle.[3] The higher the valence of the ion, the greater is its effectiveness in producing flocculation, but this rule, known as the Hardy-Schulze rule, is modified by the adsorbability of the ion. In the case of lyophile colloids, *e.g.*, gelatin and albumin, stability is due less to the presence of an electric charge than to adsorption of the dispersion medium. Accordingly, such colloids are not readily precipitated by electrolytes, unless the adsorbed liquid is first removed;[4] and the presence of such colloids protects lyophobe colloids and renders them less readily precipitated by electrolytes. Although lyophile colloids are not readily precipitated by electrolytes, addition of electrolytes alters the water content of the colloid, with varying degrees of effectiveness, according to the nature of the ions.[5] This fact is of much importance in many biological processes and also in agriculture. In the latter case, the water-retaining power of the soil may be altered and, in the former case, a greater adaptability to change is given to the living organism, which consists largely of hydrophile colloids.

Colloids and colloid properties play a very important part not only in the plant and animal organism, but also in very many industrial and other operations—in agriculture, in breadmaking, in porcelain manufacture, in tanning and dyeing, in the manufacture of artificial silk, in the production of rubber, etc.[6]

During the past twenty years notable advances have been made in the study of unimolecular films and of large molecules.[7]

When a drop of an oil or of a liquid such as oleic acid spreads on

[1] A. Tiselius, J. B. Eriksson-Quensel, *Biochem. J.*, 1939, 33, 1752; L. G. Longsworth, R. K. Cannan and D. A. MacInnes, *J. Amer. Chem. Soc.*, 1940, 62, 2580.
[2] A. Tiselius, *Trans. Faraday Soc.*, 1937, 33, 524; A. Tiselius and F. L. Horsfall, *J. Exp. Med.*, 1939, 69, 83; G. L. Foster and C. L. A. Schmidt, *J. Biol. Chem.*, 1923, 56, 545; *J. Amer. Chem. Soc.*, 1926, 48, 1709.
[3] *J. prakt. Chem.*, 1882, 25, 431; 1883, 27, 320.
[4] H. R. Kruyt, *Z. physikal. Chem.*, 1922, 100, 250.
[5] Franz Hofmeister. *Arch. f. experim. Path. und Pharm.*, 1888, 24, 247.
[6] See W. D. Bancroft, *Applied Colloid Chemistry*; A. E. Alexander and P. Johnon, *Colloid Science*.
[7] *Recent Advances in Colloid Science*, edited by E. O. Kraemer.

the surface of water it forms, if the surface area is sufficiently lar ge a film of only one molecule in thickness (Lord Rayleigh, 1899); and in 1917 it was pointed out by William D. Harkins (1873–1951), of the University of Chicago, that substances which spread on water contain a polar group (OH, COOH, etc.), between which and the water molecules there exists a strong dipole attraction, whereas there is little attraction between the hydrocarbon part of the molecule and water. (Sir) William Hardy (1864–1934), of the University of Cambridge, and Irving Langmuir in America, therefore regarded the molecules of the unimolecular film as being arranged side by side and orientated in such a way that the polar group of the molecule was attached to the water surface while the hydrocarbon chains stood more or less vertically away from the water surface and parallel to one another. From a study of such films Langmuir[1] and N. K. Adam,[2] formerly Professor of Chemistry at Southampton, were able to calculate the thickness of the film and the average cross-section and average length of the molecules in the film. Later the behaviour of unimolecular layers of organic substances was intensively studied by (Sir) Eric K. Rideal[3] and his co-workers in the Department of Colloid Science, University of Cambridge, and much light has been thrown on many problems of biological importance, such as the nature of cell walls, the varying toxicity of similar compounds, staining of living tissues,[4] etc.

The importance of natural polymers needs no stressing, for they form an essential part of all living cells. In recent years, the theory of polymers has been greatly stimulated by the industrial development of a very wide range of artificial polymers. They include such materials as polyethylene and the related polypropylene, polystyrene, polymethylmethacrylate, and polyvinyl chloride which are encountered in everyday life in a variety of moulded forms. In addition, fibrous polymers have been developed to rival the traditional textiles, such as wool, cotton, silk and linen; outstanding among the man-made fibres are nylon and Terylene, both of which are now used in enormous quantities in a variety of applications.

Because of their exceptional theoretical and practical importance,

[1] *J. Amer. Chem. Soc.*, 1917, **39**, 1869.
[2] *Proc. Roy. Soc.*, 1921, A, **99**, 336; 1922, A, **101**, 452.
[3] Visiting Professor of Physical Chemistry, University of Illinois, 1919–20; Professor of Colloid Science, Cambridge University, 1930–46; Fullerian Professor of Chemistry, Royal Institution, London, 1946–49; Professor of Physical Chemistry King's College, London, 1950–55. Created Knight, 1951.
[4] *Endeavour*, 1945, **4**, 83.

the structures of polymers have been closely investigated, both by conventional chemical methods of degradation and synthesis and by X-ray crystallography. Allusion has earlier been made to these (p. 169) in connection with one particularly important group of natural polymers, the proteins.

When only one monomer is polymerized, the product formed is called a homopolymer; it is chemically homogeneous, although the sizes of the individual units may lie within a wide range of molecular weight. If two monomers (say, X and Y) are polymerized, the result is a copolymer. In these the order of the monomer units may be random—say, –X–X–Y–X–Y–Y–X– or, if special catalysts are used, may follow a regular pattern (isotactic polymer), such as

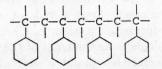

The field of polymer chemistry, both pure and applied, is now so great that no single reference adequately covers the field for the general reader, but Fred W. Billmeyer, junr's *Textbook of Polymer Chemistry* (Interscience, 1957) is a good introduction.

Attention was drawn, at an earlier point (p. 106), to the striking analogy which exists between dilute solutions and gases and which led to the statement of the general law that the osmotic pressure of a solution is proportional to the concentration expressed in gram-molecules per unit volume of solution.[1] Great, however, as was the practical importance of the van't Hoff theory of solutions, it was explicitly stated as valid only for very dilute solutions; and the determinations of osmotic pressure, indirect as well as direct, which had at that time been carried out, were too few in number and insufficient in accuracy to define the limits of concentration within which the theory could, in any given case, claim validity. The extensive series of direct determinations of the osmotic pressure of solutions, more especially of cane sugar and of glucose, which were carried out in the opening decade of this century, were therefore of the highest interest and importance. These determinations we owe mainly to the American chemists, Harmon Northrop Morse (1848–1920), Joseph Christie Whitney Frazer and co-workers at

[1] See A. Findlay, *Osmotic Pressure* (Longmans).

Johns Hopkins University,[1] and to the Earl of Berkeley (1865–1942) and Ernald George Justinian Hartley[2] in England.

From the experimental results obtained by these workers it became clear that the validity of the van't Hoff theory extends scarcely up to decimolar concentration, and that the divergence of the experimental value from the value calculated according to the theory of van't Hoff becomes increasingly great with increasing concentration. The range of validity of the van't Hoff theory, therefore, is comparatively small, in the case at least of aqueous solutions of glucose and cane sugar.

Since the van't Hoff relationship between osmotic pressure and concentration was thus shown to possess no general validity (to which, as a matter of fact, it had made no claim), the question arose as to whether any general expression could be obtained which would be applicable at all concentrations. In view of the analogy between gases and dilute solutions, the assumption, really rather unwarrantable, was made by a number of physical chemists that the relationship between osmotic pressure and concentration should take a form similar to the van der Waals equation for gases,[3] but although several such expressions were suggested, they were found to have only a restricted validity. A better way was followed, first by the Dutch physicists van der Waals,[4] Johannes Jacobus van Laar,[5] and later by others.[6] Instead of introducing into the thermodynamic deduction, as van't Hoff had done, the simplifying assumption of an infinitely dilute solution, the assumption was introduced that the solution was *ideal*. For such a solution, consisting of two completely miscible constituents which undergo neither association nor dissociation and which do not interact with each other, the relation was deduced:

$$P = \frac{RT}{V_0} \left[\, - \log_e (1 - x) \, \right] - \tfrac{1}{2}aP^2$$

where P is the osmotic pressure, V_0 the molecular volume of the

[1] *Amer. Chem. J.*, 1901, 26, 80, *et seq.* Morse, *The Osmotic Pressure of Aqueous Solutions* (Carnegie Institution, Washington).

[2] *Phil. Trans.*, 1906, A, 206, 486; *Proc. Roy. Soc.*, 1909, A, 82, 271; 1916, A, 92, 477.

[3] "The dynamical condition of molecules in solution is essentially and utterly different from that of a molecule of a gas" (Fitzgerald, *J. Chem. Soc.*, 1896, 69, 885).

[4] *Z. physikal. Chem.*, 1890, 5, 1631.

[5] *Ibid.*, 1894, 15, 457; *Sechs Vorträge über das thermodynamische Potential*.

[6] G. N. Lewis, *J. Amer. Chem. Soc.*, 1908, 30, 668; E. W. Washburn, *Z. physikal. Chem.*, 1910, 74, 385.

solvent, x the molar fraction of the solute and a the coefficient of compressibility of the solution. This equation has been tested, indirectly, for mixtures of two normal, non-associated liquids and good agreement found between the experimental and the calculated values. In the case of aqueous solutions of cane sugar, however, no agreement is found, for in these cases one is evidently not dealing with ideal solutions; and the deviations from the theoretical behaviour have been attributed mainly to hydration of the solute.

Although the Arrhenius theory of electrolytic dissociation (p. 109) adequately accounted for the behaviour of weak or slightly ionized electrolytes, it failed to give an equally satisfactory account of the behaviour of strong electrolytes, such as the ordinary salts and the mineral acids and alkalis; and it was not until the electronic theory of atomic constitution had been established and an explanation of valency, based on this theory, had been given, that a satisfactory explanation of the behaviour of strong electrolytes could be offered.

According to modern views, the metallic and acid radicals of a salt are united by electrovalency, so that the existence of ions not merely in solution but even in the crystalline state may be assumed.[1] From this one is led to the conclusion that salts are completely ionized in solution, but the electrolyte may not behave as if it were completely ionized owing to the existence of interionic forces. In such solutions, as S. Roslington Milner, Professor of Physics in the University of Sheffield, first showed,[2] there can be no random distribution of the ions, owing to electrical attractions and repulsions; but although Milner was able to deduce a relation between the depression of the freezing point of a solution of a strong electrolyte and the dilution, he failed to obtain a satisfactory expression for the electrical conductivity. Stimulated by the publication of an erroneous theory of conductivity put forward by the Indian chemist, Jnanendra C. Ghosh, in 1918, Peter Debye, Professor of Physics at the High School of Technology, Zurich, took up the investigation of the problem and, along with his assistant, E. Hückel, succeeded in obtaining an expression for the equivalent conductivity of a solution of a strong electrolyte.[3]

[1] This view is supported also by the fact that salts, even in the solid state, have been found to conduct the electric current when heated to a suitable temperature (C. Tubandt and S. Eggert, Z. anorg. Chem., 1920, 110, 196), and conduction readily takes place when the salts are fused.

[2] Phil. Mag., 1912, 23, 551; 1913, 25, 742.

[3] Physikal. Z., 1923, 24, 305; 1924, 25, 145; Trans. Faraday Soc., 1927, 23, 334.

According to this theory, an ion, owing to interionic forces, will build up an "atmosphere" of ions of opposite sign, and when an ion moves under the influence of an applied electromotive force the ionic atmosphere will be unsymmetrical; for the ion has to build up a new ionic atmosphere in front, while the atmosphere behind the moving ion will be dispersed. This dispersal, however, requires a certain amount of time ("time of relaxation"), and there will therefore always be, in the rear of a moving ion, an excess of ions of opposite sign, whereby the mobility of the ion will be diminished. Moreover, the movement of an ion will be subject to a retarding electrical drag owing to the movement of oppositely charged ions in the reverse direction, and the total reduction of the mobility of an ion was found to be proportional to the square root of the concentration. By taking into account the factors influencing the mobility of an ion, Debye and Hückel deduced an expression for the variation of the equivalent conductivity with the concentration. The expression obtained by Debye and Hückel was, in 1926, improved by the Norwegian physicist, Lars Onsager, by taking into account the Brownian movement of the ion;[1] and the equation

$$\Lambda_c = \Lambda_0 - \left[\frac{0 \cdot 986 \times 10^6}{(\text{D.T})^{\frac{3}{2}}} \cdot (2 - \sqrt{2}) \cdot z^2 \cdot \Lambda_0 + \frac{58 \cdot 0}{(\text{D.T})^{\frac{1}{2}} \eta} \cdot z \right] \cdot \sqrt{2zc}$$

which applies to a z-valent binary electrolyte, was obtained. This equation, which involves the dielectric constant, D, and the viscosity, η, of the solvent, is of the same general form as that deduced empirically, in 1907, by Kohlrausch, namely, $\Lambda_c = \Lambda_0 - k \sqrt{c}$. Deviations from the values calculated by the Debye-Hückel-Onsager equation may be regarded as due to some form of ionic association, such as the formation of ionic doublets which simulate unionized molecules. The extent to which this will occur will depend on the interionic forces, and these, in turn, will depend on the dielectric constant of the solvent (Nernst-Thomson rule).[2] It will depend also on the extent to which the ions undergo solvation, i.e., on the nature of the ions and of the solvent. In a complete theory of the behaviour of a strong electrolyte, therefore, chemical as well as purely physical properties must be taken into account.

In the application of the law of mass action to homogeneous equilibrium, the concentration in gram-molecules per litre was

[1] Physikal. Z., 1926, 27, 388; 1927, 28, 277.
[2] W. Nernst, Z. physikal Chem., 1894, 13, 531; J. J. Thomson, Phil. Mag., 1893, 36, 320.

regarded by Guldberg and Waage, and by the physical chemists of the nineteenth century, as a measure of the "active mass" (p. 88). As Professor G. N. Lewis, however, pointed out,[1] this can be valid only in the case of ideal gases and ideal solutions; and in order to preserve the law of mass action Lewis replaced concentrations by *activities*, or quantities which are proportional to the active mass of the reacting substances and by the use of which concordance with the law of mass action is obtained.

In dealing with reactions between strong electrolytes, more especially, the concept of activity was proved of much importance. In the case of weak electrolytes, the degree of ionization was obtained from the ratio of equivalent conductivities, $\Lambda_v / \Lambda_\infty = \alpha$ (p. 110); and the concentration of an ion is represented by $\alpha.c$, where c is the concentration of electrolyte. In the case of strong electrolytes, however, this no longer holds, and the ionic activity or mean ion activity will be given by $\gamma.c$, where γ is an activity coefficient which can be determined, for example, by measurements of the electromotive force of concentration cells and in other ways. It is this quantity $\gamma.c$, therefore, which had to be used in mass law equilibrium equations involving reactions between salts.[2] For a general account of the modern theory of solutions of electrolytes, see a recently revised book by H. S. Harned and B. B. Owen.[3]

During the second half of the nineteenth century, as we have seen, chemists were able, on the basis of the theories of Couper and Kekulé, van't Hoff and Le Bel, to interpret in a satisfactory manner the chemical behaviour of large classes of artificially produced and naturally-occurring organic compounds, and to build up the extraordinarily imposing edifice of organic chemistry with which we are acquainted at the present day. Although, however, great advances were made in the development of the practical methods of synthetic chemistry and in the elucidation of molecular structure, by physico-chemical as well as by chemical methods, little or nothing essentially new had been contributed to chemical theory. Then at the end of the nineteenth century there came the discovery of X-rays and of radioactivity, the investigation of which led to the development of new views regarding the constitution of the atoms of matter and to an explanation of valency; and the introduction

[1] *Proc. Amer. Acad.*, 1901, **37**, 45; 1907, **43**, 259.
[2] Lewis and Randall, *J. Amer. Chem. Soc.*, 1921, **43**, 1112; *Thermodynamics*.
[3] *The Physical Chemistry of Electrolytic Solutions* (3rd edn.), 1958.

at a somewhat later time of special mathematico-physical conceptions (*e.g.*, quantum theory, wave mechanics) and experimental methods gave an entirely different outlook and motive to chemistry. Whereas the classical physical chemistry of the nineteenth century dealt with the properties of substances, regarded as collections of molecules, the newer physical chemistry seeks to investigate the individual molecule; to gain a knowledge not only of the mutual arrangement of the atoms within the molecule but also of the forces which hold them together. It seeks, moreover, to ascertain the dimensions of the molecules, the distances apart of their constituent atoms and the angles at which their valencies are oriented in space, etc. These are some of the aims of modern theoretical chemistry, and in the history of physical science nothing perhaps is more remarkable than the way in which chemists and physicists, in intensely active co-operation, have, in recent years, applied a whole series of physical methods of different kinds to the elucidation of the problems of molecular constitution and the nature of the linkages which unite the atoms in the molecule.

As it was on crystals that the first determinations of interatomic distances were carried out, we may discuss them first.

When X-rays were discovered in 1895 much discussion took place regarding their nature; and although the view was held that they are of the same nature as the rays of light, with a very short wave-length, there was no means of testing this view. In 1912, however, Max von Laue, Professor of Physics in the University of Zurich, and Nobel Prizeman for Physics (1914), suggested that if X-rays are of the same nature as ordinary light, diffraction effects should be obtained if X-rays are passed through a regular arrangement of particles sufficiently small in size to act as a diffraction grating. Such a grating, Laue thought, might be formed by the regular arrangement of atoms or atomic groups in a crystal; and when the suggestion was put to the test by the German physicists, Walter Friedrich and Paul Knipping,[1] of the University of Munich, Laue's expectation was confirmed.

The production of a diffraction pattern when a fine pencil of X-rays was passed through a crystal not only proved that X-rays are of the same nature as ordinary light, but it suggested a means whereby the atoms and groups in crystals might be located in space and the dimensions of the crystal unit or space lattice be determined.

[1] Friedrich, Knipping and Laue, *Ber. K. Bayer. Akad.*, 1912, p. 303; *Le Radium*, 1913, **10**, 47.

A simpler method for the determination of the internal structure of crystals was introduced in 1912 by (Sir) William Henry Bragg (1862–1942), at that time Professor of Physics in the University of Leeds, and his son (Sir) William Lawrence Bragg.[1] This method depends on the determination of the angle at which a beam of X-rays of definite wave-length is reflected by the successive layers of atoms lying in planes parallel to the crystal face.[2] Without discussing the details of the measurements and necessary calculations, it may be said that it has now become possible to determine the position of the atoms in a crystal and their distance apart. Thus, in the case of sodium chloride, the molecules of which, according to the modern view, are made up of sodium ions and chloride ions held together by electrical forces, the crystals have the form of a cube. Each sodium ion is surrounded by six chloride ions and each chloride ion is similarly surrounded by six sodium ions, and the distance apart of the ions is $2 \cdot 817 \times 10^{-8}$ cm., or $2 \cdot 817$ Ångström units (A.U.).

By X-ray analysis of diamond it was found that the carbon atoms are arranged in such a way that each is at the centre of a regular tetrahedron and is joined to four other atoms which lie at the corners of the tetrahedron. The atoms are spaced at equal distances[3] from one another, namely $1 \cdot 54$ A.U., so that each atom is bound with equal strength to four symmetrically arranged atoms; an arrangement which serves to account for the hardness of diamond. Moreover, the linking of the atoms in the manner described leads to the formation of "rings" of six carbon atoms, but the carbon atoms forming a hexagon are not all in one plane.[4]

[1] See W. H. Bragg, *An Introduction to Crystal Analysis* (Bell).

William Henry Bragg, Professor of Physics, successively, at the Universities of Adelaide (Australia) and Leeds, was Fullerian Professor of Chemistry, Royal Institution, London, 1923–42. He was awarded, jointly with his son, the Nobel Prize for Physics in 1915. He was created Knight Commander of the Order of the British Empire (K.B.E.) in 1920, and had the Order of Merit conferred on him in 1925.

William Lawrence Bragg was Professor of Physics, University of Manchester, 1919–37; Director of the National Physical Laboratory, 1937–38; Cavendish Professor of Experimental Physics, Cambridge, 1938–53. He was appointed Fullerian Professor of Chemistry, Royal Institution, London, in 1953. He shared with his father the Nobel Prize for Physics in 1915 and was created a Knight in 1941.

[2] The X-rays may also be passed through a small quantity of crystalline powder and measurements made by a method due to P. Debye and P. Scherrer (*Phys. Z.*, 1916, **17**, 277) and to A. W. Hull (*Proc. Nat. Acad. Sci.*, 1917, **3**, 470).

[3] The interatomic distances are the distances between the centres of adjacent atoms.

[4] W. H. Bragg and W. L. Bragg, *Proc. Roy. Soc.*, 1913, A, **89**, 277; W. Ehrenberg, *Z. Krist.*, 1926, **63**, 320.

In the case of graphite, X-ray examination shows that the crystal consists of layers of co-planar carbon atoms forming a hexagonal network,[1] the distance between the centres of adjacent carbon atoms being 1·42 A.U. The distance between the layers is relatively large and equal to 3·41 A.U., so that the force between the different layers of carbon atoms is much less than between the atoms of a given layer. The soft, flaky nature of graphite thus finds an explanation.

The arrangement of the carbon atoms in diamond and in graphite is of much interest in connection with the structure of benzene and its derivatives, in the molecules of which Kekulé, as we have seen, inferred the presence of a hexagonal ring of six carbon atoms. Since benzene is not crystalline at the ordinary temperature, the X-ray investigation of the benzene ring had to be carried out on certain of its derivatives. From the X-ray analysis of crystals of diphenyl, hexamethylbenzene, etc., the presence of a ring of six carbon atoms was confirmed, and the distance between the carbon atoms in the ring was found to be 1·42 A.U., as in the case of graphite;[2] and the same value was found in the case of naphthalene and anthracene,[3] in harmony with the conclusions of chemists that these hydrocarbons are made up of two and of three benzene rings respectively. It was thus shown that the crystal structures of benzene derivatives are more closely related to that of graphite than to that of diamond; and that the benzene ring is to be regarded as a *flat* ring of six carbon atoms.

In the case of long-chain aliphatic compounds, X-ray analysis indicates that the carbon atoms have a zigzag arrangement in the chain.

The method of the Fourier series as applied to crystal analysis—suggested by W. H. Bragg[4] in 1915 and later employed by J. M. Robertson, Professor of Physical Chemistry, University of Glasgow, in his investigation of naphthalene and anthracene—was later extended to the direct determination of the relative orientation of the constituent atoms and to the determination of the nature of the chemical bonds.

The methods of X-ray analysis have in recent years been

[1] J. D. Bernal, *Proc. Roy. Soc.*, 1924, A, **106**, 749; O. Hassel and H. Mark, *Z. Phys.*, 1924, **25**, 317.
[2] J. Dhar, *Indian J. Phys.*, 1932, 7, 43; K. Lonsdale, *Proc. Roy. Soc.*, 1929, A, **123**, 494.
[3] J. M. Robertson, *Proc. Roy. Soc.*, 1933, A, **140**, 79; **142**, 674.
[4] *Phil. Trans.*, 1915, A, **215**, 253.

greatly extended, notably by use of computers, and have been applied to the investigation of metals and their alloys, and to many inorganic and organic compounds;[1] and in the case of the organic compounds they have yielded an amazing verification of the structural formulae which had been adopted on purely chemical grounds. These formulae, therefore, gain a degree of reality which they did not formerly possess, a reality which is emphasized by the introduction of exact metrical representation.

The analysis of crystal structure may be effected not only by means of X-rays but also by means of electron diffraction. According to a theory put forward in 1923 by the French physicist and Nobel Prizeman, Prince Louis de Broglie,[2] a moving electron is associated with waves which, like light waves, can show interference, diffraction effects, etc.; and the predictions from this theory were realized experimentally in 1927 by (Sir) George Paget Thomson,[3] at that time Professor of Natural Philosophy in the University of Aberdeen, and by C. Davisson and L. H. Germer,[4] of the Bell Telephone Laboratories, New York. This discovery has led to the use of electron beams in place of X-rays for the investigation of crystals, and the results agree with those obtained by means of X-rays.

The voltages used in the X-ray diffraction of crystals vary enormously, depending on the wave-length required. If the electron falls through a potential of 40 volts the wave-length corresponds to approximately 2 A.U.; at 100,000 volts, it corresponds to less than 0·04 A.U. Because these wave-lengths are much less than those of visible light their use in the electron microscope, now a standard research instrument, enables very high resolving powers to be attained.

Although the determination of molecular dimensions was first effected by means of an X-ray analysis of substances in the crystalline state, a number of other physical methods have since been introduced by means of which molecular dimensions and configurations of substances in the gaseous or liquid state can be

[1] See S. B. Hendricks, *Chem. Reviews*, 1930, **7**, 431; J. M. Robertson, *ibid.*, 1935, **16**, 417; H. Mark, *Z. Elektrochem.*, 1934, **40**, 413; *Annual Reports of the Chemical Society*, 1935, **32**, 193.

[2] *Compt. rend.*, 1923, **177**, 507, 548.

[3] *Nature*, 1927, **119**, 890; **120**, 802; *Proc. Roy. Soc.*, 1928, A, **117**, 600.
George Paget Thomson, son of Sir J. J. Thomson, was Professor of Physics, Imperial College of Science, London, 1930–52; Master of Corpus Christi College, Cambridge, 1952–62; awarded Nobel Prize for Physics, 1937; created a Knight, 1943.

[4] *Nature*, 1927, **119**, 558; *Phys. Rev.*, 1927 [ii], **30**, 705.

ascertained and light thrown on the nature of the valency bonds, etc.[1] In discussing this rapidly developing and highly specialized field of chemical physics, brevity of treatment is necessarily imposed.

In 1929, Debye,[2] by a technique similar to that used in the examination of crystalline powders (p. 256), found that X-ray interference patterns could be obtained with gases as well as with liquids, although in the case of the latter the results, owing to the scattering of the rays by the molecules as well as by the atoms, could not easily be interpreted. With gases, however, it was possible from the interference patterns to determine interatomic distances, as in the case of crystals; and a number of such determinations were carried out, more especially by pupils of Debye.

This method was very soon displaced. In 1930 the German physicist, R. Wierl,[3] showed that a beam of electrons could be used with advantage in place of X-rays for the investigation of gases and vapours, because, owing to the greater intensity of the electron beam, the period of exposure necessary to get the photographic image is reduced from a few hours (required with the X-ray method) to a fraction of a second. The method, moreover, can be applied to gases and vapours under reduced pressure. An experimental requirement is that the electron beam (30,000–70,000 volts) be maintained at a very constant voltage.

Examination of benzene and of diphenyl, etc., by the electron diffraction method has shown that the length of the $C-C$ bond is 1·39 and not 1·42 A.U. as found by X-ray measurements.[4] The shortening of the $C-C$ bond in benzene as compared with the value found in graphite may be due to mesomerism (p. 263).

Information regarding interatomic distances and molecular configuration can also be obtained from an investigation of molecular spectra. Whereas comparatively simple line spectra, due to energy transitions, are emitted by atoms, molecules yield band spectra,

[1] See, for example, A. Weissberger (ed.), *Physical Methods of Organic Chemistry* (3rd edn.), 1959 (Interscience).

[2] P. Debye, L. Bewilogua and F. Ehrgardt, *Physikal. Z.*, 1929, **30**, 84. The mathematical treatment of the subject had previously been worked out by Debye (*Ann. Physik*, 1915, **46**, 809) and by P. Ehrenfest (*Proc. K. Akad. Amsterdam*, 1915, **17**, 1184).

[3] *Physikal. Z.*, 1930, **31**, 366, 1028; *Ann. Physik*, 1931, **8**, 521; 1932, **13**, 453. See J. T. Randall, *Diffraction of X-rays and Electrons by Amorphous Solids, Liquids and Gases* (Chapman and Hall).

[4] L. Pauling and L. O. Brockway, *J. Amer. Chem. Soc.*, 1937, **59**, 1223; Miss I. L. Karle and L. O. Brockway, *ibid.*, 1944, **66**, 1974.

a band being formed by a large number of fine lines crowded together. Although lines due to electronic transition are found, spectra are also observed in the infra-red, due, on the one hand, to changes in the energy of rotation of the molecule, and, on the other hand, to changes in the energy of vibration of the constituent atomic nuclei.[1] Of these the rotation-vibration spectrum is, at present, the most important and has been most widely studied—generally as an absorption spectrum—owing to the fact that it lies mainly in that part of the infra-red which is nearest to the visible spectrum, while the pure rotation spectrum lies in the far infra-red.[2] From a study of the fine structure of the rotation-vibration spectrum one can not only calculate the angles between the bonds linking the constituent atoms, but also deduce the moment of inertia of the molecule. In this way it is possible to distinguish between a linear molecule and, say, a triangular one. Heats of dissociation of linkages may also be calculated. Owing, however, to difficulties in interpreting the band spectra, the method was at first applied only to the simpler molecules. In recent years improvements in the methods of detecting infra-red radiation have been introduced and other technical developments have taken place which have made possible a rapid increase in the range of chemical applications of infra-red spectroscopy;[3] and a great advance in our knowledge of molecular structure has resulted from a study of the infra-red absorption spectra of larger molecules. Owing to the individuality of the vibration spectrum of a large molecule one is able to distinguish between normal and iso-compounds, *cis-* and *trans*-isomers, monomers and polymers, etc. The examination of the infra-red absorption spectra of hydroxylic compounds has revealed the existence of a "hydrogen bridge" of the type $-O \cdot H \cdots O-$ in the molecules of many of them. Infra-red spectroscopy has become one of the most important methods of analysis.

At the other end of the visible spectrum is the ultra-violet, which in the 2,000–4,000 A.U. range has been very useful in chemical analysis and structure determination, for both organic compounds and some inorganic ions and their complexes. Absorption in this region depends, in organic compounds, upon the presence of some degree of unsaturation within the molecule. Groups such as keto

[1] A. Kratzer, *Z. Physik*, 1920, 3, 289, 460; 1921, 4, 476; 1924, 23, 298.
[2] M. Czerny, *Z. Physik*, 1925, 34, 227; 1927, 44, 235; 1927, 45, 476; R. M. Badger and C. H. Cartwright, *Phys. Rev.*, 1929, 33, 692.
[3] H. W. Thompson, *Endeavour*, 1945, 4, 154.

(C = O), or olefinic (C = C) cause characteristic changes in the ultra-violet absorption spectrum. The absorption may be very intense, permitting detection of substances at extremely high dilutions, *e.g.*, one part in a million.

In 1923 Adolf Smekal,[1] Professor of Physics in the University of Halle, predicted, on theoretical grounds, that if a beam of monochromatic light is passed through a transparent substance a scattering of the light by the molecules of the substance will take place, and that in the scattered light there will be radiations of different frequencies from that of the incident light. These frequency differences are characteristic of the scattering medium. This alteration of the frequency was verified in 1928 by (Sir) Chandrasekhara Venkata Raman,[2] and is generally known as the *Smekal-Raman effect*. From a study of the Raman spectra valuable information regarding molecular structure and the number and nature of the linkages between atoms can be obtained, in the case at least of the simpler types of molecules.[3]

It has been pointed out (p. 231) that two atoms may combine by means of a co-valent link; that is, by the sharing of two electrons between the orbits of the two atoms. If the two electrons are shared equally between the two atoms the resulting molecule will be neutral and non-polar; but if the two shared electrons, constituting a single co-valent linkage, are not shared equally between the two atoms, then one atom will have an excess of positive electricity associated with it and the other an excess of negative electricity. The molecule, therefore, will be polar and will possess a dipole moment equal to the charge on one of the atoms multiplied by the distance between the atoms. The existence of a dipole moment, which is normally found when a co-valent bond unites two unlike atoms, is evidence that an unequal sharing of electrons has taken place. From a study of these dipole moments, which may be calculated from determinations of the dielectric constant of the substance or from the deviation of a "molecular beam" in a non-

[1] *Naturwiss.*, 1923, **11**, 873; *Z. Physik*, 1925, **32**, 241.

[2] C. V. Raman and K. S. Krishnan, *Indian J. Physics*, 1928, **2**, 399; Raman, *Trans. Faraday Soc.*, 1929, **25**, 781. C. V. Raman was Professor of Physics, University of Calcutta, 1917–33, and Director of the Indian Institute of Science, Bangalore, 1933–   . He is Director of the Raman Research Institute, Bangalore. He was created a Knight in 1929 and awarded the Nobel Prize for Physics in 1930.

[3] See D. S. Villars, *Chem. Reviews*, 1932, **11**, 369; Jevons, *Report on Band Spectra of Diatomic Molecules*, 1932; J. H. Hibben, *The Raman Effect and its Chemical Applications* (*Amer. Chem. Soc. Monographs*); G. B. B. M. Sutherland, *Infra-red and Raman Spectra*; G. Herzberg, *Infra-red and Raman Spectra* (van Nostrand).

homogeneous magnetic field,[1] much insight has been gained into molecular structure.[2]

In illustration of the great advance which has already been made —and the advance is rapidly proceeding—one may mention only a few simple cases. Elementary molecules, *e.g.*, H_2, O_2, are found to have no dipole moment, and, consequently, there is a uniform sharing of the electrons; but in the case of HCl, for example, there is a dipole moment of $1 \cdot 04 \times 10^{-18}$ electrostatic units, the distance between the atoms being $1 \cdot 28$ A.U. In the case of a triatomic molecule in which one atom is united to two atoms of another element, *e.g.*, H—O—H, one may conclude that if the molecule is linear the electric moments of the two bonds will be equal in magnitude and opposite in direction, and will thus neutralize each other. The molecule will have no dipole moment. This is found to be the case with carbon dioxide, for the molecule of which, therefore, one must assume a linear structure, O—C—O. The distance between the carbon atom and each of the oxygen atoms is $1 \cdot 15$ A.U. Unlike carbon dioxide, the molecule of water is found to have a dipole moment of $1 \cdot 84 \times 10^{-18}$ electrostatic units. The molecule, therefore, cannot be linear but the two valency bonds must be inclined at an angle to each other. This angle was found to be[3] 105° $6'$, and the distance between the oxygen atom and each of the hydrogen atoms is $0 \cdot 970$ A.U. A similar configuration is found in the case of hydrogen sulphide (H_2S) and sulphur dioxide (SO_2). The molecule of ammonia is pyramidal, the height of the pyramid being $0 \cdot 52$ A.U., while the atoms of carbon tetrachloride are arranged as a regular tetrahedron, the interatomic C—Cl distance being $1 \cdot 83$ A.U., and the Cl—Cl distance $2 \cdot 98$ A.U. Other molecules, *e.g.*, $O-C\diagup\diagdown{}^H_H$, are shown to have a Y-shape.

From a study of the dipole moments of molecules, also, it has been found possible to calculate the moments associated with particular linkages and to calculate the residual charges on the atoms.

The development of wave mechanics has proved of great importance not only in physics but also in chemistry, for, on the basis of that concept, it is possible to calculate the properties of some of the simpler molecules at least. Of importance for organic

[1] See E. Wrede, *Z. Physik*, 1927, **44**, 261; I. Eastermann and M. Wohlwill, *Z. physikal. Chem.*, 1928, B, **1**, 161; 1929, B, **2**, 287; 1933, B, **20**, 195.

[2] See P. Debye, *Dipole Moment and Chemical Structure* (Blackie); R. J. W. Le Fèvre, *Dipole Moments* (Methuen).

[3] R. Mecke, *Z. Physik*, 1933, **81**, 313.

chemistry, more particularly, is a principle of structure which has been deduced on the basis of wave mechanics, the principle of *resonance*, or *mesomerism*, as N. V. Sidgwick proposed it should be called, which has been applied more especially by Linus Pauling, of the California Institute of Technology, Pasadena, and his pupils.[1] On the basis of wave mechanics it can be shown that if two electronic structures are possible for the same molecule, the normal state of the molecule is neither the one nor the other of the two separate states, but a linear combination of the two. The molecule, in other words, is to be regarded as passing from one state to the other with very great frequency, or as having a structure intermediate between the two and not expressible by the ordinary structural formulae. The molecule, therefore, exhibits some of the properties of each of the two possible electronic structures.

This principle of mesomerism affords, according to some, an explanation of the views expressed by A. Lapworth, R. Robinson, C. K. Ingold, and others,[2] according to which, changes in the reactivity of organic compounds are due to a drift of electrons from their normal positions. The occurrence of mesomerism, also, produces a characteristic shortening of linkages or a reduction of the "atomic radius,"[3] and it may be that the shortening of the C—C link in benzene to $1\cdot39$ A.U., in place of $1\cdot42$ A.U. found in graphite, is due to mesomerism; and the fact that the benzene molecule has no dipole moment indicates that it consists of a flat ring of carbon atoms with the electrons regularly distributed as a result of mesomerism. The theory of mesomerism, therefore, leads to the view that in benzene the bonds between adjacent atoms of carbon are identical, as its chemical properties indicate, and not alternately single and double. The hypothesis of mesomerism, although of great interest, has not been universally accepted.[4]

. . . . .

[1] *J. Amer. Chem. Soc.*, 1931, **53**, 3225 onwards.
Linus Carl Pauling, Professor of Chemistry since 1931 and Director of Gates and Crellin Laboratories of Chemistry, California Institute of Technology since 1937, was Fisher Baker Lecturer in Chemistry, Cornell University, 1937–38, and George Eastman Professor, Oxford University, 1948. He was awarded Nobel Prize for Chemistry in 1954, and Nobel Peace Prize, 1962.
[2] See R. Robinson, *Outline of an Electrochemical (Electronic) Theory of the Course of Organic Reactions* (Inst. of Chem.); Fry, *The Electronic Conception of Valence and the Constitution of Benzene* (Longmans).
[3] Since the interatomic distance is the distance between the centres of two linked atoms, the "atomic radius" of an atom may be taken as half the interatomic distance between two like atoms.
[4] See A. Burawoy, *Chem. and Ind.*, **1944**, 434; G. W. Wheland, *Resonance in Organic Chemistry* (1955).

In the field of stereochemistry great advances have been made during the present century through the application of the newer physical ideas and methods.[1] Thus, from a study of the dipole moment, for example, it is possible to distinguish between *cis*- and *trans*-isomerides (*e.g.*, maleic and fumaric acids), since the *cis*-form will be polar and have a dipole moment, whereas the *trans*-form will be non-polar.[2]

Further, determinations of atomic size and of valency angles are of much importance in the study of an optical activity which is dependent on restricted rotation within the molecule. Thus, by means of the physical methods already discussed, diphenyl is shown to consist of two flat benzene nuclei joined co-axially, and capable of free rotation about the common axis. Such a molecule is symmetrical and cannot be obtained in different optically active forms. If, however, the hydrogen atoms of the two benzene rings are substituted in the ortho-positions to the connecting bond, optically active molecules can be obtained if the dimensions (atomic radii) of the substituting groups are sufficiently large to interfere with each other and prevent free rotation about the common axis. In this case the molecule cannot be wholly planar, and it therefore becomes asymmetric and capable of existing in optically active forms.[3] These spatial or steric effects are of much importance.[4]

In 1895 Paul Walden (1863–1957), Professor of Chemistry, Rostock, observed that a reversal of the sign of rotation of an optically active compound may take place when one of the atoms or groups attached to the asymmetric carbon atom is replaced by another atom or group. Thus, when dextro-chlorosuccinic acid is converted into malic acid by means of silver hydroxide, dextro-rotatory malic acid is obtained; but when the conversion is effected by means of potassium hydroxide, laevo-rotatory malic acid is formed. To this phenomenon the term *Walden inversion* was applied.

From 1907 onwards the phenomenon of optical inversion was studied by many chemists, but although a great and rather bewildering mass of facts was established, no theory was put forward which accounts satisfactorily for the phenomenon and makes possible the prediction of the results of substitution (including influence of structure, of solvent and of reagent).

---

[1] See "A Stereochemical Centenary", *Endeavour*, 1953, **12**, 171.
[2] J. Errera, *Physikal. Z.*, 1926, **27**, 764.
[3] F. Bell and J. Kenyon, *Chem. and Ind.*, 1926, **4**, 864; W. H. Mills, *ibid.*, p. 884; W. H. Mills and K. A. C. Elliott, *J. Chem. Soc.*, **1928**, 1291.
[4] E. E. Turner, *Science Progress*, 1936, **31**, 29.

The nearest approach to a satisfactory theory was made on the basis of certain views put forward by A. Werner (1911), T. M. Lowry (1925) and others. By these workers it was postulated that the substituting element first forms an addition complex with the radical or element to be substituted. As a result a change in the distortion of the molecule takes place, so that the substituting group does not enter the molecule in the position of the radical or atom substituted, and thus brings about an inversion of the optical activity. Thus:

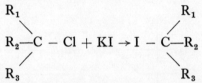

These views were supported by M. Polanyi (1936), of the University of Manchester, and E. D. Hughes and co-workers (1935) at University College, London. A. R. Olson (1936), of the University of California, expressed the view that every reaction of substitution is accompanied by inversion, and that if there is apparently no inversion the reaction must take place in two stages, with an inversion in each stage.

That the occurrence or non-occurrence of inversion is dependent on the mechanism of the reaction was borne out by the experiments of C. K. Ingold (1935).

The success which attended the van't Hoff-Le Bel theory of the asymmetric carbon atom in accounting for the optical activity of carbon compounds led chemists to apply considerations of a similar kind to the atoms of other elements; and attempts were soon made, by Le Bel and others, to prepare optically active nitrogen compounds. These attempts, however, remained unsuccessful until 1899 when W. J. Pope and S. J. Peachey (1877–1936),[1] by the use of camphor sulphonic acids and of improved methods of working (p. 62), proved the occurrence of optical activity due to an asymmetric nitrogen atom by preparing d- and l-phenylbenzylmethylallylammonium salts; and other active compounds were later obtained in which the optical activity was associated with the presence, as centre of asymmetry, of an atom of sulphur, selenium, tellurium, silicon, tin, phosphorus, etc.[2]

[1] J. Chem. Soc., 1899, 75, 1127.
[2] W. J. Pope and S. J. Peachey, J. Chem. Soc., 1900, 77, 1972; W. J. Pope and H. A. D. Neville, ibid., 1902, 81, 1552; T. M. Lowry and F. L. Gilbert, ibid.,

The application of stereochemical considerations to the spatial arrangement of ammonia and other groups about the central metal atom in the co-ordination complexes suggested by Werner (p. 232), made it possible to predict the existence of stereo-isomers; and the guidance thus given led to their preparation. In the case of the co-ordination complex $[Co,(NH_3)_4Cl_2]$, two isomeric forms are known, and Werner showed that these could be accounted for if it is assumed that the six radicals have an octahedral arrangement round the central metal atom, thus:

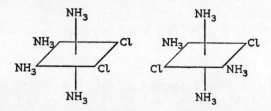

Optical isomerism, moreover, should be possible when six different atoms or groups are present in the complex. The difficulties attending the preparation of such complexes were avoided when Werner, in 1911, found that three molecules of ethylene diamine, $CH_2(NH_2) \cdot CH_2(NH_2)$ (contracted to *en*), could replace six molecules of ammonia, giving rise to the two enantiomorphous, asymmetric forms:

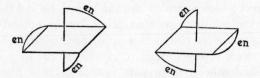

These forms are non-superposable mirror-images, one of the other, and it has been found that salts such as $[Co, en_3]Cl_3$ can be resolved into optically active isomers. Optically active compounds have now been prepared from no fewer than twenty-one elements.[1]

In nature most asymmetric compounds are found occurring in *one* of the optically active forms only, but when it is attempted to prepare an asymmetric compound in the laboratory from symmetric substances alone, from substances, that is to say, which are

1929, p. 2867; F. S. Kipping, *ibid.*, 1907, **91**, 209; 717 W. J. Pope and S. J. Peachey, *Proc. Chem. Soc.*, 1900, **16**, 42, 116; J. Meisenheimer and L. Lichtenstadt, *Ber.*, 1911, **44**, 356.
[1] T. M. Lowry, *Optical Rotatory Power* (Longmans).

not themselves optically active, it is always found that the product obtained is inactive. How then do these single optically active forms arise in the living organism? For many years it was considered that the formation of these compounds could take place only in association with living matter and that it was the prerogative of life; a view which was also held by Pasteur in 1860. "Artificial products," wrote Pasteur, "have no molecular asymmetry; and I could not point out the existence of any more profound distinction between the products formed under the influence of life and all others." It is true that the inactive, racemic form which is obtained as the result of artificial syntheses can be separated into two optically active forms with the help of asymmetric, optically active compounds or with the help of living organisms, but these separations are due, directly or indirectly, to living matter. Moreover, when sodium ammonium racemate is resolved by crystallization (p. 60), both active forms separate together and have to be separated by the living operator; and this act of separation was held by Francis Robert Japp (1848–1925), Professor of Chemistry in the University of Aberdeen, to be not altogether dissimilar to the selection made by the mould *Penicillium glaucum*.[1] Japp, therefore, argued strongly for the view that "at the moment when life first arose, a directive force came into play—a force of precisely the same character as that which enables the intelligent operator, by the exercise of his will, to select one crystallized enantiomorph and reject its asymmetric opposite." Very many attempts have been made to obtain an answer to the question of the origin of optical activity in nature.

As early as 1894 Emil Fischer had pointed out that if a compound containing an asymmetric carbon atom is synthesized from a symmetrical compound in presence of an optically active substance or group, the latter exerts a directive action and excess of one of the optically active forms may be obtained. The first undoubted asymmetric synthesis of this kind was effected in 1904 by Alexander McKenzie;[2] and since then many other asymmetric syntheses of an analogous kind have been carried out.[3] Thereby it has been fully proved that new asymmetric molecules can be

---

[1] *Stereo-chemistry and Vitalism* (*British Association Reports*, 1898).

[2] *J. Chem. Soc.*, 1904, **85**, 1249. A. McKenzie was Head of the Chemistry Department, Birkbeck College, London, 1905–13, and Professor of Chemistry, University College, Dundee, 1914–38.

[3] See Patrick D. Ritchie, *Asymmetric Synthesis and Asymmetric Induction* (Oxford University Press).

formed under the directing influence of previously existing asymmetric molecules; and one can conceive the possibility that once even a relatively simple asymmetric molecule has been formed, more and more complex asymmetric molecules might in the course of ages be built up. But what force directed the first asymmetric synthesis? How did the first asymmetric, optically active molecule arise?

While Pasteur recognized the necessity for the existence of asymmetric forces "at the moment of the elaboration of natural organic products," he conceived the possibility that such asymmetric forces might lie outside the living organism and "reside in light, in electricity, in magnetism, or in heat"; and van't Hoff, also, in 1894, made the suggestion that the formation of optically active compounds might take place under the directive action of right or left circularly polarized light.[1] This suggestion was eagerly seized upon, and during the following years many attempts were made to effect an asymmetric synthesis under the action of circularly polarized light. All, however, met with failure until, in 1929, Werner Kuhn and E. Braun,[2] and later Kuhn and E. Knopf,[3] of the University of Heidelberg, attained success. By the photochemical decomposition of a complex amide an optically active compound was obtained, the rotatory power of which varied in sign according as the light was dextro- or laevo-circularly polarized. In 1930, also, an analogous asymmetric photochemical decomposition was effected by Stotherd Mitchell,[4] of the University of Glasgow; and, in 1938, M. Betti and E. Lucchi announced that, by the action of circularly polarized light on a mixture of $CH_2{:}CHMe$ and chlorine or bromine (X), optically active $CH_2X{\cdot}CHXMe$ was obtained.

Such experiments are of the highest interest and importance for they demonstrate the possibility of an asymmetric synthesis under the directive action of a purely physical force. Moreover, since it was shown by Alfred Byk in 1904, that dextro-circularly polarized light predominates at the earth's surface in reflected sunlight,[5] it would appear that an unsymmetrical form of photochemical energy exists in nature under the directive action of which an asymmetric synthesis could take place. The possibility that the

---

[1] Van't Hoff, *Die Lagerung der Atome im Raume.*
[2] *Naturwiss.*, 1929, **17**, 227.
[3] *Naturwiss.*, 1930, **18**, 183; *Z. physikal. Chem.*, 1930, B, **7**, 292.
[4] *J. Chem. Soc.*, **1930**, 1829.
[5] *Z. physikal. Chem.*, 1904, **49**, 641; *Ber.*, 1904, **37**, 4696.

formation of the first asymmetric molecule took place under such directive influence is thereby established; the proof that it did in fact so take place cannot of course be given.

The occurrence of optical activity in an organic molecule, as was pointed out by Pasteur, is associated with a molecular asymmetry; but the classical theory does not enable one to state which of the two mirror images represents the dextro- and which the laevo-rotatory form. For a long time the problem of determining the absolute configuration of optical antipodes seemed insoluble but, although it originally seemed incapable of doing so, X-ray analysis has enabled this problem to be solved.[1]

An interesting new development in stereochemistry has been the discovery of the so-called clathrate compounds.[2] These are seemingly crystalline molecular compounds, corresponding to a definite formula—e.g., three molecules quinol to one molecule methyl alcohol—in which there is, however, no chemical combination between them. The constancy of the formulae depends on the fact that one molecule fits snugly—though without chemical combination—into the interstices of the crystalline structure of the other.

It has long been known that light energy may be converted into chemical energy, and as a result of the earlier investigations of the influence of light on chemical reactions it was recognized even in 1818 by Grotthuss and more fully in 1841 by John William Draper (1811–82), Professor of Chemistry in the Medical School of the University of New York, that only such rays as are absorbed are effective in producing chemical change.[3] The systematic investigation of photochemical change was begun in 1854 by Bunsen and Roscoe; and from their investigation of the influence of light on the combination of hydrogen and chlorine they stated the law that the amount of photochemical change is proportional to the amount of light energy absorbed.[4] On the basis of this law, from which, however, deviations were later frequently found, *actinometers*, or instruments for measuring the intensity of light, were constructed.

During the last quarter of the nineteenth century not a little attention was paid to the investigation of reactions brought about or catalytically accelerated by light, and much information of a

[1] J. M. Bijvoet, *Endeavour*, 1955, **14**, 71. [2] H. M. Powell, *Endeavour*, 1950, **9**, 154.
[3] *Phil. Mag.*, 1841, **19**, 195 *et seq.*
[4] *Ann. Physik*, 1855, **96**, 373; 1857, **100**, 43, 481; **101**, 235; 1859, **108**, 193; Ostwald's *Klassiker*, Nos. 34 and 38.

varied kind collected; but it was not until about 1912 that any notable advance was made in the general theory of photochemical reactions. In that year a new era in photochemical investigation was inaugurated by Albert Einstein (1879–1955), at that time Professor of Physics in the University of Prague and, later, Professor of Theoretical Physics, Princeton University, who, on the basis of Planck's quantum concept, deduced the *law of photochemical equivalence*. According to this law, when a substance undergoes photochemical reaction each molecule absorbs one quantum of radiation, represented by $h\nu$ where h is Planck's elementary quantum of action ($6 \cdot 62 \times 10^{-27}$ erg-seconds), and ν is the frequency characteristic of the absorbing molecule. If, therefore, the total energy of radiation absorbed by 1 gram-molecule is represented by U, the law of photochemical equivalence can be expressed by the equation, $U = N.h\nu = N.hc/\lambda$, where N is the Avogadro number, c is the velocity of light ($2 \cdot 998 \times 10^{10}$ cm. per sec.), and λ is the wavelength of the absorbed radiation in Angström units.[1] From this it follows that the energy of light of short wave-length is greater than that of long wave-length. This deduction is borne out by the fact that in many photochemical reactions ultra-violet light is more effective than visible light of longer wave-length.

According to the Einstein generalization, the first stage in a photochemical reaction is the absorption of one quantum of energy by one molecule, and if the process is purely photochemical the change will be in accordance with the Einstein law; but secondary reactions which are independent of the action of light may cause deviations from this law. The Einstein law, therefore, although not adequate in itself, has been found to form a useful basis for the interpretation of photochemical phenomena.

In the case of photochemical reactions in which there is an increase of free energy, there must be an absorption of considerable amounts of energy of radiation. Such reactions, since they make possible the conversion of light energy into chemical energy, are of very great importance. Thus the photosynthetic processes which take place in the green leaves of plants make possible the conversion of carbon dioxide and water into sugars, starch and other complex organic compounds through absorption of the energy of sunlight.[2]

Elucidation of the mechanism of photosynthesis is essentially

---

[1] The value of λ in A.U. must be multiplied by 10^{-8} to obtain the value in cm.

[2] R. Willstätter and A. Stoll, *Untersuchungen über die Assimilation der Kohlensäure*, 1918.

a matter for the biochemist and it is in any case too complex to explore in any detail here. Nevertheless, the process is of such fundamental importance to all life as we know it—it is estimated that 10^{11} tons of organic carbohydrate are manufactured by plant cells every year—that some mention must be made of recent research.[1]

In the study of photosynthesis much progress has been made since radioactive carbon (C^{14}) became generally available as a tracer. It appears that the first stage is a combination between a pentose sugar (ribulose) and carbon dioxide. The product is a carboxy-acid which is then reduced by hydrogen derived from water. This eventually leads, through a complex series of enzymatic reactions, partly to hexose sugars and partly to regeneration of ribulose, which maintains the cycle. The overall reaction requires about 112 kcal/mole and this is derived from sunlight—absorbed mainly by chlorophyll.

$$CO_2 + H_2O \xrightarrow[\text{112 kcal}]{\text{sunlight}} O_2 + (CH_2O)$$

Until recently, it was supposed that chlorophyll was the only pigment involved in photosynthesis, but it now appears that other plant pigments, such as yellow carotenoids, also play a minor part in the necessary energy transfer. Within the cell, the site of photosynthesis appears to be lamellated structures known as chloroplasts.

For reactions which, like the combination of hydrogen and chlorine, are accompanied by a decrease of free energy, light acts only as a catalyst. In such cases a small amount of light energy may produce a great chemical effect.

In recent years the study of photochemical reactions has been actively prosecuted.[2]

Although, according to the Grotthuss-Draper law only such rays as are absorbed are effective in producing chemical change, the light rays need not necessarily be absorbed by the reacting substance; a "sensitizer," which may not take part in the reaction at all, may act as absorbent. This fact was discovered[3] in 1873 by Hermann Wilhelm Vogel (1834–98), Professor of Photography in

[1] R. Hill and C. P. Whittingham, *Photosynthesis*, 1955; E. I. Rabinowitch, *Photosynthesis and related processes* (3 vols.), 1945–56; M. Calvin, *J. Chem. Soc.*, **1956**, 1895.

[2] Griffith and McKeown, *Photoprocesses in Gaseous and Liquid Systems* (Longmans); G. R. Rollefson and M. Burton, *Photochemistry and the Mechanism of Chemical Reactions* (Prentice-Hall).

[3] *Ber.*, 1873, **6**, 1302.

the Technical High School, Berlin, and has received important application in the production of panchromatic photographic plates. By the addition of suitable dyes the silver halide, which is sensitive chiefly to rays of short wave-length, may be made sensitive also to light of long wave-length. In recent years, the photochemistry of silver halides has been very closely studied,[1] largely because of its fundamental importance to the economically very important photographic industry.

Finally, mention must be made of electron spin resonance and nuclear magnetic resonance, both of which are not only of fundamental importance in atomic theory but have proved of outstanding value in structure investigations. Neither the original quantum theory, nor the theory of wave mechanics which followed it, could explain the multiple structure of spectral lines, such as the doublets in the alkali metal spectra. In 1925, Uhlenbeck and Goudsmit attributed this to electron spin. On this basis a new quantum number s is introduced, corresponding to clockwise or anticlockwise spin and having two possible values, namely $+\frac{1}{2}$ or $-\frac{1}{2}$. Hyperfine spectral structure can be attributed also to the spin of protons and neutrons within the nucleus; the nuclear and electron spins can be opposed or supplementary. In 1946, Purcell and Bloch introduced the method of nuclear magnetic resonance which has since proved an immensely powerful tool for elucidating structural information.

[1] J. W. Mitchell and N. F. Mott, *Phil. Mag.*, 1957, 1149.

CHAPTER THIRTEEN

THE DEVELOPMENT OF INDUSTRIAL CHEMISTRY

IN the preceding pages the aim has been mainly to indicate how the general form and design of the edifice of chemical science, as revealed by the hypotheses and theories which give co-ordination and meaning to the data of experiment, have been developed; and to trace the main paths along which chemists have advanced to a knowledge of the nature of matter and of the changes which it undergoes. From the very dawn of civilization, however, man has sought to turn his knowledge to practical account and to make use of the different forms of matter by which he is surrounded and of the changes which he is able to bring about in matter, in order to increase the comfort and to improve the amenities of everyday life. The extraction and working of metals, the production of alcohol by fermentation, the manufacture of leather, faience, glass and enamel, the extraction of vegetable oils and alkaloids, and the dyeing of fabrics, are among the arts and crafts known even to the early Egyptians; but, until near the end of the eighteenth century, the various processes by which man sought to utilize and transform the natural products had almost all been developed empirically, on a basis of knowledge gained by unorganized experiment or casual observation, uncontrolled by scientific theory.

During the past hundred years, however, the rapid development of chemical science, both theoretical and practical, has not only brought about a marked change in many industrial processes but has led to the development of many important new industries and the production of a multitude of new and useful materials. This industrial development has been especially rapid in all countries during the present century and in the years following the First World War. At the end of our survey of the growth of chemical theory, therefore, an attempt will be made to indicate briefly how some, at least, of the discoveries of the scientific laboratory have found application to the advance of industry and the enhancement of the social and economic wellbeing of man.[1] Some of the

[1] See A. Findlay, *Chemistry in the Service of Man.*

273

great organic chemical industries—production of dyes, etc.—have already been discussed.

Perhaps the oldest and most important of the chemical industries is that dealing with the extraction of metals from their ores, and this industry has grown and developed to an extraordinary extent during the past hundred years. It is, moreover, during that time that the processes of extraction and alloying of metals, processes which frequently involve problems in chemistry and chemical equilibrium of a very complex character, have been made the subject of scientific investigation and of exact control. Such investigations have led not only to a more complete understanding but also to the improvement of the processes of manufacture.

While it is not possible to discuss in detail methods of extracting metals from their ores,[1] attention may be drawn to the fact that whereas the metals commonly in use down to the nineteenth century numbered only some seven or eight—metals known even in ancient times—not a few other metals, as a result of scientific investigation during the past hundred years, either have been discovered or have acquired industrial importance. Such metals are, for example, aluminium, magnesium, nickel, manganese, chromium, molybdenum, tungsten, vanadium, beryllium, uranium, titanium, zirconium, and tantalum.

Aluminium, although isolated by Wöhler in 1827, remained a rare and costly metal until, in 1886, Charles Martin Hall (1863–1914), a graduate student of Oberlin College, Ohio,[2] and soon after him Paul L. V. Héroult (1863–1914), a French electro-chemist, showed that the metal could be obtained by the electrolysis of a solution of purified bauxite (aluminium oxide) in molten cryolite. It is by this means that the entire world's production of the metal is now obtained.

It was also by an electrolytic process, the electrolysis of fused magnesium chloride, that magnesium was first isolated by Sir Humphry Davy in 1808, and is still produced. Until recently carnallite from the Stassfurt salt deposits was the main source of magnesium chloride, but processes have been developed in America

---

[1] One of the most important advances in the winning of metals was the introduction in 1887 of the MacArthur-Forrest cyanide process for the extraction of gold, which was worked out by the industrial chemist, J. S. MacArthur, and two brothers Forrest, medical practitioners in Glasgow. By this process the output of gold was greatly increased and the life of many mines prolonged.

[2] Later Vice-President of the Aluminium Company of America.

whereby magnesium hydroxide, which can then be converted into the chloride, is successfully produced from sea-water. The metal is also produced by reducing the oxide, obtained from the carbonate (magnesite), by means of ferrosilicon. Magnesium has found rapidly expanding use as an industrial metal, the lightest of all industrial metals, with a specfic gravity of only 1·74. Sodium, used in certain chemical industries, is produced by the electrolysis of fused sodium hydroxide.

The production and application of the metal nickel were greatly stimulated by the introduction of a novel process discovered, in 1890, by Dr. Ludwig Mond[1] of Brunner, Mond and Co., in England. The nickel ore, after conversion to the oxide, is heated with producer gas in order to reduce the oxide; and carbon monoxide, under pressure, is passed over the gently heated reduced metal. Volatile nickel carbonyl is thus formed, and is passed into a chamber where it is decomposed by heat with deposition of pure metal and liberation of carbon monoxide.

Whereas many metal oxides can be reduced fairly readily by means of carbon or hydrogen, other more stable oxides are best reduced by means of aluminium, the affinity of which for oxygen is very great. In the process introduced by Hans Goldschmidt (1861–1923), an industrial chemist of Essen, in 1905, a mixture of the metal oxide and aluminium powder is heated locally by means of burning magnesium ribbon in contact with barium peroxide. The aluminium then reduces the metal oxide with the evolution of a very large amount of heat. This process finds application, more especially, in the thermit process for the welding of steel rails and for the reduction of the oxides of manganese and chromium, so that these metals, formerly obtainable with difficulty, are now produced economically and with ease.

Probably the most notable development in metallurgy during the past hundred years has taken place in the scientific investigation, manufacture and application of alloys, more especially nonferrous alloys. On the scientific side, the introduction of the microscopic examination of the structure of metals led to the recognition of the fact that the properties of a metal or alloy depend very greatly on the crystalline structure or texture; and it became possible to correlate the changes which take place in the physical properties (ductility, tensile strength, etc.), of metals and alloys with change in the crystalline structure. For this development we are indebted

[1] Mond, Langer and Quincke, *J. Chem. Soc.*, 1890, **57**, 749.

mainly to Henry Clifton Sorby (1826–1908), of Sheffield, a scientific worker of private means who, in 1863, devised the technique of the preparation and microscopic examination of metal sections, the crystalline structure being revealed by treatment of the polished surface of the metal with a suitable etching liquid. In recent years pure metals and alloys have also been subjected to examination by means of X-rays and the electron microscope, and a fuller knowledge of their crystalline structure and of the spacing of the atoms in the crystals has been obtained.

The more systematic and comprehensive investigation of alloys was also profoundly stimulated by the work of Hendrik Willem Bakhuis Roozeboom, who, in 1887, introduced to chemists the great generalization known as the Phase Rule (p. 100), and made its practical applicability to the study of alloys apparent. Under the guidance of the Phase Rule, the nature of the solid phase (alloy) separating out from molten mixtures of metals has been very thoroughly and systematically investigated.[1] In this way very many alloys, possessing properties of the highest value, have been discovered; and investigation has shown that the crystalline texture and therefore the physical properties of an alloy may be markedly altered not only by additions (sometimes very small) of other substances, but also by mechanical and thermal treatment.

A notable development on the practical side of alloy manufacture was the introduction by the English ironmaster (Sir) Henry Bessemer, in 1855, of his process of steel manufacture, a process in which the oxidation of the impurities present in cast iron (silicon, sulphur, phosphorus) was effected by blowing air through the molten metal. This was followed in 1863 by the introduction by (Sir) William Siemens of the open-hearth process, in which oxidation was effected by haematite (ferric oxide). Of scarcely less importance was the lining of the converter or hearth with a basic material, magnesite or dolomite, with which the oxidized phosphorus combined to form a slag, known as *basic slag*, a material for which an outlet was found as a valuable phosphatic fertilizer in agriculture. The invention of the basic lining, which came into use about 1879 and which we owe to Sidney Gilchrist Thomas (1850–1885), a magistrate's clerk in London, has made available for steel manufacture hundreds of millions of tons of phosphorus-containing

---

[1] See, for example, C. H. Desch, *Metallography* (Longmans) and *Intermetallic Compounds* (Longmans); A. Findlay, *The Phase Rule and its Applications* (Longmans).

ore. Initially, this was of particular value for exploiting the phosphoric ores of Loraine, acquired by Germany after the Franco-Prussian war.

An important recent development in iron smelting has been the replacement of the air blast with an oxygen-enriched blast of air and steam. This is an important new industrial use of oxygen.

Through the scientific investigation of metals and their alloys, the ever more exacting demands of the engineer have been met; and developments in engineering practice have been made possible which not very long ago would have been impracticable. Thus, steels have been rendered non-corrodible or "stainless" by additions of chromium and nickel (H. Brearley, 1912), or made suitable for rock crushing and resistance to wear by the addition of manganese (Sir Robert A. Hadfield, 1882); and steels which retain their "temper" at high temperatures and are suitable for the high-speed cutting of metals are obtained by the addition of tungsten and molybdenum to a chrome steel (L. Aitchison, 1900). Steels (*e.g.*, invar, platinite) containing 36–46 per cent of nickel have a very small coefficient of expansion with heat (C. E. Guillaume, 1896), and are used for measuring-rods, surveyors' tapes and pendulums of clocks, and for sealing into glass. The alloy, duralumin, also, which combines lightness with great strength and has proved of great importance in the construction of aeroplanes, is obtained by alloying aluminium with copper (4 per cent), with small additions of magnesium and manganese; and magnalium, an alloy of aluminium and magnesium (1–2 per cent), equals brass in strength. Many other alloys of practical utility have been invented (such as the "ultra-light" alloys, containing 90 per cent or more of magnesium), and have exercised a great influence on industrial development in many directions, especially in the aircraft industry.

Among the oldest and most important of the chemical-manufacturing industries are those which are concerned with the manufacture of acids, alkalis, bleaching materials, etc.; and at the centre of these is the manufacture of sulphuric acid, which, even yet, is the most important chemical reagent in industry. It is by far the most important acid known and is used in the production of hydrochloric acid, nitric acid and most other acids; in the production of ammonium sulphate and of superphosphate, for use in agriculture; in the refining of petroleum; in electric storage batteries; in the manufacture of glucose; in the dyeing, calico printing and tanning industries; and in the manufacture of many other industrial products. In fact, there is scarcely an industry or trade in which it does not find application. Demand for sulphuric acid has long been accepted as a useful index of industrial activity.

Sulphuric acid was, even in the eighteenth century, being produced by a process essentially the same as that in use at the present day. In the lead-chamber process, sulphur dioxide is oxidized by atmospheric oxygen under the catalytic action of nitric oxide which is formed by the reduction of nitric acid or by the oxidation of ammonia (p. 286). According to the explanation put forward by the French chemist, Péligot,[1] in 1844, and which may still in essence be accepted, the nitric oxide combines with oxygen to form nitrogen dioxide, and this oxide reacts with sulphur dioxide in presence of water to form sulphuric acid, with regeneration of nitric oxide. In order to avoid the loss of the valuable oxides of nitrogen, Gay-Lussac, in 1827, suggested an absorption tower into which the gases issuing from the lead chamber should be passed, and this tower came into use in 1835. In the Gay-Lussac tower the nitrogen dioxide is absorbed in sulphuric acid which trickles over the coke with which the tower is filled. In order to recover the oxides of nitrogen, the "nitrated" sulphuric acid formed in the Gay-Lussac tower is allowed to trickle over flints contained in another tower introduced at Newcastle-on-Tyne in 1861 by the English chemical manufacturer, John Glover (1817–1902). Into this tower are passed the hot gases from the pyrites burners, and these not only carry the oxides of nitrogen with them into the lead reaction chamber, but serve also to concentrate the acid in the tower.

Although the sulphur dioxide required for the manufacture of sulphuric acid was at first obtained by burning sulphur (produced mainly in Sicily), the increase in the price of sulphur following on the introduction in 1835 of the monopoly for the exportation of Sicilian sulphur led to the increasing use of iron pyrites and other sulphur-containing ores; but, in America, sulphur again came into use in the nineties, when the deposits of sulphur in Louisiana and Texas were exploited by the process introduced by the American engineer, Herman Frasch, in 1891. In more recent years, also, the catalytic oxidation of ammonia has come increasingly into use for the production of the nitric oxide in place of the nitric acid produced by the action of sulphuric acid on sodium nitrate.

As far back as 1817, it was suggested by Sir Humphry Davy that

[1] *Ann. Chim.*, 1844 [3], **12**, 263.

platinum might be employed to accelerate the oxidation of sulphur dioxide by oxygen, and in 1831 Peregrine Phillips,[1] a vinegar manufacturer of Bristol, attempted to make practical application of this suggestion. From 1875 onwards sulphur trioxide was produced by various manufacturers in England and Germany by this so-called "contact process," for the production of fuming sulphuric acid or "oleum," formerly produced at Nordhausen by heating ferrous sulphate (Nordhausen sulphuric acid). As, however, sulphur or sulphuric acid itself, decomposed on hot brick surfaces, was used as the source of sulphur dioxide, the fuming sulphuric acid formed was too costly except for special purposes. It was only after 1897, when a cheap supply of "oleum" became necessary for the manufacture of synthetic indigotin (p. 144), that the contact process was finally put on an economic basis, through the scientific investigation of the process and the elimination of the "poisoning" of the catalyst by arsenic.[2] The process is now largely used, platinum and, more recently, vanadium compounds[3] being employed as catalyst.

The production of sulphuric acid received a powerful stimulus through the introduction of the Leblanc soda process. Previous to the nineteenth century the carbonate of sodium (soda), required more especially for the ancient industries of glass and soap manufacture, was obtained mainly from the soda lakes of Egypt and elsewhere, or from the ash of the saltwort. This plant was grown very largely along the coast of Alicante, having been introduced into Spain by the Saracens. The ash obtained by burning the saltwort, and sold under the name of *barilla*, contained up to 15–20 per cent of sodium carbonate. The increasing demand for soda, however, led the French Academy in 1775 to offer a prize for a process of converting sodium chloride into sodium carbonate; and although the prize was never awarded, one solution of the problem was reached by the French chemist, Nicolas Leblanc (1742–1806),[4] who, in 1791, under the patronage of the Duke of Orleans, opened a factory at St Denis, near Paris, for the manufacture of soda.

Owing to the disturbed conditions prevailing in France after the Revolution, it was in England, after the abolition of the salt tax in

[1] See E. Cook, *Nature*, 1926, **117**, 419.

[2] R. Knietsch, *Ber.*, 1901, **34**, 4069. The poisoning of the catalyst by impurities had been discussed by Squire and Messel in 1876 (see Miall, *A History of British Chemical Industry*, p. 32).

[3] See J. Alexander, *J. Soc. Chem. Ind.*, 1929, **48**, 871; H. N. Holmes and A. L. Elder, *J. Ind. Eng. Chem.*, 1930, **22**, 471.

[4] T. S. Patterson, *Chem. and Ind.*, **1943**, 174.

1823, that the Leblanc process was mainly developed. The real founder of the soda industry was James Muspratt (1793–1885) who, born in Ireland, erected a factory in Liverpool and began the manufacture of sulphuric acid and soda in 1823. The industry had many difficulties, such as excessive production of hydrochloric acid and the accumulation of noxious alkali waste (calcium sulphide) to contend with; but these difficulties were finally overcome.

The handicap imposed by the excessive production of hydrochloric acid was surmounted in 1868 by Henry W. Deacon (1822–1877) by the atmospheric oxidation of hydrogen chloride in presence of copper chloride as catalyst (Deacon's process). With the passing of the Leblanc process, Deacon's process became obsolete. It is interesting to note that with by-product hydrogen chloride once again plentiful Deacon's process, using modern techniques of catalysis, has been revived. The chlorine so produced was used in the manufacture of bleaching powder (p. 282) for which there was a considerable demand. Manganese dioxide, also, had been used for the oxidation of hydrochloric acid, but the loss of expensive manganese made this process uneconomical. In 1869–70, however, Walter Weldon (1832–85), a journalist and amateur of chemistry, showed that by treating the liquors containing manganese chloride with calcium carbonate and excess of milk of lime and then blowing air through the hot mixture, manganese dioxide was regenerated and could be used for the oxidation of a further quantity of hydrochloric acid.[1] By the introduction of this recovery process the cost of bleaching powder and consequently the cost of paper and textiles were reduced, and the Leblanc process was given a new lease of life.

For the recovery of sulphur from alkali waste (calcium sulphide), Alexander Macomb Chance (1844–1917),[2] a chemical manufacturer at Oldbury, in 1887, improved a process suggested by C. F. Claus in 1883. By passing carbon dioxide (flue gases) through a suspension of alkali waste in water, hydrogen sulphide was formed; and when this was passed with a small amount of air into a kiln containing oxide of iron as catalyst, the hydrogen of the hydrogen sulphide was burned to water and the sulphur was liberated. This sulphur was then used for the manufacture of sulphuric acid, itself required in the first stage of the Leblanc process. By this process a material which was a useless waste product and the accumulation of which was a public nuisance became a source of profit.

[1] *Chem. News*, 1870, **22**, 145.   [2] *J. Soc. Chem. Ind.*, 1888, **7**, 162.

Although the Leblanc process grew and flourished during the first sixty or seventy years of the nineteenth century, a rival process, the ammonia-soda process, was thereafter developed, under the attacks of which the Leblanc process finally succumbed. Even before 1811 Augustin Jean Fresnel (1788–1827), better known as a physicist than as a chemist, discovered that sodium chloride reacts with ammonium bicarbonate with production of sodium bicarbonate and ammonium chloride; and in 1838 H. G. Dyar and J. Hemming, in England, attempted unsuccessfully to apply the reaction on an industrial scale. The fundamental problem was to limit the loss of volatile expensive ammonia. In 1863 the Belgian industrial chemist, Ernest Solvay (1838–1922), successfully developed the process and showed that soda could thereby be produced more cheaply and in a purer form than by the Leblanc process. Solvay's principal contribution was that of the carbonating tower by which the process was made continuous. It is by the ammonia-soda process that sodium carbonate is now almost entirely manufactured, although a small amount is produced by the action of carbon dioxide on caustic soda.

The conversion of sodium carbonate into sodium hydroxide or caustic soda by the action of slaked lime has been known from earliest times, and until about 1890 all the caustic soda was produced by that process. Even at the present day most of it is still so produced. As early as 1851, however, it had been suggested by Charles Watt, in England, that sodium hydroxide might be produced industrially by the electrolysis of brine,[1] but it was only after the invention of the electric dynamo that such a process could hope to become economically practicable. Various difficulties had to be surmounted, so that it was not till 1890 that the electrolytic process was definitely introduced as an industrial process by Griesheim-Elektron Works in Germany. Since then various types of electrolytic cell—e.g., Hargreaves-Bird cell (1892), Castner-Kellner cell (1892)—have been introduced and the process improved. Owing, however, to the fact that the demand for caustic soda is greater than that for chlorine, which is also produced by the electrolysis of sodium chloride, a certain limit is set to the extension of that process; and in spite of the discovery of new uses for chlorine (e.g., in the manufacture of dyes and insecticides), a substantial part of

---

[1] That an alkaline solution is produced at the cathode when a solution of sodium chloride is electrolysed had been observed by William Cruickshank in 1800 (*Nicholson's J.*, 1801, **4**, 187).

the caustic soda is still produced by the older chemical process. In order to utilize the electrolytic chlorine, it has been found economical in some cases to combine it with hydrogen, which is also a by-product of electrolysis, to form pure hydrochloric acid.

It has been pointed out that one of the difficulties with which the Leblanc process had to contend was the over-production of hydrogen chloride, and the process was saved largely through the introduction of the Deacon process for the conversion of hydrogen chloride to chlorine, for which a growing demand existed, and by the Weldon process for the recovery of manganese. As early as 1785 the French chemist, Berthollet, introduced chlorine as a bleaching agent, and thereby revolutionized the textile industry. The bleaching liquor, *eau de Javelle*—a liquor containing potassium hypochlorite—which Berthollet manufactured near Paris, was superseded by the cheaper and more convenient bleaching powder, introduced by Charles Tennant, of Glasgow, in 1798. During the nineteenth century the manufacture of bleaching powder, increasingly required by the textile industry, developed into a large and prosperous industry.

The production and commercial supply of liquid chlorine in cylinders and tank wagons has led in large measure to the supersession of bleaching powder by the element itself, for purposes of bleaching, purification of drinking-water, treatment of sewage, etc.

It is to Justus von Liebig that we owe the first clear perception of the important relations between chemistry and agriculture, and it was he who, in 1840, first pointed out that plants do not subsist merely on the water and carbon dioxide which they abstract from the air and the soil, but that they have to be fed with the elements necessary for the building up of their structures in a form in which they can be assimilated by the plants. Although no fewer than about thirteen elements are required by plants, the most important apart from the carbon, hydrogen and oxygen which are derived from water and atmospheric carbon dioxide, are nitrogen, phosphorus and potassium.

Until well after the opening of the nineteenth century, agriculture depended for its supply of nitrogenous fertilizers on waste animal and vegetable matter, but after about 1830 a supply—small at first but, after the eighties, rapidly increasing—of Chile saltpetre became available, and this, together with the ammonium sulphate derived from the distillation of coal to make gas, was the main source of combined nitrogen used in agriculture. And the supply appeared to be adequate to the demand. In 1898 nearly 300,000 tons of nitrogen in a combined state was used for fertilizer purposes, about two-thirds of this being in the form of Chile saltpetre. In that year Sir William Crookes pointed out the danger of a wheat shortage unless the yield of the soil could be greatly increased by intensive cultivation, which, in turn, would apparently exhaust in a few years all the known sources of combined nitrogen; and he urged that chemists should turn their attention to devising methods for the "fixation" or bringing into combination of atmospheric nitrogen. So successful have chemists been in doing this that it is now with difficulty that the suppliers of Chile saltpetre can compete with the manufacturers of synthetic nitrogen compounds.

The first successfully to solve the problem of the fixation of atmospheric nitrogen on an industrial scale were two Norwegians, Kristian Birkeland (1867–1917), Professor of Physics, University of Christiania (Oslo), and the engineer, Samuel Eyde.[1] Applying the observation made by the Hon. Henry Cavendish in 1784, that when an electric discharge is passed through air oxides of nitrogen are formed, Birkeland and Eyde in 1903 made use of an electric arc in order to obtain the high temperature necessary for the combination of nitrogen and oxygen to form nitric oxide.[2] This oxide was then caused to combine with oxygen to form nitrogen dioxide which, when absorbed in water, gives nitric acid. For convenience of transport and handling, the nitric acid was neutralized with limestone, and the calcium nitrate obtained was used as a fertilizer under the name of Norwegain saltpetre. This direct production of nitric acid and nitrates from the air depended for its success on cheap electric power and was carried out mainly, although not exclusively, in southern Norway. After about twenty-five years the Birkeland and Eyde process was, in 1928, forced to surrender to a more efficient rival—the direct synthesis of ammonia.

In 1892 the French chemist, Moissan, had found that when a mixture of lime and carbon is heated to a high temperature in an electric furnace, a compound calcium carbide is formed; and the same discovery was made in the following year by T. L. Willson, in America. When water acts on calcium carbide, acetylene is

---

[1] *J. Ind. Eng. Chem.*, 1912, **4**, 771.
[2] The reaction $N_2 + O_2 = 2NO$ is an endothermic reaction, and the proportion of nitric oxide in the equilibrium mixture increases with rise of temperature.

formed; and works for the manufacture of carbide were erected
in various countries where cheap power was obtainable, in the
expectation, not realized, that acetylene had a great future before it
as an illuminant. Considerable quantities of acetylene, nevertheless, are used for the cutting and welding of metals by means of the
oxyacetylene blowpipe, and acetylene is also used for the production of acetic acid (p. 293), ethyl alcohol, butyl alcohol, butadiene
(p. 180), and other substances.

In 1898 the German chemist, F. Rothe, working as assistant to
Professor Adolf Frank and Dr. N. Caro, of the Technical High
School, Charlottenburg, found[1] that when nitrogen is passed over
calcium carbide at a temperature of about 1,000° C., combination
takes place with formation of calcium cyanamide, $CaCN_2$, a compound which, it was found, could be used as a nitrogenous fertilizer.
The process was patented by Frank and Caro, and the material
was put on the market under the name of *nitrolim* or *lime nitrogen*.
Moreover, when calcium cyanamide is treated with superheated
steam it is decomposed with formation of ammonia. A means was
thus given of obtaining an abundant supply of ammonia from
atmospheric nitrogen for use as a fertilizer or for conversion into
nitric acid (p. 285).

Since the success of both the preceding processes depended on
cheap electric power, their development was restricted. In 1912,
however, as the result of ably directed and painstaking endeavour,
a process was devised whereby ammonia could be synthesized from
nitrogen and hydrogen; and this process, not being dependent on
exceptionally cheap electric power, was capable of unlimited
expansion and development in different countries.

Various attempts had been made from the time of Regnault in
1840 onwards to produce ammonia by the combination of nitrogen
and hydrogen; but no advance was made towards a practicable
synthesis of ammonia until, in Germany, Walter Nernst and F.
Jost,[2] and later, Fritz Haber and his pupil Robert Le Rossignol,[3]
made a systematic investigation of the equilibrium between ammonia, nitrogen and hydrogen at different temperatures and under
different pressures. In accordance with the theorem of Le Chatelier, it was found that the percentage of ammonia in the equilibrium
mixture is increased by increasing the pressure under which the

[1] Frank, *Z. angew, Chem..* 1906, **19**, 835; 1909, **22**, 1178.
[2] Nernst, *Z. Elektrochem.*, 1907, **13**, 521; Jost, *Z. anorgan. Chem.*, 1908, **57**, 414;
Z. Elektrochem., 1908, **14**, 373.
[3] *Z. Elektrochem.*, 1913, **19**, 53.

gases react, and by lowering the temperature. To enable this synthetic process to take place sufficiently rapidly at the lower temperatures, a suitable catalyst had to be found,[1] and precautions taken to remove impurities which poisoned the catalyst. On the basis of these investigations a laboratory process was worked out about 1905, and by 1912 it had been developed into an industrially successful process. In this process, generally known as the Haber process,[2] the nitrogen and hydrogen are circulated over a catalyst under a pressure of 150–200 atmospheres, and at a temperature of about 500° C. In the modifications of the Haber process, introduced by Georges Claude in France in 1918, and by Luigi Casale (1882–1927), in Italy, pressures of 800–1,000 atm. are employed. For use as a nitrogenous fertilizer the ammonia is converted into ammonium sulphate, either by neutralization with sulphuric acid or by interaction with gypsum (calcium sulphate). The latter method, now very largely employed, was devised by the Badische Anilin- und Sodafabrik at Ludwigshafen, as the result of a shortage of sulphuric acid during the First World War, and consists in passing carbon dioxide through a suspension of powdered gypsum or anhydrite in a solution of ammonia. Calcium carbonate is precipitated and ammonium sulphate remains in solution. The introduction of this process materially lowered the cost of ammonium sulphate.

As a consequence of the development of the synthetic ammonia process, the amount of nitrogen compounds produced by the industrial fixation of atmospheric nitrogen now greatly exceeds the amount contributed by Chile saltpetre and by the distillation of coal.

The direct synthesis of ammonia has gained great importance owing not only to the demand for nitrogenous fertilizers and for large amounts of liquid ammonia for purposes of refrigeration, but also to the fact that ammonia can, by oxidation, be converted into nitric acid. As early as 1788 it had been found by Isaac Milner (1750–1820), First Jacksonian Professor of Natural Philosophy and President of Queen's College, Cambridge, that ammonia is oxidized to nitric acid by passing it over heated manganese dioxide,[3] and this process is stated to have been used in France

[1] Various substances were found to have a catalytic action. It is probable that the catalyst used is oxide of iron to which potassium oxide, molybdenum, aluminium oxide or other substances are added to increase the efficiency.
[2] Also called the Haber-Bosch process owing to the participation of Dr. Bosch, of the Badische Anilin- und Sodafabrik, in its industrial development.
[3] *Phil. Trans.*, 1789, **79**, 300.

for the manufacture of saltpetre during the Napoleonic wars.[1] In 1839 Frédéric Kuhlmann, of Lille, observed[2] that oxides of nitrogen are formed when a mixture of air and ammonia is passed over platinum sponge at a temperature of 300° C., and during the nineteenth century various workers studied the action of other substances in catalytically accelerating the oxidation of ammonia.

It was not, however, till 1900 that the conditions necessary for the industrial application of the catalytic oxidation of ammonia in presence of platinum were investigated in Germany by Wilhelm Ostwald and his assistant, Eberhard Brauer.[3] Based on these investigations, a plant was erected in Germany in 1909 for the production of nitric acid from ammonia—obtained at that time as a by-product of the distillation of coal—and by this process Germany was able to provide herself, during the First World War, with the nitric acid and ammonium nitrate required for use in agriculture and in the manufacture of explosives.[4] Besides platinum various substances have been found effective as catalysts for the oxidation of ammonia, and this process is employed at the present day not only for the manufacture of nitric acid but also for the production of oxides of nitrogen required for the production of sulphuric acid by the lead-chamber process.

The development of the synthetic ammonia process called for the supply of large quantities of nitrogen and hydrogen. The former gas may be obtained by the distillation of liquid air—the production of which was, in consequence, greatly increased.[5] For the industrial production of hydrogen various processes are employed, according to local conditions and circumstances. In some cases hydrogen is produced by the electrolysis of dilute solutions of caustic alkali, and in other cases by the passage of steam over red-hot iron. In 1912, also, the Badische Anilin- und Sodafabrik developed a process which later acquired great importance. On passing water-gas (essentially a mixture of hydrogen and carbon monoxide) and steam over a suitable catalyst (e.g., a mixture of

[1] Joseph Black, Lectures on Chemistry, 1803, II, 455; J. R. Partington, Nature, 1922, 109, 137.

[2] Annalen, 1839, 29, 280; Compt. rend., 1838, 7, 1107.

[3] Chem. Zeit., 1903, 27, 100.

[4] "Even if, under present conditions," wrote Kuhlmann in 1838, "the transformation of ammonia into nitric acid by means of platinum sponge and air offers no adequate advantage, still the time may come when that transformation may become economically possible."

[5] There has also resulted therefrom an increased supply of argon and of neon, two gases which are now very largely used in filling electric light bulbs and for electric signs, etc.

iron and chromium oxides with the addition of 0·5 per cent of thorium oxide), at a temperature of about 500° C., the carbon monoxide reacts with the steam, forming hydrogen and carbon dioxide. The latter gas can then be used in connection with the production of ammonium sulphate (p. 285), and for many other purposes. Today, the necessary hydrogen is being increasingly obtained by decomposition of hydrocarbons with steam.

Before 1840 the phosphorus which is essential for healthy plant growth was added to the soil chiefly in the form of crushed bones, the phosphorus being present in these as calcium phosphate. Between 1840 and 1850 (Sir) John Bennet Lawes (1814–1900)[1] showed that, by treatment with sulphuric acid, the phosphate is converted into acid calcium phosphate and is thereby rendered more readily soluble in the soil, and therefore more readily available as a plant food. Such treatment had, in fact, been previously recommended by Liebig in 1839–40, and had been suggested even in 1835 by a German, Escher. The first to use this process commercially was James Murray (1788–1871), a Dublin physician, who took out a patent for the manufacture of superphosphate in 1842. This treatment, moreover, as Murray showed, could be applied with excellent results to mineral phosphates which, ordinarily too hard and insoluble, are thereby rendered available as fertilizers. Since that time the production of superphosphate (a mixture of acid calcium phosphate and calcium sulphate) from mineral phosphates has developed very greatly and the process of manufacture has been greatly improved; and superphosphate now constitutes the chief phosphatic fertilizer. As phosphatic fertilizer there is, however, also used finely ground basic slag (p. 276), which has proved specially valuable on pasture land. Later, also, the introduction of the Liljenrot process, whereby, in presence of a catalyst (e.g., nickel), steam reacts with phosphorus with production of phosphoric acid and hydrogen, led to the production of ammonium phosphate as a fertilizer.

The salts of potassium which are used in agriculture and which were formerly mainly derived from the wood ashes of lumber camps are now nearly all of mineral origin; and the chief sources of these were, for many years, the great deposits of salts near Stassfurt, in Germany, and the deposits which were discovered in Alsace in

[1] Landowner, chemical manufacturer and founder, in 1843, of the Agricultural Experimental Station at Rothamstead, Hertfordshire. For his services to agricultural science he was created a baronet by Queen Victoria in 1882.

1904. In the deposits at Stassfurt, which have been worked since about 1860, a number of different salts occur, and the extraction and purification of the potassium salts by crystallization from solution were, in later years, guided by the long and laborious investigations of the solubilities of the salts in presence of one another carried out, more especially, by van't Hoff and his pupils from 1897 onwards.

Very large quantities of potassium salts are now being produced in America from underground deposits of sylvinite in Texas and near Carlsbad, New Mexico, as well as from the brines of Searles Lake, California, and at Bonneville, Utah. Considerable quantities of potash salts are also being extracted from the waters of the Dead Sea, and large deposits of potassium salts have been discovered in Russia. Other minor sources of potassium salts are also known, *e.g.*, the flue dust from blast furnaces and cement works, seaweeds, minerals, etc.

The progress of civilization has depended very largely on man's ability to utilize, control and direct energy, and on his ability to obtain this energy in a concentrated form. Hence the importance of *explosives* in the operations of mining and quarrying, and in the removal of material obstructions to the constructive labours of the engineer. Almost all the explosives in use require for their production nitrates or nitric acid.

The first explosive to be used—black gunpowder—remained the only one in use until near the middle of the nineteenth century, when the first great advance in the chemistry and manufacture of explosives was made. In 1846 Christian Friedrich Schönbein (1799–1868)—a Swabian by birth and Professor of Chemistry in the University of Basle from 1829 till his death—discovered that by the action of a mixture of nitric and sulphuric acids on cotton (cellulose), gun-cotton, a nitrate of cellulose, is obtained.[1] When loose, gun-cotton burns with great rapidity, and when subjected to a shock, as when a little fulminate of mercury is caused to detonate near it, it suddenly decomposes and gives rise to a large volume of gaseous products. This explosive decomposition, moreover, can

---

[1] The production of a readily inflammable substance by the action of concentrated nitric acid on wood fibre was first observed by H. Braconnot (*Ann. Physik*, 1833, **29**, 176), and the reaction was also studied by Pelouze (*J. prakt. Chem.*, 1839, **16**, 168).
Recognition of the technological importance of the reaction and the production of gun-cotton as an explosive are due to Schönbein.

take place even when the gun-cotton is wet. The disruptive effect of gun-cotton is very great, so that while it is useful as a high explosive, *e.g.*, in torpedoes, it cannot be used as a low or propulsive explosive. In 1886, however, the French physicist, Paul Vieille, of the École Polytechnique, Paris, showed that if a mixture of gun-cotton and soluble nitrocotton is made into a paste with ether and alcohol, the gelatinized, horn-like material left after the removal of the ether and alcohol, burns smoothly and rapidly and can be used as a propellant. This was the first smokeless powder to be used for military purposes.

A further advance in the chemistry of explosives was made in 1846 by the Italian chemist, Ascanio Sobrero (1812–88), Professor of Chemistry at the Institute of Technology, Turin, who obtained the very highly explosive liquid, glyceryl nitrate ("nitroglycerine") by the action of nitric acid on glycerine. This substance was first manufactured on a commercial scale in 1862 by the Swedish chemist and engineer, Alfred Bernhard Nobel (1833–96), but was found to be difficult to handle on account of its great sensitiveness to shock. In 1866, however, Nobel discovered that if the liquid nitroglycerine was mixed with kieselguhr it could be transported and handled with comparative freedom from danger. This mixture was introduced commercially in 1867 under the name of *dynamite*. Its explosion is initiated by means of a detonator.

In 1875 a new type of explosive was produced by Nobel by mixing nitrocotton with nitroglycerine to form a tough jelly-like mass. This was known as *blasting gelatine*, and is one of the most powerful blasting explosives known. By mixing blasting gelatine with varying amounts of such materials as potassium nitrate, sodium nitrate, ammonium nitrate, wood meal, chalk, etc., various *gelignites* are produced. These are the most commonly used blasting agents in present-day use.

After the invention of blasting gelatine, attempts were successfully made by Nobel to utilize nitroglycerine as a gelatinizing agent for nitrocotton in the production of a smokeless propellant, *ballistite*; and in 1889 a mixture of gun-cotton and nitroglycerine, with a small proportion of vaseline, was introduced in England under the name of *cordite*. The invention of this smokeless powder was due to (Sir) Frederick Abel (1827–1902), Professor of Chemistry at the Royal Military Academy, and (Sir) James Dewar.

In 1885 trinitrophenol was introduced in France as a high explosive, as a filling for shells, etc. This compound, first prepared in 1771, was, on account of its bitter taste, christened by Dumas picric acid. When it came to be used as an explosive for shells, it was called by the French *mélinite*, on account of the quince-yellow colour of the molten acid. *Lyddite* and various other names have also been applied to picric acid when used as an explosive.

As a high explosive for military purposes, *trinitrotoluene* (T.N.T.) came into prominence during the First World War. A mixture of 20 per cent of trinitrotoluene and 80 per cent of ammonium nitrate was used as a filling for shells under the name of *amatol*.

The compound *cyclonite* (*cyclo*-trimethylenetrinitramine) is an explosive which in *brisance* and blasting power is greatly superior to any other explosive hitherto used for bursting charges. It was first prepared in 1899 by G. F. Henning, and is produced by the action of concentrated nitric acid on hexamethylenetetramine (or hexamine), which is, in turn, obtained by the reaction between ammonia and formaldehyde. It has the constitution

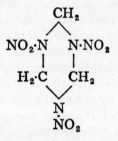

This explosive came into use during the war of 1939–45. The powerful high explosive, *torpex*, is a mixture of cyclonite, trinitrotoluene and aluminium powder.

Besides those mentioned, other explosives have been made and used for different special purposes. In recent years, of course, explosives depending for their action on physical rather than chemical processes have become increasingly important.

It has already been pointed out (p. 93) that the phenomena of catalysis, first observed more than a hundred years ago, have been intensively studied during the present century owing, largely, to the importance of catalysts in industry; and so extensive and important have been the applications of catalysts that this may almost be called the "catalytic age" in chemical industry. To a number of these applications which have in recent times acquired great importance, such as the contact process for the production of sulphuric acid, the synthesis of ammonia and the oxidation of ammonia to nitric acid, and the oxidation of the carbon monoxide of water-gas by steam, reference has already been made. There remain, however, a number of other processes of great industrial and economic importance to which attention must be given.[1]

The discovery by Sir Humphry Davy of the action of platinum in promoting oxidation reactions led to the investigation of the action of that metal in reduction processes; and in 1838 Kuhlmann showed[2] that oxides of nitrogen can be reduced to ammonia in presence of platinum sponge. Later the catalytic acceleration of other processes of reduction by hydrogen were investigated by B. Corenwinder,[3] Heinrich Debus[4] and M. P. de Wilde;[5] but it is to the scientific investigations of the Nobel Prizeman Paul Sabatier (1854–1941), Professor of Chemistry in the University of Toulouse, and of his co-workers, of whom Jean Baptiste Senderens (1856–1937) may be specially mentioned, that we owe the development of the catalytic hydrogenation processes which are of such great economic importance at the present day. In the course of investigations, carried out from 1897 onwards,[6] Sabatier showed that finely divided metallic nickel is a very efficient catalyst of hydrogenation. Many cases of hydrogenation of unsaturated organic compounds were studied, and of these the conversion of unsaturated vegetable oils to saturated fats by addition of hydrogen has found an important industrial application. By passing hydrogen through a liquid oil (olive oil, linseed oil, whale oil, etc.) at a temperature of about 250° C., in presence of finely divided nickel, hydrogenation of the oil takes place and a saturated compound, a fat, is obtained. This process was developed industrially under various patents from 1903 onwards, and is now carried on in various countries on a large scale.[7]

Not only does this process of hydrogenation of oils (hardening of oils) secure a fresh source of raw material for the manufacture of soap and of margarine, but by stimulating the cultivation of oil-producing plants it exercises a profound influence on the economic development of those countries which are suitable for such cultivation.

[1] See E. K. Rideal and H. S. Taylor, *Catalysis in Theory and Practice*; G. G. Henderson, *Catalysis in Industrial Chemistry*; T. P. Hilditch, *Catalytic Processes in Applied Chemistry*.
[2] *Compt. rend.*, 1838, **17**, 1107.   [3] *Ann. Chim.*, 1852 [3], **34**, 77.
[4] *Annalen*, 1863, **128**, 200.   [5] *Ber.*, 1874, **7**, 352.
[6] Sabatier, *Le Catalyse en Chimie Organique*; *Ann. Chim.*, 1905 [8], **4**, 319.
[7] See Carleton Ellis, *The Hydrogenation of Oils*.

Another historically important process of hydrogenation which must be mentioned is that known as *berginization,* or hydrogenation of coal and coal tar for the production of motor spirit and lubricating oils. Such processes are economic, however, only in special circumstances, such as a wartime shortage of petrol due to import difficulties, or the local availability of very cheap coal. The problem of converting coal into oil by removing the oxygen and causing the coal compounds to add on hydrogen was first solved industrially about 1935 by the German chemist, Friedrich Bergius, of Mannheim, to whom, however, the idea of such conversion had suggested itself as early as 1914.

The process of hydrogenation may be carried out in the liquid or in the vapour phase. In the liquid phase hydrogen under a pressure of about 250 atm. is passed into a paste of finely powdered coal and coal tar or heavy (creosote) oil at a temperature of about 450° C. In the presence of a suitable catalyst, hydrogenation takes place with production of gaseous hydrocarbons (methane, ethane, propane, butane[1]), and a liquid which, on being distilled, yields motor spirit, a middle oil which is used for further treatment, and a heavy oil which is used for mixing with coal.

Hydrogenation of the middle oil, just referred to, and of the creosote oils produced especially by the low-temperature distillation of coal, can be carried out in the vapour phase by passing a mixture of the vapours and hydrogen over a suitable catalyst (*e.g.,* tin) at a temperature of about 450° C. and under pressure. The product consists mainly of motor spirit, but by the choice of suitable catalysts the reactions may be directed and the nature of the products controlled.

In 1925 F. Fischer and H. Tropsch, in Germany, found that water-gas in presence of a suitable catalyst (*e.g.,* cobalt and thorium) and at a temperature of about 400° F. is converted into a mixture of liquid and gaseous hydrocarbons, from which motor spirit can be obtained. High-grade lubricating oils can also be produced from water-gas with the help of aluminium chloride as catalyst.

Such processes were of great value during the war of 1939–45, but are of no great interest—save where coal is exceptionally cheap—when natural petroleum is freely available.

Of related interest is the so-called Oxo process, for the production

---

[1] The propane and butane are separated and compressed into cylinders for use as a gaseous fuel ("Calor gas").

of alcohols, aldehydes and other oxygen-containing compounds. It consists in passing olefines over cobalt-containing catalysts in the presence of carbon monoxide and hydrogen, at a pressure of some 200 atm. The product is an aldehyde containing one carbon atom more than the original olefine, but under normal operating conditions this is reduced to the corresponding alcohol.

From the time of Boyle until 1925, methyl alcohol (methanol) or wood spirit, which is used in large quantity in the manufacture of dyes, perfumes, etc., and for the preparation of formaldehyde, was produced by the distillation of wood. This wood-distillation industry, however, received, in 1924, a serious blow when, mainly through the researches of French and German chemists, it was shown that methyl alcohol can be readily produced by passing a mixture of carbon monoxide and hydrogen (water-gas), or a mixture of carbon dioxide and hydrogen, over a suitable catalyst (zinc oxide or a mixture of zinc oxide, chromium oxide and reduced copper) under regulated conditions of temperature and pressure. The industrial production of methanol by this process is now well established in various countries. Methanol is also now manufactured by the partial oxidation of natural hydrocarbons.

Not only can one obtain hydrocarbons and methanol from a mixture of carbon monoxide and hydrogen, but by a proper choice of catalyst and by regulation of the conditions of reaction other substances (e.g., butyl alcohol) may also be obtained.

The aqueous liquor obtained by the distillation of wood contains not only methyl alcohol but also acetic acid and acetone; and for the production of these substances were catalytic processes long ago introduced. Thus in 1881 it was found by M. Kutscheroff that when acetylene is passed into a dilute solution of sulphuric acid in presence of mercuric sulphate, acetaldehyde is formed; and if air or oxygen is passed through liquid acetaldehyde in presence of manganese acetate, oxidation of the aldehyde to acetic acid readily takes place. Acetaldehyde, moreover, can not only be oxidized to acetic acid but it can also be reduced by hydrogen to ethyl alcohol in presence of a catalyst, e.g., nickel; and this process has been developed on an industrial scale.

Acetone, one of the products of distillation of wood, is produced by heating calcium acetate and by the fermentation of starch by means of a special bacterial culture, Clostridium acetobutylicum, discovered in 1910 by Auguste Fernbach (1860–1939) of the Pasteur Institute. In this process butyl alcohol is also formed. Acetone can now also be produced industrially by passing alcohol vapour and steam over an iron catalyst; by passing the vapour of isopropyl alcohol (produced from propylene which is formed in large quantities by the "cracking" of petroleum hydrocarbons) over a copper catalyst at 500° C., and by passing acetylene and steam at a high temperature over a catalyst of zinc oxide and ferrous oxide. Acetaldehyde can be made by the catalytic oxidation of ethyl alcohol or ethylene.

During the present century two great industries have developed, to which reference must be made here—the production of synthetic fibres and filaments and of synthetic resins and plastics.

Although it is stated that threads "resembling silk" were known even in the seventeenth century, George Audemars, of Lausanne, was the first to obtain a patent for the production of artificial silk from nitrocellulose in 1855; and in 1883 (Sir) Joseph Wilson Swan was the first to produce filaments by squirting solutions of nitrocellulose and the first to denitrate the filaments produced. Swan's process was adapted commercially by the French chemist, Count Hilaire de Chardonnet. In 1897 Hermann Pauly, who later became Professor of Chemistry at Würzburg, produced an artificial silk by dissolving cellulose in ammoniacal copper hydroxide and regenerating cellulose by means of sulphuric acid;[1] but the process by which most of the artificial silk is made at the present time is the viscose process introduced by the English consulting chemists, Charles Frederick Cross (1855–1935) and Edward John Bevan (1856–1921). In this process cellulose is converted into a thick vaseline-like material, known as *viscose*, by treatment with caustic soda and carbon disulphide, and the viscose is then forced through fine spinnerets into a bath containing sulphuric acid and other substances, and is thereby reconverted into cellulose. When sheets of viscose are treated in this way, *Cellophane* is obtained. The various artificial silk-like materials derived from cellulose have received the name *Rayon*.

The use of cellulose acetate for the production of artificial silk was introduced in the early years of this century. It goes by the name of *Celanese* or acetate rayon.

A filament of greatly increased tensile strength can be obtained

---

[1] The process, although patented in the name of Pauly, had been worked out by the industrialist, Max Fremery, and the engineer, Johann Urban, of the Rheinisch Glühlampenfabrik.

by hydrolysing the cellulose acetate and stretching the filament of reconstituted cellulose so as to cause the molecules to orientate themselves parallel to the axis of the filament. High-strength fibres have proved particularly valuable in the manufacture of motor-car tyres.

In the production of rayon one starts with a naturally-occurring polymer, cellulose, which is built up of a large number of smaller units (p. 168), and although it is superior to silk in lustre, it is inferior to it in elasticity, tensile strength and heat-insulating properties.

Since natural silk is a protein and has, therefore, a polypeptide structure (p. 173), W. H. Carothers (1896–1937), of the Du Pont Company in America, along with a team of collaborators, attacked the problem of producing a synthetic fibre or filament which would resemble natural silk not only superficially but also in its chemical composition and constitution, by preparing high polymers of a polypeptide nature. Many compounds were investigated and, in 1935, success was attained. A new fibre, *Nylon*, became an industrial product.

Nylon, as first produced, was obtained as a condensation product of adipic acid, $COOH \cdot (CH_2)_4 \cdot COOH$, and hexamethylenediamine, $NH_2 \cdot (CH_2)_6 \cdot NH_2$, which could then by a process of condensation be built up into a macromolecule of the structure:

$$- [CO \cdot (CH_2)_4 \cdot CO \cdot NH \cdot (CH_2)_6 \cdot NH]_n -$$

with a molecular weight of over 10,000. The compound in a molten state is forced through a small jet and the filament so formed is "cold drawn" to about four times its original length. By this means an increased degree of lengthwise orientation of the macromolecules is brought about and very fine filaments are obtained which are transparent, lustrous, highly elastic and of great tensile strength. Polymerization of amino-caproic lactam (caprolactam) gives what is called Nylon 6.

Investigations which may be regarded as a logical extension of the work of Carothers have been carried out in England by J. R. Whinfield and Dixon of the Calico Printers' Association, and have led (1946) to the production of another synthetic fibre, to which the name *Terylene* has been given. This is obtained by the condensation of terephthalic acid ($COOH \cdot C_6H_4 \cdot COOH$) with ethylene glycol ($HO \cdot CH_2 - CH_2 \cdot OH$), and polymerization of the product. Macromolecules of the constitution,

$$- [O \cdot (CH_2)_2 \cdot O \cdot OC \cdot C_6H_4 \cdot CO]_n -$$

are thus obtained. Today, Terylene is one of the most important of all the man-made fibres.

In recent years, acrylic fibres (*e.g.*, Acrilan, Orlon) have become of considerable importance, especially in knitwear and blankets. Chemically these are fibrous polymers in which the repeating unit is $-CH_2 \cdot CH(CN)-$.

One of the most remarkable developments in modern chemical industry has taken place in the production of *plastics*—synthetic solid materials of very diverse kinds which, at some stage in their production, exhibit the property of plasticity, or are capable of being moulded under pressure. Some of these materials become plastic when heated and can be moulded and remoulded by the application of heat and pressure. They are said to be *thermoplastic*. Others are *thermosetting*. They also can be moulded by heat and pressure, but, under the action of heat, they undergo chemical change and pass into a hard material which cannot be softened and remoulded by further application of heat and pressure.

To whichever group the plastic belongs, it is a compound with a large molecule. In some cases the large molecules have been built up by nature and exist in the raw material (*e.g.*, cellulose) used in the manufacture of the plastic; but, in most cases, the large molecules are formed during manufacture, either by the polymerization of smaller molecules or by a process of condensation or chemical reaction in which a large number of molecules combine with elimination of a molecule of water between each pair of molecules.

Although the earliest plastics were introduced as substitutes for naturally-occurring materials—ivory, gums and resins—they now take their place as new materials possessing their own distinctive and valuable properties. For many purposes they are superior to and have displaced wood, metal and stoneware, and have found widespread and varied use as new constructive materials.

The earliest synthetic plastics to be introduced were those derived from cellulose. In 1865 Alexander Parkes, of Birmingham, found that if nitrocellulose was mixed with alcohol and a certain amount of camphor a horn-like mass was obtained which, on warming, could be moulded by pressure. Later this was produced commercially in England under the name *xylonite*. Meanwhile, in 1869, the same process was adopted by two brothers, J. W. and J. S. Hyatt, of Newark, New Jersey, who called the material *celluloid*.

A similar thermoplastic material was produced at a later time by using cellulose acetate in place of nitrocellulose. This had the great advantage over celluloid of being non-inflammable.

In 1897, it was discovered by W. Krische, a lithographer in Hanover, and Adolf Spitteler, a chemist in Prien, Bavaria, that when casein (obtained by curdling milk with rennet) is treated with formaldehyde, a hard bone- or horn-like material, known now as *galalith*, *erinoid*, etc., is formed; and since 1900, when this material was first produced commercially in Germany, its manufacture has greatly increased. Casein plastics are widely used where great dimensional stability is not important, for example in button manufacture.

In 1872 Professor A. von Baeyer obersved that when phenol (carbolic acid) is mixed with formaldehyde a resinous material is formed, but it was not till 1908 that industrial use was made of this reaction. In that year Leo H. Baekeland (1863–1944), Hon. Professor of Chemical Engineering, Columbia University, an American chemist of Belgian birth and the inventor of "velox" photographic paper, showed that when the resinous material is heated in presence of an alkaline catalyst, it first softens and then, after a time, becomes quite hard. This artificial resin, or thermosetting plastic, known as *bakelite*, and a number of other materials of a similar kind were soon widely used for making gramophone records, for moulding articles used for a great variety of purposes, and for use in varnishes, enamels, etc.

Besides the three classes of plastics to which reference has been made, plastics derived from cellulose, from casein and the phenol-formaldehyde plastics, not a few other plastics are produced.[1] To a few only of these attention may be drawn.

Vinyl compounds, especially vinyl chloride, yield a wide range of useful plastic materials. Polyvinyl chloride is familiar in everyday life in such varied forms as floor covering, electrical insulation, leathercloth, and waterproof sheeting. Polyvinyl acetate is widely used in emulsion paints.

The amino-plastics are obtained by the condensation of urea or thiourea or their derivatives with formaldehyde and polymerization of the product. Of these beetle-ware is well known. Of greater importance are the polymerized acrylic esters, *e.g.*, methyl methacrylate, which forms a highly transparent, glass-like material

[1] V. E. Yarsley and E. G. Couzens, *Plastics* (Pelican Books); H. R. Fleck, *Plastics, Scientific and Technological*.

called *Perspex* or Lucite. This has found a widespread use, particularly for replacing glass in aeroplanes, and is also extensively employed for making an immense range of articles of everyday use.

In the presence of a trace of oxygen and when subjected to high pressure, the unsaturated hydrocarbon, ethylene, undergoes polymerization and forms a chain consisting of 1,000–2,000 — CH_2 — groups joined together. This plastic, polythene, was first produced in England in 1939. It is one of the lightest of all the plastics and floats on water. It is characterized by its toughness and flexibility, its impermeability to water, its chemical inertness and its unique electrical properties. Its tensile strength can be increased by cold-drawing. More recently, polythene has been made by the Ziegler process, in which ethylene is polymerized at atmospheric pressure in the presence of a stereospecific catalyst. From propylene has been made the corresponding polypropylene, which is cheap and light and, apart from use in mouldings, has considerable possibilities as a fibre. Fluorinated polythene has found useful applications because of its exceptional resistance to chemical attack and to heat.

Reference has been made (p. 116) to the production of silicols (or silanols) and to the condensation of these to form compounds known as silicones. Only the disilicols and trisilicols polymerize to form long chains which also exhibit cross-linkages. One of the most outstanding developments of the past few years has been the industrial production of silicones, as liquids, as semi-solids and as solids, for use as heat-resisting resins, as insulating oils for use in electrical transformers, as lubricants (the viscosity of which alters only slightly with the temperature), and as a water-repelling plastic material.[1]

One of the greatest changes in industrial chemistry in recent years is in the source of raw materials. The great majority of industrial chemicals are produced by ringing the changes on a relatively small number of raw materials—air, water, coal, limestone, salt, sulphuric acid. Of these, coal is steadily losing its position, which it has held since the earliest days of synthetic dyestuffs, as a primary source of carbon. In its place, petroleum is becoming increasingly important and already its consumption by the chemical industry is measured in millions of tons annually. So decisive is the change that is it true to say that for many processes manufacturers would

[1] See W. R. Collings, *Chem. and Eng. News*, **1945**, 1616; E. G. Rochow, *Introduction to the Chemistry of the Silicones*.

prefer petroleum to coal even if the latter were available free of charge. In almost every respect petroleum is superior as a source of carbon for industrial chemicals, hence the extraordinarily rapid growth of the petrochemical industry.

In the preceding pages it has been possible to indicate only some of the more important developments in industrial chemistry during the past hundred years; and it must constantly be borne in mind that in many other industries, including some of the most ancient industries known—fermentation industries, manufacture of glass, porcelain, soap, leather, etc.—many improvements have been introduced as a result of new discoveries in chemistry and of exact scientific control. Nothing, indeed, is more characteristic of the industrial life of all countries during the present century, and more especially since the First World War which was a stern but effective teacher, than the recognition of the futility of rule-of-thumb methods and of the vital importance for industrial progress of scientific investigation and exact control of manufacturing operations.

BIOGRAPHICAL NOTES

ANDREWS, THOMAS, was born at Belfast on December 19th, 1813. After studying chemistry under Thomas Thomson at Glasgow and under Dumas in Paris, Andrews graduated as Doctor of Medicine at Edinburgh in 1835. Returning to Belfast, Andrews engaged in medical practice and also carried out investigations in chemistry and physics. In 1845 he became Professor of Chemistry in the newly founded Queen's College, Belfast, and from this Chair he retired in 1879. He died on November 26th, 1885. Andrews carried out investigations in thermochemistry and in other departments of chemistry, and is best known for his work on the continuity of the gaseous and liquid states. He was a man of simple and unpretending manner, of remarkable skill in manipulation and of unwearying patience.

ARMSTRONG, HENRY EDWARD, born at Lewisham, Kent, on May 6th, 1848, entered in 1865 the Royal College of Chemistry, London, where Edward Frankland had just succeeded von Hofmann as Professor of Chemistry. After a year and a half of study he assisted Frankland in his work on the purification of public water-supplies and, in 1867, proceeded to Leipzig to work in Kolbe's laboratory. There he began his studies of the aromatic compounds which occupied a large part of his working life. Returning to London in 1870, Armstrong assisted Dr. Matthiessen at St. Bartholomew's Hospital Medical School for twelve years, and taught in the evening at the London Institution in Finsbury Circus. From 1884 till 1911, when he retired, he was Professor of Chemistry at the Central Technical College, South Kensington. His contributions to chemistry were concerned mainly with orientation in aromatic compounds, theory of colour in organic compounds, structure of the terpenes and nature of enzyme action. A striking personality with a disinterested love of truth, combative, impatient and self-willed, a forceful critic of new theories and an ardent propagandist of the teaching of science, he exercised a powerful influence in the world of chemistry in England for many years. He died on June 13th, 1937.

ARRHENIUS, SVANTE AUGUST, was born in the neighbourhood of Uppsala on February 19th, 1859. Educated at the Universities of Uppsala and Stockholm, Arrhenius carried out, at the latter University, investigations on the electrical conductivity of solutions, for which he was awarded the Doctor's degree. In 1886 and 1887 he worked in the laboratories of Ostwald at Riga, Kohlrausch at Würzburg, Boltzmann at Graz, and van't Hoff at Amsterdam, and thereafter became an assistant under Ostwald in the Physical Chemical Institute at Leipzig. He returned to Sweden in 1891 and became lecturer in physics and later Professor at the Technical High School there. In 1903 he was awarded the Nobel Prize for Chemistry, and in 1905 was appointed Director of the Nobel Institute at Stockholm. His genial and attractive personality made him a welcome guest everywhere. He died on October 2nd, 1927.

BAEYER, JOHANN FRIEDRICH WILHELM ADOLF VON, was born in Berlin on October 31st, 1835, and studied chemistry under Bunsen and Kekulé at Heidelberg. He later followed Kekulé to Ghent. In 1860 he returned to Berlin where, for twelve years, he taught organic chemistry at the Technical Institute, the forerunner of the Charlottenburg Technical High School. In 1872 he was appointed Director of the chemical laboratories in the newly founded University of Strasbourg and in 1875 succeeded Liebig at Munich. There, devoting himself to teaching and to the steady and enthusiastic but unhurried pursuit of knowledge, he built up his great reputation as one of the greatest of Germany's organic chemists. He died on August 20th, 1917.

BERTHELOT, MARCELIN PIERRE EUGÈNE, one of the most outstanding of the French chemists of the nineteenth century, was born on October 27th, 1827. He was appointed Professor of Organic Chemistry at the *École Supérieure de Pharmacie* in 1859, and Professor of Organic Chemistry at the *École de France* in 1865. This Chair, specially created for him, he occupied till his death on March 27th, 1907. In 1881 he was elected a Senator for life, and he held at various times the portfolio of Minister of Public Instruction and Minister for Foreign Affairs. Berthelot was the greatest and most prolific chemist of his age, "the successor of the encyclopaedists of the eighteenth century," and had a mind possessed of great generalizing power. His contributions to chemistry range over the fields of organic synthesis, rates of reaction, thermochemistry, vegetable chemistry, and the history of chemistry and alchemy. On several of these subjects he wrote comprehensive works.

BODENSTEIN, ERNST AUGUST MAX, born at Magdeburg on July 15th, 1871, studied under Victor Meyer at the University of Heidelberg. His doctoral research was an investigation of the decomposition of hydrogen iodide, which became a textbook example of chemical equilibrium. He gained the *venia legendi* at the University of Heidelberg in 1899, and in the following year joined Ostwald's staff at Leipzig. In 1908 he was appointed

Professor of Chemistry at the Technical High School at Hanover and in 1923 became Professor of Physical Chemistry in the University of Berlin. In 1929 he was a Visiting Professor at Johns Hopkins University, Baltimore. He died on September 3rd, 1942.

BOISBAUDRAN, PAUL ÉMILE LECOQ DE, was born in Cognac on April 18th, 1838, and died on May 28th, 1912. Freed from the necessity of earning a livelihood, he devoted himself to scientific investigations. Besides gallium, he was the discoverer of samarium and dysprosium (Ramsay, *J. Chem. Soc.*, 1913, **103**, 742).

BUNSEN, ROBERT WILHELM, was born at Göttingen, March 31st, 1811. After graduating at the University of Göttingen in 1831, Bunsen visited Berlin, Paris and Vienna. In 1833, he became Privat-Dozent at Göttingen and in 1836 succeeded Wöhler at the Technical School at Cassel. In 1839 he was appointed Associate Professor, and in 1841 Professor of Chemistry at the University of Marburg. In 1846 he took part in a scientific expedition to Iceland. After a year at Breslau, Bunsen became, in 1852, Professor of Chemistry at Heidelberg, and retired from the Chair in 1889. He died August 16th, 1899. Although Bunsen first gained distinction for his work on the cacodyl compounds, his later work lay in the domain of inorganic and physical chemistry, the practical side of which he greatly developed. He had no interest in theoretical speculations. He introduced a new voltaic cell, developed the methods of gas analysis, invented the Bunsen burner and improved the methods of mineralogical analysis by means of dry tests and flame colorations. In 1860 his book on *Chemical Analysis by Spectral Analysis*, which was the result of work carried out in collaboration with the physicist, G. R. Kirchhoff, introduced a new and powerful instrument of discovery into practical chemistry, and led to the discovery, by Bunsen, of the elements caesium and rubidium. Bunsen's work and character are summed up by his pupil, H. E. Roscoe, in the words: "As an investigator he was great, as a teacher he was greater, as a man and friend he was greatest" (*J. Chem. Soc.*, 1900, **77**, 513).

CANNIZZARO, STANISLAO, was born at Palermo, Sicily, July 13th, 1826. After studying medicine at Palermo, Cannizzaro, in 1845, studied chemistry at Pisa under Raffaele Piria (1813–65), the most distinguished Italian chemist of his time. A patriot as well as a chemist, Cannizzaro took part in the rebellion of 1847, and again, in 1860, while Professor of Chemistry in the University of Genoa, he joined Garibaldi and his famous Thousand in Palermo. In 1861, Cannizzaro was called to the Chair of Chemistry in the University of Palermo and, in 1872, to the Chair of Chemistry at Rome, and was also made a Senator of the Kingdom. Cannizzaro made many contributions to organic chemistry but the most important work of his life was the full and clear exposition of the meaning and importance of Avogadro's hypothesis as a basis for the determination of molecular and of atomic weights. He died May 10th, 1910.

COUPER, ARCHIBALD SCOTT, was born at Kirkintilloch, near Glasgow, on March 31st, 1831. After studying classics and philosophy at the Universities of Glasgow and Edinburgh, he took up the study of chemistry, first in Berlin and then in Paris. In 1858 he became assistant in the chemistry department of the University of Edinburgh, under Lyon Playfair; but a nervous breakdown, followed after a few months by sunstroke, brought an early termination to his scientific career. He died at Kirkintilloch on March 11th, 1892.

Couper's theory of molecular constitution, which is essentially the same as that of Kekulé and was expressed with greater precision, was, owing to the failure of Wurtz to submit it timeously to the French Academy of Sciences, not made public until a month after Kekulé had published his theory. Through this stroke of ill fortune, which contributed to his breakdown in health, Couper failed to obtain, in his lifetime, that recognition which was his due; and the theory was for long associated only with the name of Kekulé.

CROOKES, WILLIAM, one of the great Victorians in science and a man of courteous and conciliatory manners, was born in London on June 17th, 1832. Although he had studied under Hofmann at the Royal College of Chemistry, Crookes was dedicated to no single branch of science, but pursued the investigation of natural phenomena from whatever direction they might appear. He remained through life a private practitioner of chemistry, and his extraordinary energy and experimental skill, his flashes of intuition and scientific insight led to discoveries which gained for him the highest recognition. He became distinguished, more especially, for his researches in spectroscopy, discharge of electricity through rarefied gases and radioactivity. He invented the radiometer, the spinthariscope and a special glass which is, in large measure, opaque to heat rays and to ultra-violet light. Crookes, also, exercised a not inconsiderable influence on the advancement of science through the weekly journal, *Chemical News*, which he founded in 1859 and edited until 1906. He was created a Knight in 1897, and the Order of Merit was conferred on him in 1910. He died on April 4th, 1919.

CURIE, MARIE SKLODOVSKA, was born on November 7th, 1867, at Warsaw, where her father Vladislav Sklodovski taught physics at the High School for Boys. Keenly interested in science, she went in 1891 to Paris to study physics, and in 1895 married Pierre Curie, a young physicist of distinction. In 1897, attracted by Becquerel's discovery of radioactivity, she resolved to study the phenomenon and in 1898 discovered radium and polonium. In 1906 she succeeded her husband as Professor of Physics at the Sorbonne and in 1919 was made Honorary Professor of Radiology in the University of Warsaw. She was awarded, along with her husband and Becquerel, the Nobel Prize for Physics in 1903, and was later, in 1911, awarded the Nobel Prize for Chemistry. Quiet and unassuming, she was a tireless worker in science and actively promoted the application of radioactivity to the alleviation of suffering and disease. She died on July 4th, 1934.

DEWAR, JAMES, was born at Kincardine-on-Forth, Scotland, on September 20th, 1842, and died in London on March 27th, 1923. He studied chemistry under Lyon Playfair at Edinburgh and under Kekulé at Ghent. In 1875 he was appointed Jacksonian Professor of Natural Experimental Philosophy in the University of Cambridge, and two years later became also Fullerian Professor of Chemistry at the Royal Institution, London. He was pre-eminent as an experimentalist, but was not great as a teacher for his mind was of too original and impatient a type. "He lacked the gift of ready literary expression and was often an incoherent lecturer, yet his lectures were the most masterly and fascinating displays ever witnessed" (*Nature*, 1923, **111**, 472). "He was a great man, vigorous, kindly and combative." He received the honour of knighthood in 1904.

DOBBIE, JAMES JOHNSTON, was born in Glasgow on August 4th, 1852. After graduating Master of Arts at Glasgow, he studied science for four years at Edinburgh and then in Germany under Kolbe and Wiedemann. In 1884 Dobbie was appointed Professor of Chemistry in University College of North Wales, Bangor, and there he worked enthusiastically to develop the application of chemistry to agriculture. A man of unfailing tact and personal charm, his administrative gifts were recognized by his appointment in 1903 as Director of the Royal Scottish Museum and in 1909 as Head of the Government Laboratory. He was knighted in 1915, retired from service in 1920, and died on June 19th, 1924.

DRAPER, JOHN WILLIAM, was born at St. Helens on May 5th, 1811, and died in New York on January 4th, 1882. After studying chemistry at University College, London, he emigrated to the United States and in 1836 graduated M.D. at the University of Pennsylvania. In 1839 he became Professor of Chemistry in the University of New York. He published numerous researches on photochemistry and was also known as a writer on historical subjects.

DUMAS, JEAN BAPTISTE ANDRÉ, was born at Alais, in the south of France, on July 16th, 1800. After a year's apprenticeship with an apothecary in his native town, Dumas, in 1816, proceeded to Geneva where he continued his apprenticeship as apothecary, widened the range of his studies and carried out research in physiology and physiological chemistry with Dr. J. L. Prévost. The scientific reputation which he thereby made for himself ensured for him a welcome by the leading men of science in Paris, whither he went in 1823. Soon after his arrival Dumas was appointed tutorial assistant (*répétiteur*) to Thenard at the *École Polytechnique* and also lecturer at the *Athénaeum*, an institution for evening lectures supported by private subscription. In later years he was Professor of Chemistry at the Sorbonne, the *École Polytechnique* and the *École de Médecine*. A man of great energy, Dumas made many contributions of the highest importance, not only in the domain of organic but also in that of inorganic chemistry. In his books and writings Dumas showed himself a master of literary style and of lucid exposition. In 1848 he was elected a member of the National Assembly and soon became Minister of Commerce and Agriculture; and, on the restoration of the Empire, he was made a Senator. After holding the offices of Vice-president and President of the Municipal Council of Paris and Master of the Mint, Dumas retired from public life in 1870. He died at Cannes, April 11th, 1884. An investigator of exceptional ability, Dumas became one of the most influential leaders of French chemistry in the first half of the nineteenth century. Greatly ambitious, he also understood how to realize his ambitions. Favoured by fortune, and doubtless also by his personal qualities, he rose from comparatively humble circumstances to a position of distinction attained only by a few.

FARADAY, MICHAEL, was born at Newington Butts (now incorporated in the south-eastern district of London) on September 22nd, 1791. Son of a blacksmith, Faraday became, at the age of fourteen, a bookbinder's apprentice and eagerly read the books which passed through his hands. A youth of alert mind, eager, original and inquiring, he had the desire, after hearing lectures on chemistry by Sir Humphry Davy, to enter into the service of science. In 1813 he became laboratory assistant to Davy at the Royal Institution, and there he received his first systematic training and passed the whole of his scientific life. He became Professor of Chemistry in 1827. In 1825 Faraday discovered benzene, and he also carried out investigations on the composition and properties of glass and of alloys. It is, however, on his investigations and discoveries in the domain of physics and physical chemistry that his fame mainly rests—on his researches on voltaic and frictional electricity, on the magnetic rotation of the plane of polarized light, on the electrolytic decomposition of solutions and on his discovery of electric induction and magneto-electric induction. He was acknowledged as the foremost investigator in physical science of his time and one of the most attractive lecturers. He died at Hampton Court on August 25th, 1867.

FISCHER, EMIL, one of the foremost organic chemists of the nineteenth century, distinguished as an experimentalist rather than as a theorist, was born at Euskirchen, in the Rhineland, on October 9th, 1852. His predilection for mathematics and physics having, according to his father, a successful business man, too little hope of bringing material gain, chemistry was chosen for his further study. Although Fischer spent a short time in the laboratory of Kekulé at Bonn, it was only when he came under the influence of Baeyer, at Strasbourg, in 1872, that his enthusiasm for organic chemistry was aroused and the future course of his brilliant career was set. Having followed Baeyer to Munich in 1874, Fischer worked there for eight years, and then in rapid succession filled the Chairs of Chemistry at Erlangen (1882), Würzburg (1885), and Berlin (1892). It was in Berlin that he died on July 15th, 1919.

FISCHER, OTTO PHILIPP, a cousin of Emil, was born in 1852, and became Professor of Chemistry in the University of Erlangen in 1885. From this Chair he retired in 1925. Died 1932.

FRANKLAND, EDWARD, was born at Churchtown, near Lancaster, on January 18th, 1825, and died while on a visit to Norway on August 9th, 1899. After serving as druggist's apprentice in Lancaster he studied chemistry under Playfair, Liebig and Bunsen, and in 1851 became Professor of Chemistry at Owens College, Manchester. In 1863 he was appointed to a special chair of chemistry, and later succeeded Michael Faraday at the Royal Institution; and in 1865 he succeeded Hofmann at the Royal College of Chemistry and Royal School of Mines, London. From this Chair he retired in 1885. He was made the recipient of many honours and was created Knight Commander of the Bath (K.C.B.) in 1897. For many years he devoted his attention to the problems of water-supply and purification.

GERHARDT, CHARLES FRÉDÉRIC, was born at Strasbourg on August 21st, 1816. He studied chemistry at Karlsruhe and Leipzig, and in 1836–7 under Liebig at Giessen. In 1838 he went to Paris, where he became lecture assistant to Dumas. He was Professor of Chemistry at Montpellier from 1841 to 1848, when he returned to Paris and opened, with Laurent, a private laboratory for instruction in chemistry. In 1855 he was appointed Professor of Chemistry and Pharmacy at Strasbourg, where he died on August 15th, 1856. Gerhardt, who was more distinguished as a generalizer and systematizer than as an experimenter, made valuable contributions, both theoretical and practical, to organic chemistry, and his *Précis de Chimie Organique* (1844, 1845) greatly enhanced his reputation. His *Traité de Chimie Organique* was published 1853–6. The failure of Gerhardt, as of Laurent, to win that measure of material success which their work merited, must, it is apparent, be attributed largely to personal characteristics. "Gerhardt," writes M. Bloch, "never learned to bow and scrape, his spine stiffened whenever he met any of the great ones of the earth, and throughout his whole life he retained something of the character of a high-principled and capable but callow youth. To him, the respect which one pays to age and merit seemed like toadyism. He was a young man whom one could not but like on account of his integrity, but who always spoilt everything by his boorishness" (Bugge, *Das Buch der grossen Chemiker*, II, 98).

GLADSTONE, JOHN HALL, was born at Hackney, London, on March 7th, 1827. After studying chemistry under Graham at University College, London, and under Liebig at Giessen, he returned to London and was in 1850 appointed Lecturer in Chemistry at St. Thomas's Hospital. This appointment he held for two years. From 1874 to 1877 he held the Fullerian Professorship of Chemistry at the Royal Institution. A man of private means, he was able to follow freely the bent of his mind, and he took an active part in religious work, in political and public life and on scientific committees. His most important scientific work lay in the domain of physical chemistry. He died on October 6th, 1902.

GRAHAM, THOMAS, was born in Glasgow on December 21st, 1805. After graduating Master of Arts at the University of Glasgow, where he had come under the influence of Professor Thomas Thomson, Graham decided to devote himself to the study and advancement of chemistry. He proceeded to Edinburgh where he worked under Professor Hope and earned a precarious livelihood by teaching and writing. In 1829 he was appointed lecturer in the Mechanics' Institution, and in the following year Professor of Chemistry in the Andersonian University (now the Royal Technical College), Glasgow. In 1837, on the death of Edward Turner, Graham was appointed Professor of Chemistry at University College, London, where his enthusiasm for his subject soon gained him fame as a teacher; and the publication of his *Elements of Chemistry* in 1841 made his name known over the whole world. Although his lectures were characterized by accuracy and breadth of knowledge, his manner was nervous and hesitating, and he had but little fluency of speech. In 1854 Graham was appointed Master of the Mint. A bachelor of simple tastes and of not too robust a constitution, Graham led a quiet life devoted entirely to scientific pursuits. He died on September 16th, 1869.

GRIGNARD, FRANÇOIS AUGUSTE VICTOR, was born at Cherbourg on May 6th, 1871. He studied at the College of Science at Lyons under Louis Bouveault and Philippe Barbier. For a time he was a *Préparateur* at Lyons and then became Lecturer, first at Besançon and then at Lyons. From

1909 to 1919 he was Professor of Chemistry at Nancy, and in the latter year returned to Lyons as successor to Barbier. He was awarded the Nobel Prize for Chemistry in 1912 for his important contributions to synthetic chemistry through the use of the "Grignard reagent" which arose out of an investigation suggested to him by Professor Barbier. He died on December 13th, 1935.

HABER, FRITZ, was born at Breslau on December 9th, 1868, and died at Basle, while on holiday, on January 29th, 1934. An outstanding personality in academic and industrial research, he had, at the age of 26, become Assistant to Bunte at the Institute of Chemical Technology at Karlsruhe, and remained there for seventeen years. In 1906 he succeeded Karl Engler as Professor of Physical Chemistry at Karlsruhe, and five years later became Director of the newly founded Kaiser Wilhelm Institute of Physical Chemistry and Electrochemistry at Berlin-Dahlem. Owing to the political conditions in Germany he resigned this post in 1933 and accepted an invitation to work in the University of Cambridge, England. Haber had an exceptionally wide range of knowledge and a superb gift of expression and was an attractive conversationalist.

HARCOURT, AUGUSTUS GEORGE VERNON, was born on December 24th, 1834, and died on August 23rd, 1919. After studying chemistry at Balliol College, Oxford, he was in 1859 elected Lee's Reader in chemistry, and appointed in 1872 one of the Metropolitan Gas referees to report on London gas. He introduced the pentane lamp as standard of illuminating power. He was an excellent manipulator and an inspiring and lucid teacher. "If he had an ambition, it was the desire to serve others and to feel that he was loved by his friends."

HAWORTH, WALTER NORMAN, one of the most distinguished of English chemists, was born at Chorley, Lancashire, on March 19th, 1883. He studied chemistry under W. H. Perkin, junr., at the University of Manchester and, as an Exhibition (1851) Scholar, under O. Wallach at Göttingen. In 1912 he was appointed a Lecturer and later a Reader in Chemistry in the University of St. Andrews, where he joined Irvine in the investigation of the carbohydrates. He was appointed Professor of Organic Chemistry at Armstrong College (later, King's College), Newcastle-upon-Tyne, in 1920, and Professor of Chemistry, University of Birmingham, in 1925. Although of a somewhat austere and reticent nature, Haworth had wide interests in literature and art. He had, moreover, the gift of inspiring in his students enthusiasm for research, and a constant and rapid stream of memoirs poured from his laboratories. The outstanding importance of his investigations was recognized by his being awarded (jointly with P. Karrer), the Nobel Prize for Chemistry in 1937, and by his receiving the honour of Knighthood in 1948. In 1928, Haworth spent a year at Basle as an Exchange Professor, and after retiring from the Chair at Birmingham in 1948, he visited and delivered lectures in Australia and New Zealand. His death came with unexpected suddenness on March 19th, 1950.

HEILBRON, IAN MORRIS, was born in Glasgow on November 6th, 1886. After completing his course of training at the Royal Technical College, Glasgow, Heilbron was awarded a Carnegie Fellowship and studied for two years, 1907-9, under Hantzsch at Leipzig. During World War I he became Assistant Director of Supplies in Salonika, with the rank of Lieutenant-Colonel, and, in 1918, was awarded the Distinguished Service Order, the Greek Order of the Redeemer and the Medaille d'Honneur. In 1947 he received the American Medal of Freedom.

Heilbron occupied successively the Chairs of Organic Chemistry at the Royal Technical College, Glasgow (1919-20), the University of Liverpool (1920-33), the University of Manchester (1933-38), and the Imperial College of Science and Technology, London (1938-49). Deliberate in speech and precise in expression, Heilbron exercised an inspiring influence on a numerous body of pupils, and his creativeness in research contributed greatly to the advancement of chemical science. His scientific achievements no less than his services on many Councils and Committees earned for him, in 1948, the honour of Knighthood. In 1949, Heilbron retired from academic work and became the first Director of the Brewing Industry Research Foundation. To his scientific interests he added also a love of music, of painting, and of all things beautiful. He died on September 14th, 1959.

HOFF, JACOBUS HENRICUS VAN'T, who became pre-eminent for his contributions to theoretical chemistry, was born at Rotterdam on August 30th, 1852. In 1872 van't Hoff entered the laboratory of Kekulé at Bonn and there his first experimental investigations were carried out. In 1874 he studied in Paris under Adolphe Wurtz at the *École de Médecine*, where he became acquainted with Le Bel. After a short period, 1876-7, spent as a lecturer in physics at the Veterinary School at Utrecht, van't Hoff was appointed in 1877 as Lecturer and in 1878 as Professor of Chemistry at the University of Amsterdam and remained there till 1896. In this year he accepted election to the Academy of Berlin, and appointment as Honorary Professor in the University; and in Berlin he remained for the rest of his life, unravelling, with the help of graduate students from different countries of the world, the solubility relations and the conditions of formation and separation of the salts of the Stassfurt deposits. Van't Hoff's was a generalizing and speculative mind, interested in fundamental principles rather than in the accumulation of experimental facts; and so, even at the age of twenty-two, the first fruit of his habit of speculating and day-dreaming is found in the epoch-making theory of the "asymmetric carbon atom," which he published in 1874. Ten years later there was published the still more important work, *Études de Dynamique Chimique*, one of the classics of chemical science which led van't Hoff on to the enunciation of the osmotic theory of solutions in 1886. Acclaimed during his life as "the greatest living physical chemist," van't Hoff retained a kindliness of heart and quiet simplicity of character which added a glory to his genius. He died from an attack of tuberculosis on March 1st, 1911.

HOFMANN, AUGUST WILHELM VON, was born at Giessen on April 8th, 1818. Entering the university of his native town in 1836 for the purpose of studying law, Hofmann came under the influence of Liebig and forsook jurisprudence for chemistry. In 1845, at the instance of the Prince Consort, Hofmann accepted the invitation to fill the Chair of Chemistry at the newly founded College of Chemistry in London. Here he taught and worked for twenty years, training students who were to become the leading English chemists of the Victorian era, and delivering lectures to working men. In 1865, "seized with a profound homesickness for the spiritual heights of a German University," he accepted the invitation of the Prussian Government to fill the Chair of Chemistry in the University of Berlin, rendered vacant by the death of Eilhard Mitscherlich. Here Hofmann exercised an extraordinarily great influence on the development of organic chemistry, and more especially on the development of the dye industry; and as a tribute to his memory the "Hofmann-Haus" was opened in Berlin in 1900, to be the home and centre of German chemical science. He was raised to the rank of a nobleman of Prussia on his seventieth birthday, and died on May 5th, 1892. "As a teacher he was singularly interesting and lucid. . . . Enthusiastic as an investigator of scientific problems, he could impart his enthusiasm, if not his genius, to others. . . . His genial and charming manner, high flow of spirits and originality in conversation and correspondence secured for him devoted friends." (See Hofmann Memorial Lecture, *J. Chem. Soc.*, 1896, **68**, 575.)

HOPKINS, FREDERICK GOWLAND, was born at Eastbourne, Sussex, June 20th, 1861. He began his scientific education at the Royal School of Mines, London, and became private assistant to P. F. Frankland there, and he also attended lectures at University College, London. He obtained his first professional qualification by passing the examination for the Associateship of the Royal Institute of Chemistry in 1883. In 1888 he joined the medical school at Guy's Hospital and graduated M.B. in the University of London in 1894. In 1898 he joined the department of Sir Michael Foster, at Cambridge, and developed there teaching and research on the chemical aspects of physiology. In 1901 he was elected Reader in

Chemical Physiology, in 1910 Praelector in Physiological Chemistry, in 1914 Professor of Biochemistry, and in 1921 first Sir William Dunn Professor of Biochemistry, from which Chair he retired in 1943.

In 1901, in collaboration with S. W. Cole, Hopkins isolated and identified tryptophan, the first of a series of researches which led in 1906 to the announcement of the dietary accessory factors or vitamins. The leading biochemist of his time, his numerous valuable contributions to science gained for him the esteem of scientists and the award of many academic and other honours. He received the honour of knighthood in 1925 and the Nobel Prize for Medicine in 1929, and the Order of Merit was conferred upon him in 1935. He died May 16th, 1947. A man of great modesty and charm of manner, he attained through his outstanding ability and persistence a position of the highest distinction in science.

IRVINE, JAMES COLQUHOUN, was born in Glasgow, May 9th, 1877. He entered the University of St. Andrews in 1895, and later studied under Wislicenus in Leipzig. In 1902 he returned to St. Andrews and through his researches there inaugurated a new era in carbohydrate chemistry. In 1909 he was appointed Professor of Chemistry and in 1921 Principal and Vice-Chancellor of the University, an office which he held until his death on June 12th, 1952. He was an inspiring teacher and leader of research, and as Principal of the University he exhibited great administrative ability. A man of vision and of great energy, dedicated to promoting the welfare of his University, he was able, through tact and gifts of persuasion, to increase greatly the financial resources of the University and to enhance both the prosperity and prestige of that ancient foundation. Noted both for scholarship and statesmanship he rendered many services to higher education not only in Britain but also in countries oversea. He was appointed Commander of the Order of the British Empire (C.B.E.) in 1920; created a Knight in 1925, and a Knight Commander of the Order of the British Empire (K.B.E.) in 1948.

KEKULÉ, FRIEDRICH AUGUST, was born at Darmstadt on September 7th, 1829. At the age of eighteen he matriculated in the University of Giessen as a student of architecture; but Liebig's lectures proved so attractive that he decided to abandon the study of architecture in favour of the study of chemistry. His training in chemistry was completed by a year's study in Paris, and thereafter he held private research posts in Switzerland and in London. In 1856 Kekulé returned to Germany as Privat-Dozent at Heidelberg, and in 1858 was appointed to the Chair of Chemistry at Ghent. In 1867 he became Professor of Chemistry at Bonn where, his health undermined by overwork, he died on July 13th, 1896. Although the author or inspirer of a large amount of experimental work, Kekulé's greatest service to chemistry consisted in the development of the theories of molecular structure, a work to which the architectural bent of his mind doubtless contributed. One of the most brilliant and imaginative, if not one of the most profound, of the German chemists of the nineteenth century, Kekulé received many conspicuous honours and was ennobled by the German Emperor. He then adopted the name of Kekule von Stradonitz and dropped the accent on the final *e* of his name.

KOLBE, HERMANN, was born at Elliehausen, near Göttingen, on September 27th, 1818, and died at Leipzig on November 25th, 1884. A pupil of Wöhler, he became assistant to Bunsen, and he worked also under Playfair in London. In 1847 he returned to Germany to edit the *Dictionary of Chemistry* which had been started by Liebig, whom he succeeded as Professor of Chemistry at Marburg in 1851. In 1865 he accepted an invitation to the University of Leipzig. Kolbe was pre-eminent as a teacher and critic, but his influence in the latter capacity was frequently destroyed through the excessive sharpness and lack of balance of his language.

KOPP, HERMANN, was born at Hanau on October 30th, 1817, and studied under Liebig at Giessen. In 1849 he founded, along with Liebig, the *Jahresbericht der Chemie*. In 1863 he became Professor of Chemistry at Heidelberg, where he died on February 20th, 1892. Kopp's experimental investigations lay chiefly in the domain of physical chemistry, of which he may in some respects be regarded as the founder, but his fame rests mainly on his writings dealing with the history of chemistry. Of these one may mention: *Geschichte der Chemie* (1843–7); *Entwickelung der Chemie in der neueren Zeit*; *Beiträge zur Geschichte der Chemie*; *Die Alchemie in älterer und neuerer Zeit*.

LANDOLT, HANS HEINRICH, was born at Zurich on December 5th, 1831. After studying at Zurich, Breslau, Berlin and under Bunsen at Heidelberg, he became a Dozent at Breslau in 1856 and Professor of Chemistry at Bonn in 1867. In 1869 he was appointed Professor of Chemistry in the newly founded Technical College at Aix-la-Chapelle. In 1882 he became a member of the Berlin Academy and in 1891 Professor of Chemistry in the University of Berlin. From this Chair he retired in 1905, and died on March 15th, 1910. A man of serious character but of constant good humour, and never without a cigar between his lips, his contributions to science were many and valuable in the fields of refractivity and optical activity, and he was noted for his careful investigation of the law of conservation of mass.

LANGMUIR, IRVING, was born at Brooklyn, N.Y., on January 31st, 1881. He received his early education at Paris, Philadelphia and Brooklyn. After graduating at the School of Mines, Columbia University, in 1903, Langmuir engaged in postgraduate study and research under Nernst at Göttingen, where he obtained the degree of Ph.D. in 1906. For a short period he was Instructor in Chemistry at the Stevens Institute of Technology in Hoboken, N.J. In 1909 he entered the Research Laboratory of the General Electric Company at Schenectady, N.Y., of which he later became Chemical Director; and from the time of his retirement in 1950 he remained Associated Director until his death on August 16th, 1957. Perhaps the most outstanding scientific investigator connected with industry in America, Langmuir was unrivalled in the width of his scientific interests and powers of observation; and his researches, everywhere opening up new vistas, ranged over a wide field of chemistry, both fundamental and applied. For his achievements in science and invention he was made the recipient of a large number of honours, including the award, in 1925, of the Cannizzaro Prize by the Royal Academy of the Lincei in Rome, and of the Nobel Prize for Chemistry in 1932. The U.S. Army and the U.S. Navy presented him with the Medal for Merit in 1948. From his life of scientific endeavour he found relaxation in mountaineering, winter sports, music and flying.

LAURENT, AUGUSTE, was born at La Folie, near Langres, in France, on November 14th, 1807. After qualifying as a mining engineer, he became tutorial assistant to Dumas at the *École Centrale des Arts et Métiers* in Paris, where he carried out his first chemical investigation. In 1838 he was appointed Professor of Chemistry at Bordeaux, but resigned in 1845 and returned to Paris. In 1848 he succeeded Péligot as assayer at the Mint. Laurent was distinguished mainly for his experimental and theoretical studies in organic chemistry, and during the later years of his life was closely associated with his fellow-countryman, Charles Gerhardt. Of an enthusiastic and artistic temperament, Laurent was perhaps oversensitive to the sharp criticisms—not always unjustified—which were levelled against his experimental work and theoretical speculations; and he was embittered by the want of appreciation which he considered to be his due. He died on April 23rd, 1853. His *Méthode de Chimie* in which he had collected his views on general chemical questions was published posthumously in 1854.

LE BEL, JOSEPH ACHILLE, a nephew of the agricultural chemist, Jean Baptiste Boussingault, was born at Pechelbron, Alsace, on January 21st, 1847. He became assistant to Balard and later to Wurtz at the *École de Médecine*. After the sale of a petroleum property which he inherited, he devoted himself to scientific investigations. His few publications were permeated by a philosophic spirit. A man of originality of thought and of an individualistic temperament, he was somewhat difficult of access, but was a delightful companion to those whom he admitted to his friendship. He died in Paris on August 6th, 1930.

LEWIS, GILBERT NEWTON, one of the most distinguished of American physical chemists, was born in Weymouth, Massachusetts, October 23rd, 1875. After six years' study at the University of Nebraska and at Harvard, he engaged in research at the latter University and graduated Ph.D. in 1899. In 1900 and 1901 he worked in the laboratories of Wilhelm Ostwald at Leipzig, and of Nernst at Göttingen. After a period spent in the Philippine Islands in charge of weights and measures, he became in 1907 a member of A. A. Noyes's staff at the Massachusetts Institute of Technology and was promoted to full professorial rank there in 1911. From 1912 until his death on March 26th, 1946, he was Professor of Chemistry in the University of California, Berkeley. Quiet and unassuming in manner, he had a wide and philosophic outlook and a most engaging and attractive personality. Through his many experimental investigations and his books he made a valuable contribution to the development of physical chemistry and chemical theory and exercised a wide influence on the teaching of the applications of thermodynamics in chemistry.

LIEBIG, JUSTUS, who was ennobled and became VON LIEBIG in 1845, was born at Darmstadt on May 12th, 1803. From playing with the chemicals in his father's laboratory, where pigments and varnishes were prepared, Liebig acquired an interest in chemistry; and although at first apprenticed to an apothecary, he soon gave up this work in order to study chemistry in Bonn and Erlangen. After two years (1822–4) spent in Paris, Liebig was appointed Associate Professor, and in 1825 Professor of Chemistry in the University of Giessen. In 1852 he accepted an invitation to the University of Munich, where he remained till he died on April 18th, 1873.

Liebig was a pioneer in chemical education, and the laboratory which he designed at Giessen and the courses of instruction which he organized served as models elsewhere. In later years he turned his attention more and more to the investigation of problems in agricultural and physiological chemistry; and the two books which he wrote, *Chemistry in its Application to Agriculture and Physiology* (1840) and *Organic Chemistry in its Application to Physiology and Pathology* (1842), may be said to have laid the foundations of these important branches of study. These works and his popular *Chemical Letters* made Liebig's name and reputation very widely known and gave him a supreme position of influence in scientific and economic chemistry. Although possessed of a warm heart and charming manner, Liebig had the temperament of a reformer. His impatience with what he regarded as looseness of thought or inaccuracy of experiment and his enthusiasm for truth gave a regrettable brusqueness and violence to his controversial writings. Still, as one of the greatest of his pupils wrote, "no other man of learning, in his passage through the centuries, has ever left a more valuable legacy to mankind" (A. W. Hofmann, *Berichte*, 1873, **6**, 470).

LOWRY, THOMAS MARTIN, son of a Methodist minister, was born at Low Moor, Bradford, on October 26th, 1874. In 1896 he became assistant to H. E. Armstrong, under whom he had studied at the Central Technical College of the City and Guilds Institute, South Kensington, London. In 1913 he was appointed Lecturer and later Professor in Guy's Hospital Medical School, and in 1920 was chosen to fill the newly founded Chair of Physical Chemistry in the University of Cambridge. His chief contributions to chemistry were in the field of stereochemistry and dynamic isomerism. He was the author of several textbooks of which the most important are his *Inorganic Chemistry* and *Optical Rotatory Power*. He died on November 2nd, 1936.

MENDELÉEFF, DMITRI IVANOVITSCH, one of the foremost chemical philosophers of the nineteenth century, was born at Tobolsk in Siberia on January 27th, 1834 (old style). In 1866 he became Professor of Chemistry in the University of St. Petersburg (Leningrad), where he carried out many important investigations into the properties of solutions. He resigned his professorship in 1890 and was appointed in 1893 Director of the Bureau of Weights and Measures. He died on January 20th, 1907.

MEYER, JULIUS LOTHAR, was born on August 19th, 1830, at Varel, Oldenburg. After studying medicine at Zurich and at Würzburg, where he graduated Doctor of Medicine in 1854, Meyer devoted himself to physiological investigations, and only in 1859 became a Privat-Dozent for Physics and Chemistry at Breslau. Here, in 1864, he published his *Modern Theories of Chemistry* and won a name for himself. In 1866 he became a lecturer at the Forestry Academy at Neustadt-Eberswalde and in 1868 Professor of Chemistry at the Polytechnic at Karlsruhe. In 1876 Meyer accepted the Chair of Chemistry at Tübingen, where he died on April 11th, 1895.

MOISSAN, FERDINAND FRÉDÉRIC HENRI, was born in Paris on September 28th, 1852. Apprenticed at first to a pharmacist, Moissan, in 1872, began the serious study of chemistry under Edmond Frémy at the *Musée d'Histoire Naturelle*. In 1880 he became lecture-assistant and demonstrator in chemistry at the *École Supérieure de Pharmacie*, and while there succeeded in isolating fluorine. In 1886 he was elected to the professorship of toxicology, and in 1899 to the Chair of Inorganic Chemistry at the School of Pharmacy. In 1900 he became Professor of Chemistry in the University of Paris. Moissan gained distinction not only through his investigations of fluorine and its compounds but also through his study of reactions at high temperatures. In 1906 he was awarded the Nobel Prize for Chemistry. A man of charming personality, the key of his character is to be found in his own words, "Nous devons tous placer notre idéal assez haut pour ne pouvoir jamais l'atteindre" (Ramsay, *J. Chem. Soc.*, 1912, **101**, 477). He died in Paris, February 20th, 1907.

NERNST, HERMANN WALTHER, was born on June 25th, 1864, at Briesen in West Prussia. Educated at Graudenz Gymnasium, where he became Head of the School, his main interests were at first centred in classics and literature, with a strong leaning towards poetry. Becoming interested in physics, he studied at the Universities of Zurich, Würzburg and Graz, and graduated Ph.D. at Würzburg in 1886. After a period spent as assistant in Ostwald's laboratory at Leipzig, Nernst became, in 1891, Privat-Dozent in Physics and in 1895 Professor of Physical Chemistry in the University of Göttingen. From 1905 to 1922 and again from 1924 to 1934, when he retired, he was Professor of Physical Chemistry in the University of Berlin. From 1922 to 1924 he was President of the *Physikalisch-Technische Reichsanstalt* at Berlin-Charlottenburg. He died on November 18th, 1941. A man of great energy, fond of travel and keen on shooting, Nernst had a versatile and original mind and attracted many of the younger workers in physical chemistry, especially those with a bent towards mathematics or physics, to his laboratory at Göttingen and at Berlin. His *Theoretische Chemie* had a very great influence in encouraging the study of physical chemistry to the advancement and development of which he made many important contributions. His formulation of the Nernst Theorem or Third Law of Thermodynamics was of outstanding importance in the history of physical science. He was awarded the Nobel Prize for Chemistry in 1920.

NILSON, FREDRIK, was born on May 27th, 1840, in East Gothland, and died on May 14th, 1899. He studied at the University of Uppsala and engaged in mineralogical investigations along with Sven Otto Petterson, who, in 1881, became Professor of Chemistry in the University of Stockholm. In 1878 he was appointed Professor of Analytical Chemistry in the University of Uppsala, and in 1883 was transferred to the Agricultural Academy at Stockholm. By the employment of a potash fertilizer he converted the calcareous moors of the Island of Gothland into land suitable for the growth of sugar beet.

OSTWALD, WILHELM FRIEDRICH, was born at Riga on September 2nd, 1853. From 1881 to 1887 he was Professor of Chemistry in the Polytechnic at Riga, and from 1887 to 1905 Professor of Physical Chemistry in the University of Leipzig. He was awarded the Nobel Prize in 1909. Through his experimental work Ostwald laid the foundations of many sections of physical chemistry, and as an inspiring teacher and as a lucid and enthusiastic writer and expositor in the realm of this branch of science, he was without an equal. To him, also, we owe the industrially important process for the production of nitric acid by the oxidation of ammonia. Ostwald was a man of large mind and large heart, and his freshness of mind and variety of interests gave to all intercourse with him an unforgettable charm. He died on April 4th, 1932.

PASTEUR, LOUIS, was born at Dôle in the Jura on December 27th, 1822. He studied chemistry under Balard at the *École Normale*, Paris, and under Dumas at the Sorbonne. In 1848 he became Deputy-Professor and in 1852 Professor of Chemistry in the University of Strasbourg. In 1854 he was transferred to Lille, and in 1857 to the *École Normale*, Paris. Pasteur became famous not only for his investigations on optical activity but also for his biochemical and bacteriological researches to which we owe our earliest knowledge of the bacterial origin of disease and production of immunity by vaccines. He died on September 28th, 1895.

PERKIN, WILLIAM HENRY, was born in London on March 12th, 1838. Having, while at school, shown an interest in chemical experiments, he entered the Royal College of Chemistry in 1853 as a student under Hofmann; and by the end of his second year there had carried out his first piece of research work (*J. Chem. Soc.*, 1856, **9**, 8). His zeal for chemistry is shown by the fact that he fitted up a small laboratory in his father's house, and there he carried on his experiments after the day's work at the College was over. In this laboratory, during the Easter vacation 1856, Perkin discovered the dye, mauve; and in the following year he opened a factory for its commercial production. Although he gained much success as an industrialist, Perkin never gave up his purely scientific research work; and in 1867 he published the first of a series of papers on the action of acetic anhydride on aromatic aldehydes. This work culminated in the method of synthesizing unsaturated acids known as the "Perkin reaction." In 1868 Perkin announced the synthesis of coumarin, the first vegetable perfume to be synthesized from a coal-tar product. In 1869 he patented a method for the production of the dye, alizarin, from anthraquinone. In 1874 Perkin retired from business and devoted himself to scientific research, more especially to the study of magnetic rotatory power. In 1906 the honour of knighthood was conferred upon him, and he died on July 14th, 1907.

PERKIN, WILLIAM HENRY, junr., was born at Sudbury on June 17th, 1860. After two years at the Royal College of Science (1877-9), Perkin spent five years in Germany, working under Wislicenus at Würzburg and under Baeyer at Munich. In 1886 he was appointed Professor of Chemistry at

the Heriot-Watt College, Edinburgh, and in 1892 he succeeded Schor-
lemmer at the University of Manchester. In 1912 he was appointed Professor of Chemistry in the University of Oxford. He died on September
17th, 1929. His chemical insight and experimental skill made him one
of the foremost synthetic organic chemists of the time, while his love of
music and genial manner won for him a wide circle of friends.

POPE, WILLIAM JACKSON, was born in London on March 31st, 1870. In
1890 he became assistant to H. E. Armstrong, under whom he had studied
Chemistry at Finsbury Technical College and at the Central Technical
College, South Kensington. Under Armstrong's influence he took up the
study of crystallography and lectured on this subject at the Central Technical College from 1895 to 1901, when he was appointed Professor of
Chemistry at the Municipal School (now College) of Technology, Manchester. For four years, 1897–1901, he was Head of the Chemistry
Department at the Goldsmiths' Institute, New Cross, London. In 1908
he succeeded G. D. Liveing as Professor of Chemistry in the University
of Cambridge, and occupied this Chair until he died on October 17th,
1939. He was created Knight Commander of the Order of the British
Empire in 1919. Pope made many contributions of outstanding importance to chemistry, especially in the domain of stereochemistry and molecular asymmetry. He introduced the use of camphor sulphonic acids for
the resolution of externally compensated bases and he effected the resolution into their optical antipodes of externally compensated compounds of
nitrogen, sulphur, tin and selenium. He was the first to prepare organic
compounds of platinum (with Peachey) and of gold (with Gibson).

PURDIE, THOMAS, was born on January 27th, 1843. After a period
devoted to business pursuits in the Argentine, Purdie took up the study
of chemistry, first under Edward Frankland in London and then under
Wislicenus at Würzburg. In 1884 he was appointed to the Chair of
Chemistry in the University of St. Andrews, and from this Chair he retired
in 1909. Distinguished as a chemist, Purdie was also a man who by his
tact and sympathy won the affection of all with whom he came into contact; and by his own quiet and persuasive personality he moulded the
character of his students. He died on December 14th, 1916.

RAMSAY, WILLIAM, was born at Glasgow on October 2nd, 1852. After
studying at the Universities of Glasgow, Heidelberg and Tübingen,
Ramsay held a junior post in the Chemistry Department of the University
of Glasgow. In 1880 he became Professor of Chemistry at University
College, Bristol, and in 1887 succeeded Williamson as Professor of
Chemistry at University College, London. From this latter Chair he
retired in 1912. He died on July 23rd, 1916. Perhaps the most brilliant

of British chemists since the time of Sir Humphry Davy, Ramsay earned greatest fame through his discovery of a whole group of elements (the rare gases) in the periodic system of Mendeléeff, and he was created Knight Commander of the Bath in 1902. In 1904 he was awarded the Nobel Prize for Chemistry. Lovable and sociable, musical and having the gift of tongues. Ramsay possessed personal qualities which endeared him to all.

RAOULT, FRANÇOIS MARIE, was born at Fournes, in the Département du Nord, on May 10th, 1830. The burden of earning a livelihood compelled him to relinquish prematurely his studies in Paris. For a number of years he filled various teaching-posts of a junior character, and in 1863 he obtained the degree of *Docteur ès-Sciences physiques* of the University of Paris. In 1867 he was appointed *Chargé du Cours de Chimie*, and in 1870 Professor of Chemistry in the University of Grenoble, where he remained till he died on April 1st, 1901.

ROOZEBOOM, HENDRIK WILLEM BAKHUIS, was born at Alkmaar on October 24th, 1854. After working for some time in a butter factory, he accepted, in 1878, an invitation to become assistant to Jakob Maarten van Bemmelen, Professor of Chemistry in the University of Leyden. Here he pursued his studies in chemistry and graduated in 1884. In 1896 he succeeded van't Hoff as Professor of General Chemistry in the University of Amsterdam, where he died on February 8th, 1907.

ROSCOE, HENRY ENFIELD, was born in London on January 7th, 1833, and died on December 18th, 1915. He studied chemistry under Graham at University College, under Williamson at the Birkbeck Laboratory, London, and under Bunsen at Heidelberg. It was Bunsen's influence that moulded the whole of his future scientific life. In 1857 Roscoe succeeded Edward Frankland as Professor of Chemistry at Owens College, Manchester, and the success of this recently founded College was largely due to Roscoe's educational and scientific activities and to his popularization of science. It was through his service to the teaching of chemistry rather than by his original contributions that Roscoe acquired fame. In 1885 Roscoe was elected a Member of Parliament, and retained his seat for ten years; thereafter he served on many educational and scientific committees and commissions. Roscoe was knighted in 1884, and sworn a Member of the Privy Council in 1909.

SIDGWICK, NEVIL VINCENT, a worthy member of a highly gifted family, was born at Oxford on May 8th, 1873. Elected to an open scholarship in natural science, he entered Christ Church, Oxford, in 1892. A man of outstanding ability and with great width of interests, Sidgwick, after obtaining a First Class in the Honour School of Natural Science in 1895, read Classics and obtained a First Class in "Greats" in 1897. A unique record in academic history. After two years spent at Tübingen in the laboratory of von Pechmann, Sidgwick was, in 1901, elected to a Fellowship at Lincoln College, Oxford; and this became the centre of his life and work until his death on March 15th, 1952. In 1931, Sidgwick was Fisher Baker Lecturer in Chemistry at Cornell University, and in 1933 the University of Oxford conferred on him the title of Professor of Chemistry. In 1935 he was appointed Commander of the Order of the British Empire (C.B.E.). As a tutor, Sidgwick was held in high esteem and won both the gratitude and affection of his pupils. Although not eminent as an experimenter, his acute intellect and reliable memory, and his mastery of clear and elegant language, made him one of the foremost systematizers of chemical phenomena and expositors of chemical theory of his time. Through his books and writings, of which one may mention, more especially, *The Electronic Theory of Valency* (1927) and *The Chemical Elements and their Compounds* (1950), Sidgwick exercised a very great and vitalizing influence on chemical science.

SIMPSON, MAXWELL, was born in Co. Armagh, Ireland, on March 15th, 1815, and studied under Graham at University College, London, and also, at various times, under Dumas and Wurtz, Bunsen and Kolbe. From 1847 to 1857 he held a lectureship in chemistry at Ledwich School of Medicine, Dublin, but resigned this in order to work under Wurtz in Paris. Returning to Dublin in 1860, he carried out research in a private laboratory and in 1872 was appointed Professor of Chemistry in Queen's College, Cork. He retired in 1891 and lived in London till his death on February 26th, 1902 (*J. Chem. Soc.*, 1902, **81**, 631).

SODDY, FREDERICK, was born at Eastbourne on September 2nd, 1877. After graduating at Oxford in 1898 he was appointed, in 1900, Demonstrator at McGill University, Montreal. Here he had the good fortune to meet Ernest Rutherford, Professor of Physics, and to join him in radiochemical research. It was a happy partnership and out of it came the disintegration theory of radioactivity. Soddy returned to England in 1903 and worked with Sir William Ramsay at University College, London. In 1904, before taking up his appointment as Lecturer on Physical Chemistry in the University of Glasgow, Soddy undertook a lecturing tour in Australia and New Zealand. It was at Glasgow, between the years 1904 and 1914, when he was appointed Professor of Chemistry in the University of Aberdeen, that Soddy accomplished most of his work and made himself the leader in Great Britain in radiochemical research. In 1919 he was appointed Dr. Lee's Professor of Physical Chemistry at Oxford, and he remained there until his retirement in 1936. In 1913 Soddy was awarded the Cannizzaro Prize and in 1921 the Nobel Prize for Chemistry. At

Oxford Soddy failed to obtain either the accommodation or the encouragement which would have made possible the establishment of a school of Radiochemistry, parallel with the Rutherford School of Radiophysics at Cambridge; and no further experimental work was done. A man of complex personality and somewhat unapproachable, Soddy was not dedicated solely to the advancement of chemistry. An idealist, no doubt, he sought to resolve the wrongs of the world and the weaknesses of our modern civilization, and, Paracelsus-like, he was prone to rebellion against authority and was intolerant of the views of others. He died at Brighton on September 22nd, 1956, in somewhat tragic scientific isolation.

THOMSEN, HANS PETER JÜRGEN JULIUS, was born at Copenhagen on February 16th, 1826, and became Professor of Chemistry in the University there. Although he gained for himself a high reputation in applied chemistry, it is on his numerous and valuable contributions to thermo-chemistry, published in four volumes under the title *Thermochemische Untersuchungen*, that his fame mainly rests. He died on February 13th, 1909.

THORPE, JOCELYN FIELD, son of a barrister, was born in London on December 1st, 1872, and studied chemistry at the Royal College of Science under W. A. Tilden and at the University of Heidelberg under Victor Meyer. He was Assistant Lecturer and, later, Lecturer on Organic Chemistry and Biochemistry in Owens College, Manchester, from 1896 to 1910, and a Sorby Research Fellow of the Royal Society, 1909–13. In 1914 he was appointed to the Chair of Organic Chemistry in the Imperial College of Science and Technology from which he retired in 1938. He was created a knight in 1939. Genial and debonair, highly cultured and many-sided in his interests, Thorpe had an alert and versatile mind. He is noted for his work on tautomeric phenomena and on imino-compounds, and for his investigations of various types of ring-formations. He died on June 10th, 1940.

TILDEN, WILLIAM AUGUSTUS, was born on August 15th, 1842, and died on December 11th, 1926. Apprenticed at the age of fifteen to a pharmacist, he was given the opportunity of attending Hofmann's lectures at the College of Chemistry, and was thus led to pass from pharmacy to chemistry. In 1863 he became demonstrator at the Pharmaceutical Society, London, and later, in 1872, Science Master at Clifton College, Bristol. In 1880 Tilden was appointed Professor of Chemistry at the newly founded Mason College (afterwards University), Birmingham, and in 1894 he succeeded T. E. Thorpe as Professor of Chemistry at the Royal College of Science, London. From this Chair he retired in 1909 and received the honour of a knighthood. Tilden was one of the early workers in terpene chemistry and he made many valuable contributions to this as well as to other branches of chemical science. A man of dignified and distinguished appearance, he had an interest in all aspects of chemistry; and the books which he wrote were a valuable enrichment of chemical literature.

WALKER, JAMES, was born in Dundee on April 6th, 1863. After an apprenticeship in a jute business he entered the University of Edinburgh and studied chemistry under Crum Brown. After graduation as B.Sc. in 1885 he engaged in research under T. Carnelley at University College, Dundee, and graduated D.Sc. at Edinburgh in 1886. Although mainly interested in the newly developing branch of physical chemistry, Walker was induced to work under Claisen at Munich, but in 1888 he went to Leipzig to work under Ostwald. In 1889 he returned to Edinburgh as Research Assistant to Crum Brown, and carried out his electro-synthesis of dibasic acids. In 1892 he went to University College, London, and became a member of Ramsay's staff in the following year. He was appointed Professor of Chemistry at Dundee in 1894 and in the University of Edinburgh in 1908. From this Chair he retired in 1928. He received the honour of knighthood in 1921, and died on May 6th, 1935. Through his teaching and research work and through the publication of his *Introduction to Physical Chemistry* (1899), Walker did much to establish and develop the study of physical chemistry in Great Britain.

WERNER, ALFRED, born on December 12th, 1866, at Mulhouse, in Alsace, early developed an interest in chemistry and, while undergoing his military training at Karlsruhe, took the opportunity of attending classes in chemistry at the Technical College. In 1886 he entered the Technical High School at Zurich where he was able to study under Lunge, Hantzsch and Treadwell. After obtaining the Diploma of the College in 1889, he became Assistant to Lunge. After working for a semester with Berthelot in Paris, Werner returned to Zurich in 1892 and became a Privat-Dozent. A year later he published his *Beiträge zur Konstitution anorganischer Verbindungen*. The excellence of this work led to his appointment as an Associate Professor in 1893 and as full Professor in 1895. Full of the joy of life and a tireless worker, he inspired his numerous pupils with his own enthusiasm for research; and the importance of the contribution to chemical science which he made in the formulation of his co-ordination theory and through the experimental support which he gave to it earned for him the award of a Nobel Prize in 1913. He died on November 15th, 1919.

WILLIAMSON, ALEXANDER WILLIAM, was born at Wandsworth, May 1st, 1824. Until about the age of sixteen he had very delicate health, and as a consequence lost the sight of his right eye and had but little power in his left arm. In 1840, by which time his father had retired and was living on the Continent, Williamson entered the University of Heidelberg in order to study medicine, but he later decided to take up the study of chemistry, and in 1844 proceeded to Giessen where for two years he worked under Liebig. After a further period of study in Paris, Williamson was appointed in 1849 Professor of Analytical and Practical Chemistry, and while retaining this Chair was in 1855 appointed to succeed Graham as Professor of Chemistry at University College, London. He retired in 1887 and died at Haslemere, Surrey, on May 6th, 1904. Williamson's most important work was in connection with the theory of etherification.

WINKLER, CLEMENS ALEXANDER, was born at Freiberg, in Saxony, on December 26th, 1838, and died on October 8th, 1904. Even as a boy he had shown much interest in chemistry, and in 1857 he entered the School of Mines at Freiberg, where he speedily attained much success in research. A most distinguished analyst, Winkler was appointed to the Chair of Chemical Technology and Analytical Chemistry at Freiberg. To his scientific he added great literary and musical gifts.

WISLICENUS, JOHANNES ADOLF, descended from a family of Polish refugees and the son of a Lutheran pastor of liberal views, was born at Klein-Eichstädt, near Querfurt, on June 24th, 1835. In 1853, owing to the modernist views of Wislicenus's father, the family had to find safety in America, where Wislicenus obtained a post at Harvard University and also conducted an analytical laboratory in New York. In 1855 the family returned to Europe and Wislicenus resumed his interrupted studies at Halle. In 1859 he left Halle for Zurich, where in 1870 he became Professor of Chemistry at the Polytechnic. After thirteen years (1872–85) spent at Würzburg, Wislicenus succeeded Kolbe at Leipzig, where he died on December 5th, 1902. Wislicenus was one of the leaders in organic chemistry in the nineteenth century, and a man "of extraordinary directness of purpose and splendid character." "His long beard, his fine head with its intellectual features and his majestic carriage aided in producing a sensation which in younger men was not far from veneration." (Perkin, *J. Chem. Soc.*, 1905, **87**, 508.)

WÖHLER, FRIEDRICH, was born at Eschersheim, near Frankfort, on July 31st, 1800. In 1823 he graduated Doctor of Medicine at Heidelberg, but instead of pursuing a medical career he resolved, on the advice of Leopold Gmelin, Professor of Medicine and Chemistry, to devote himself to chemistry. A year was then spent at Stockholm with Berzelius, whose methods of analysis he learned. In 1825 Wöhler became a teacher of chemistry in the Municipal Technical School in Berlin; in 1831 he was appointed professor at the Technical School at Cassel and in 1836 he became Professor of Chemistry at the University of Göttingen, where he remained till his death on September 23rd, 1882. Wöhler combined a great love of the open air, the beauties of nature and of all living things with his great passion for chemistry, and he attained a position of the highest distinction through his writings and the very numerous original contributions which he made not only to organic but also, and more especially, to inorganic chemistry. A man of quiet and peaceful mind and free from self-assertion, Wöhler pursued the even tenor of his way, taking but little part in the controversies of his time.

WURTZ, CHARLES ADOLPHE, was born at Strasbourg on November 26th, 1817. For many years he was Professor of Chemistry at the *École de Médecine* and at the Sorbonne, Paris. He was known not only for his researches in organic chemistry but also for his many literary works, of which one may mention specially *La Théorie Atomique* (1879). He was editor of a *Dictionnaire de Chimie Pure et Appliquée*, and after 1868 one of the editors of the *Annales de Chimie et de Physique*. He died in Paris on May 12th, 1884.

INDEX